# TECHNOLOGY AND THE CULTURE OF MODERNITY IN BRITAIN AND GERMANY, 1890–1945

This book examines the obsession for new technology that swept through Britain and Germany between 1890 and 1945. Drawing on a wide range of popular contemporary writings and pictorial material, it explains how, despite frequently feeling overwhelmed by innovations, Germans and Britons nurtured a long-lasting fascination for aviation, glamorous passenger liners, and film as they lived through profound social transformations and two vicious wars.

Public discussions about these "modern wonders" were torn between fears of novel risks and cultural decay on the one hand, and passionate support generated by nationalism and social fantasies on the other. While the investigation focuses on tensions between technophobia and euphoria, the book also examines the relationship between responses to technology and the differing political cultures in Britain and Germany before and after 1933.

This innovative study will prove invaluable reading to anyone interested in German and British history as well as the history of technology.

BERNHARD RIEGER teaches modern history at the International University Bremen. He has co-edited *Meanings of Modernity: Britain from the Late-Victorian Era to World War II* (2001) with Martin J. Daunton.

# NEW STUDIES IN EUROPEAN HISTORY

*Edited by*

PETER BALDWIN, University of California, Los Angeles
CHRISTOPHER CLARK, University of Cambridge
JAMES B. COLLINS, Georgetown University
MÍA RODRÍGUEZ-SALGADO, London School of Economics and Political Science
LYNDAL ROPER, University of Oxford

The aim of this series in early modern and modern European history is to publish outstanding works of research, addressed to important themes across a wide geographical range, from southern and central Europe, to Scandinavia and Russia, and from the time of the Renaissance to the Second World War. As it develops the series will comprise focused works of wide contextual range and intellectual ambition.

*For a full list of titles published in the series, please see the end of the book.*

# TECHNOLOGY AND THE CULTURE OF MODERNITY IN BRITAIN AND GERMANY, 1890–1945

BERNHARD RIEGER

*International University Bremen*

CAMBRIDGE UNIVERSITY PRESS
Cambridge, New York, Melbourne, Madrid, Cape Town, Singapore, São Paulo, Delhi

Cambridge University Press
The Edinburgh Building, Cambridge CB2 8RU, UK

Published in the United States of America by Cambridge University Press, New York

www.cambridge.org
Information on this title: www.cambridge.org/9780521845281

© Bernhard Rieger 2005

This publication is in copyright. Subject to statutory exception
and to the provisions of relevant collective licensing agreements,
no reproduction of any part may take place without the written
permission of Cambridge University Press.

First published 2005
This digitally printed version 2008

*A catalogue record for this publication is available from the British Library*

*Library of Congress Cataloguing in Publication data*
Rieger, Bernhard, 1967–
Technology and the culture of modernity in Britain and Germany, 1890–1945 / Bernhard Rieger.
p. cm. – (New studies in European history)
Includes bibliographical references and index.
ISBN 0 521 84528 9 (alk. paper)
1. Technology – Social aspects. I. Title. II. Series.
T 14.5R56 2004   303.48´3´0940904 – dc22   2004054641

ISBN 978-0-521-84528-1 hardback
ISBN 978-0-521-09314-9 paperback

*To Liz*

# Contents

| | | |
|---|---|---|
| *List of illustrations* | *page* | viii |
| *Acknowledgments* | | ix |
| 1 Introduction | | 1 |
| 2 "Modern wonders": technological innovation and public ambivalence | | 20 |
| 3 Accidents: the physical risks of technology | | 51 |
| 4 Elusive illusions: the cultural and political properties of film | | 86 |
| 5 Pilots as popular heroes: risk, gender, and the aeroplane | | 116 |
| 6 "Floating palaces": passenger liners as objects of pleasure | | 158 |
| 7 Fantasy as social practice: the rise of amateur film | | 193 |
| 8 Technology and the nation in Britain and Germany | | 224 |
| 9 Conclusion | | 276 |
| *Bibliography* | | 286 |
| *Index* | | 315 |

## Illustrations

| | | |
|---|---|---|
| 2.1 | Postcard of the liner *Bremen*. Author's collection. | page 27 |
| 2.2 | Postcard of the *Titanic*. Author's collection. | 44 |
| 3.1 | Photomontage of *L1*. Deutsches Schiffahrtsmuseum, Bremerhaven. | 63 |
| 5.1 | From Elly Beinhorn, *Ein Mädchen fliegt um die Welt*. | 143 |
| 5.2 | Star postcard of Amy Johnson. John Capstack Archive. | 145 |
| 6.1 | Drawing of riveters in a shipyard. Deutsches Schiffahrtsmuseum. | 174 |
| 6.2 | Fritz Breuhaus de Groot, *Bauten und Räume*. Verlag Ernst Wasmuth. | 179 |
| 6.3 | The *Queen Mary*'s cocktail bar. University of Liverpool Library. | 182 |
| 7.1 | Cine Nizo advertisement. Deutsches Museum, Munich. | 200 |
| 7.2 | Advertisement for a Kodak film projector. Kodak Inc. | 214 |

*Acknowledgments*

While working on this book, I have incurred innumerable debts of gratitude. For financial research support I would like to thank the German Historical Institute London, the British Academy, the Royal Historical Society, the Remarque Institute at New York University, the Research Institute at the Deutsches Museum in Munich, the National Science Foundation in Washington DC, and the Fellows of Churchill College for electing me to a Bye-Fellowship in 2002. A special "thank you" has to go to the history department at University College London, which funded my PhD and thus made possible four wonderful years. Furthermore, the historians at Iowa State University gave me that most elusive of commodities: a first proper job. Amy Sue Bix, Hamilton Cravens, Jerry Garcia, John Monroe, George McJimsey, Andrejs Plakans, and David Wilson: thanks to you all, and particularly to Jean and Alan I. Marcus.

As far as their "sources" are concerned, few historians manage to progress beyond the stage of the obsessive hunter and gatherer. I certainly did not, and the librarians and archivists of a wide range of institutions helped to satisfy my atavistic instincts. This book could not have been written without the assistance of staff at the British Library, the Staatsbibliotheken in Berlin and Munich, the Deutsches Museum Munich, Erlangen University Library, Imperial College Library, the Centre of South Asian Studies in Cambridge, Liverpool University Archives, the Deutsches Schiffahrtsmuseum Bremerhaven, the British Film Institute, and the Stiftung Deutsche Kinemathek Berlin. Every effort has been made to identify copyright-owners of sources quoted.

Companionship, conversation, meals, and, above all, good humor are what my friends shared with me. Heidi and Achim Bade, James Andrews and Maggie LaWare, James Berg and Martha Woodruff, Frank Hack, Marisa Kern, Tracy and Ethan Kleinberg, Elke Metzner and Thomas Schauer together with Moritz and Anika, Konstanze Scharring, and Jonathan Williams have all made sure that there is more to life than work.

I have the fondest memories of the time spent with my granny and her "boyfriend" Robert Winterstein, who entertained and helped Liz and me when research took us to Munich. Thanks also to my mother and father.

Many scholars have generously discussed and challenged my ideas at conferences and seminars, while others commented on chapters and offered organizational advice. Those who helped the project to become a reality include Rachel Aucott, Peter Baldwin, Geoff Crossick, David Edgerton, David French, Kevin Greenbank, Matthew Hilton, Klaus-Peter Kiedel, Frieder Kießling, Axel Körner, Jon Morris, Mary Nolan, John North, Erika Rappaport, Lyndal Roper, Sonya Rose, Bruce Seely, Willibald Steinmetz, J. Adam Tooze, Frank Trentmann, and Helmut Trischler. David Cannadine examined the dissertation and has offered encouragement in many forms since. In countless emails and phone conversations Michael Dintenfass helped me to try out fuzzy ideas, as did Johannes Paulmann, David Feldman, and Peter Mandler, who repeatedly took the time to turn my hunches into arguments. One of my former dissertation advisors, Rebecca Spang, has provided many a spark over the years. Jane Caplan and Geoff Eley took a persistent interest in this project from the outset and have supported my efforts with embarrassing generosity. Martin Daunton has shown more patience, encouragement, and enthusiasm than anyone becoming a historian could possibly begin to expect from a mentor.

The book is dedicated to Liz, who is my biggest source of inspiration.

CHAPTER I

## Introduction

"A civilisation is known by its realised dreams. If another age than ours should ask, 'What did you do with your time?' here, in the more than Roman magnificence of our engineering, is one answer we can give." It was the new liner *Queen Mary*, awaiting its launch in front of 250,000 spectators in Glasgow on 26 September 1934, that prompted this celebration in the *Manchester Guardian*. Profoundly impressed, the journalist proclaimed the new vessel to be a revelation because the *Queen Mary* turned an intangible "dream" into a material "reality," and a formidable one at that – a structure over 1,000 feet (300 meters) long and weighing more than 35,000 tons (32,000 tonnes). Although large vessels such as the *Kaiser Wilhelm der Große* (1897), the *Lusitania* (1907), the *Mauretania* (1907), the infamous *Titanic* (1912), the *Imperator* (1913), the *Bremen* (1929), and the *Europa* (1930) had galvanized popular attention on both sides of the Atlantic with almost predictable regularity since the late nineteenth century, neither the press nor its readers had grown accustomed to these giants' ever-increasing size, speed, and luxuriousness. "Never before has the launch of a ship given the popular imagination so lively a thrill in anticipation," claimed one of Britain's leading broadsheets as the *Queen Mary* became ready to take to the water. This boat, every observer agreed, put all previous naval constructions to shame. Even the pouring rain that forced many onlookers of the launch to spend hours in "flooding cornfields, [mud] slowly oozing over their ankles," could not mar the universal mood of joyful anticipation. As the hull entered the water, "one long sigh [traveled] all down the mile long line of" soaked spectators. As much as the nation as a whole, the people of Glasgow stood in awe of their "grand" work. A fitting symbol of an age that prided itself on the "conquest of nature," the *Queen Mary* struck the *Manchester Guardian*, a newspaper that usually shied away from hyperbole, as a "secular miracle."[1]

[1] *Manchester Guardian*, 25 September 1934, 8; 26 September 1934, 8; *Daily Mail*, 25 September 1934, 8. The technical data are taken from James Steele, *Queen Mary* (London, 1995), 51, 235.

By acclaiming Britain's latest transatlantic liner as a "miracle," the *Manchester Guardian* aligned the *Queen Mary* with countless other prominent technological artifacts that had struck observers as "wonders" since the onset of the Industrial Revolution. Industrial machinery, railways, aeroplanes, airships, film, photography, electricity, motor cars, the gramophone, the radio – all these and more counted among the mechanisms that, as a plethora of commentators insisted, filled the world with new, "miraculous" objects. Throughout the nineteenth century and beyond, contemporaries expressed their astonishment at technological transformations by persistently describing innovative mechanisms as the "wonders" of the modern age. Yet the rhetoric of the "modern wonder" of engineering, despite all the fervent enthusiasm it generated, did not prevent the emergence of a profound sense of ambivalence that just as persistently accompanied the appearance of technological innovations. Characteristically, the correspondent for the *Manchester Guardian*, whom we have just encountered as he marveled at the *Queen Mary*'s magnificence, also drew attention to a disconcerting discrepancy between the boat's creators and their creation. Reflecting on the ship's dimensions, he shuddered at the vessel's hubristic size, noting that mere "six-foot high human beings" were responsible for "this mountain of metal," a "sea mammoth" that dwarfed not only its makers but everything in its vicinity. Like many other spectacular new technological objects, the *Queen Mary* elated observers *and* sent shivers down their spines. While observers fervently admired many innovations, they simultaneously responded to novel mechanisms with anxiety and found spectacular innovations intellectually and emotionally confusing. Throwing contemporaries off balance, technology as a wonder contributed to a pervasive sense of dislocation that coincided with the advent of a time that struck contemporaries as a new, distinctly "modern" age.

Technological change mounted profound emotional and intellectual challenges to the popular imagination because boisterous mechanical progress created and subsequently intensified a novel dilemma in Western societies from the middle of the nineteenth century onwards. As technology played crucial roles in reshaping the external world, most people confronted this process of transformation from a position of profound ignorance.[2] Very few individuals had the expert knowledge that would have rendered

---

[2] A similar lack of understanding has been noted by works on assessments of technology in North America from the middle of the nineteenth century. See John F. Kasson, *Civilizing the Machine: Technology and Republican Values in America, 1776–1900* (New York, 1999 [1976]), 141–2; Carolyn Marvin, *When Old Technologies Were New: Thinking About Electrical Communication in the Late Nineteenth Century* (New York, 1988), 17–22.

intelligible the machines that were changing their natural and social environments. Contemporaries not only lacked an understanding of the scientific findings on which many innovations were based; they also failed to comprehend how the new mechanisms that were coming into existence functioned. Although they figured prominently as symbols of change, many inventions remained beyond the grasp of the majority of the population. In addition to transforming the environment, technological innovation also created novel modes of representing and perceiving this ever-changing external world. The first third of the twentieth century witnessed the arrival and proliferation of innovative "representational technologies," which included the tabloid press, photography, film, radio, and the gramophone. As many new media quickly acquired a wide popular following, contemporaries faced the task of determining what kinds of knowledge these novel technologies generated. Commentators found it problematic to locate innovative representational technologies within existing hierarchies of knowledge production.[3] In transforming the external world, as well as the means of representing and perceiving it, technological innovation thus created *a problem of knowledge.*

Of course, a restricted grasp of technological intricacies did not silence public debate; in fact, the opposite was true. Discussion often found ways of addressing technological issues that bypassed details of engineering but were nonetheless crucial to popular understandings of technology. This study examines the frequently passionate British and German public exchanges about technological innovation that, between the last decade of the nineteenth century and World War II, ascribed meanings to technologies as symbols of change and rendered the material transformation of the external world intelligible to technological laypersons. Such public considerations of technologies' significance went far beyond instances of popularization when scientists and engineers explained to non-experts how certain mechanisms were produced and how they worked. Public debate performed the cultural work of advancing interpretations that determined technology's place in contemporary economic, political, and social life. Discussion about technological innovation represented negotiations of ignorance, in the course of which the participants inscribed social and cultural meaning into objects on the basis of a partial technical understanding

---

[3] On these debates, see Dan LeMahieu, *A Culture for Democracy: Mass Communication and the Cultivated Mind in Britain Between the Wars* (Oxford, 1988); Jonathan Crary, *Techniques of the Observer: On Vision and Modernity in the Nineteenth Century* (Cambridge, MA, 1992). On debates about this problem in France, see Martin Jay, *Downcast Eyes: The Denigration of Vision in Twentieth-Century French Thought* (Berkeley, 1993).

of technology. Given their ambivalent nature, assessments of technology as a "modern wonder" possessed the potential to generate both public euphoria and technophobia. Public insecurity about technology, however, did not give rise to a cultural atmosphere that opposed change *per se*; instead, the broad range of the interventions that we shall analyze established a cultural climate conducive to innovation – albeit in the presence of deep ambivalence.

To show how public enthusiasm for, and unease about, innovative technology interacted to further technological change in Britain and Germany between the 1890s and World War II, this book investigates the intersecting debates about the following three technologies: aviation, transatlantic passenger shipping, and film. These examples have been selected for several reasons. The technologies in question either came into existence or embarked upon vigorous technological development during the 1890s and early 1900s. At this time, critiques of modernity increased significantly in both countries, and the debates under consideration, therefore, shed light on how contemporaries evaluated technology as anxieties about the "modern age" spread. Since public debate about technology underwent significant transformations after World War II (as the concluding pages will briefly show), the year 1945 provides our chronological endpoint. Furthermore, combining aviation with passenger shipping and film brings into view technology's military and civilian dimensions. Developments in these sectors attracted strong public attention, and discussions about these artifacts, therefore, shaped central assumptions about technological change in times of peace and war. Debates about the technologies in question not only grant opportunities to explore how British and German public interpretations ascribed meanings to a changing world of objects; incorporating discussions about film also allows for an investigation of how contemporaries judged the most successful and most controversial representational technology. Given the speed with which this medium secured a mass market, film undoubtedly, in economic terms, counts among the most prolific technological inventions of the late nineteenth century. Curiously, film, despite its evidently mechanical nature, only rarely received extended consideration as a technological phenomenon in contemporary debate, and we shall chart how cinematography lost its status as a technology and came to be conceived as a primarily cultural phenomenon. Thus this study analyzes how the British and German publics evaluated both the artifacts that transformed the external world and the novel means of representing and perceiving these spectacular metamorphoses. Moreover, an analysis of luxurious ocean liners, aeroplanes, and airships illustrates that economic profits, or expectations of such gains, alone do not explain support for innovation, since the operation of these technologies consistently generated financial losses and, consequently, required substantial subsidies. Contemporaries shared the conviction that the importance of these technologies lay beyond narrow economic definitions, a belief sustaining enthusiasm even for financially costly innovations. There was much more to the love of technology than narrow economic calculation. Finally, the examples under scrutiny guard against an inquiry that employs the concept of "technology" in a reified manner. As technologies proliferated in contemporary society, they gained a wider range of public meanings than studies of one technology or of debates about "technology" in philosophical circles can capture. The debates about heterogeneous technologies bring out the multifaceted, context-related, and frequently contradictory meanings that the term "technology" denotes.

Historical research has not systematically addressed how, given the mixed receptions accorded to a plethora of technological devices, British and German public debate supported technological innovation. Recent work on technology as systems has placed artifacts in specific social and economic contexts by identifying which actors have shaped engineering solutions and applications, as well as, more rarely, by illustrating how these solutions and applications expressed their creators' values. Focusing on the actors who shaped innovations, the approach of studying technologies as systems achieves great insights when it details the choices that lead to the adoption of certain designs and uses in engineering, but it also has a blind spot: it tacitly regards general social and cultural support for innovations beyond entrepreneurial and engineering circles as a fact that requires little explanation in itself.[4] Whereas the "systems approach" tends to examine specific engineering cultures, this book concentrates on the public cultures that arose from debates among technological laypersons upon whose consent technological change was contingent. The question posed here is different and, indeed, more fundamental: why and how did British and German societies foster a cultural climate conducive to innovation processes despite considerable public insecurity about technology between 1890 and 1945?

---

[4] Prominent examples include Thomas P. Hughes, *Networks of Power: Electrification in Western Society* (Baltimore, 1983); David E. Nye, *Electrifying America: Social Meanings of a New Technology, 1880–1940* (Cambridge, MA, 1990); Donald MacKenzie, *Inventing Accuracy: A Historical Sociology of Nuclear Missile Guidance* (Cambridge, MA, 1990); Gabrielle Hecht, *The Radiance of France: Nuclear Power and National Identity after World War II* (Cambridge, MA, 1998); Bill Luckin, *Questions of Power: Electricity and the Environment in Interwar Britain* (Manchester, 1990).

Historians of Britain have recently rejected claims that the dominance of purportedly "rural" and "gentlemanly" ideals predisposed British culture against technology and industry with grave results for the country's economic performance in the twentieth century.[5] British society, the argument runs, promoted technological innovation through military research, by training large numbers of graduates in engineering subjects and, most relevant to the topic under consideration, by fostering strong enthusiasm for new machines in general. Furthermore, ideas of scientific and technological superiority pervaded imperialist thought and played a crucial role in defining notions of British modernity.[6] Studies in design history have also alerted us to the modernist aesthetic impulses that often openly celebrated technology as an artistic and cultural inspiration.[7] While recent work has stressed Britain's overwhelmingly pro-technological cultural disposition, little work has analyzed in detail how fascination and fear interacted in the public promotion of novel artifacts. Patrick Wright's book on the cultural reception of the tank is a rare exception, as is Dan LeMahieu's examination of the reactions of the British educated elite to the development of a commercial gramophone, radio, and cinema culture.[8] Current scholarship, whether it deals with technology implicitly or explicitly, has often foregrounded innovative dimensions in British society, politics, economy, and culture in the first half the twentieth century, thereby touching on a developing field of inquiry: the study of British modernity, long thought of as a contradiction in terms.[9]

---

[5] Classic statements can be found in Martin J. Wiener, *English Culture and the Decline of the Industrial Spirit, 1850–1950* (Cambridge, 1981); Correlli Barnett, *The Collapse of British Power* (Gloucester, 1984), 19–71; Paul Warwick, "Did Britain Change? An Inquiry into the Causes of National Decline," *Journal of Contemporary History* 20 (1985), 99–134. James Winter has examined early environmentalism without reference to the debate about decline; see James Winter, *Secure from Rash Assault: Sustaining the Victorian Environment* (Berkeley, 1999).

[6] A measured inquiry is that of Michael Dintenfass, *The Decline of Industrial Britain, 1870–1980* (London, 1992). A brief list of critiques of works portraying Britain as anti-technological should include W. D. Rubinstein, *Capitalism, Culture and Decline in Britain, 1750–1990* (London, 1994); David Edgerton, *England and the Aeroplane: An Essay on a Militant and Technological Nation* (Basingstoke, 1991); idem, *Science, Technology and British Industrial "Decline"* (Cambridge, 1996); Peter Mandler, "Against 'Englishness': English Culture and the Limits of Rural Nostalgia, 1850–1940," *Transactions of the Royal Historical Society*, sixth series, 7 (1997), 155–75; Michael Adas, *Machines as the Measure of Man: Science, Technology, and Ideologies of Western Dominance* (Ithaca, 1990).

[7] Michael Saler, *The Avant-Garde in Interwar England: Medieval Modernism and the London Underground* (New York, 1999); James Peto and Donna Loveday (eds.), *Modern Britain, 1929–1939* (London, 1999); Ian Carter, *Railways and Culture in Britain: The Epitome of Modernity* (Manchester, 2001).

[8] Patrick Wright, *Tank: The Progress of a Monstrous War Machine* (London, 2000); LeMahieu, *A Culture for Democracy*. See also Sean O'Connell, *The Car and British Society: Class, Gender and Motoring, 1896–1939* (Manchester, 1998).

[9] See the following recent works: Mica Nava and Alan O'Shea (eds.), *Modern Times: Reflections on a Century of English Modernity* (London, 1996); Becky Conekin, Frank Mort, and Chris Waters (eds.),

## Introduction

In German history, inquiries into public evaluations are embedded in wider examinations of the ways anti- and pro-modern sentiments contributed to the rise, consolidation, and consequences of National Socialism. Recent work has extensively critiqued research that focused on an unfolding narrative of technophobia, claiming that hostility to technology played into the hands of National Socialism with its purportedly anti-modern, agrarian, and anti-urbanist ideology.[10] Jeffrey Herf was among the first to begin a theoretical reorientation by identifying a "reactionary modernism," during the Weimar Republic and in National Socialism, which combined anti-democratic politics with militaristic and productivist notions.[11] Because Herf's focus on intellectual history rendered it difficult to support some of his wide-ranging conceptual claims empirically, other historians have taken up related lines of inquiry in examinations of debates about industrial rationalization in early twentieth-century Germany, metaphorical descriptions of the human body as a machine, enthusiasm for aviation and military equipment, and widespread demands for an ethos of *Sachlichkeit* (sobriety or matter-of-factness) that celebrated the functionalism of the machine as an exemplary model for individual conduct in the interwar years.[12] These approaches, which have demonstrated the pervasiveness of technology in contemporary thought and culture, complement scholarship that emphasizes that the National Socialists strove to bring into existence

---

*Moments of Modernity: Reconstructing Britain, 1945–1964* (London, 1999); Martin J. Daunton and Bernhard Rieger (eds.), *Meanings of Modernity: Britain from the Late-Victorian Era to World War II* (Oxford, 2001).

[10] Rolf-Peter Sieferle, *Fortschrittsfeinde: Opposition gegen Technik und Industrie von der Romantik bis zur Gegenwart* (Munich, 1984).

[11] Jeffrey Herf, *Reactionary Modernism: Technology, Culture and Politics in Weimar and the Third Reich* (Cambridge, 1984). For an incisive critique of this work see Anson Rabinbach, "Nationalsozialismus und Moderne: Zur Technik-Interpretation im Dritten Reich," in Wolfgang Emmerich and Carl Wege (eds.), *Der Technik-Diskurs in der Hitler-Stalin-Ära* (Stuttgart, 1995), 94–113.

[12] Mary Nolan, *Visions of Modernity: American Business and the Modernization of Germany* (New York, 1994); Thomas Rohkrämer, "Antimodernism, Reactionary Modernism and National Socialism: Technocratic Tendencies in Germany, 1890–1945," *Contemporary European History* 8 (1999), 29–50; Michael T. Allen, "The Puzzle of Nazi Modernism: Modern Technology and Ideological Consensus in an SS-Factory at Auschwitz," *Technology and Culture* 37 (1996), 527–71; Anson Rabinbach, *The Human Motor: Energy, Fatigue, and the Origins of Modernity* (Berkeley, 1992); Peter Fritzsche, *A Nation of Fliers: German Aviation and the Popular Imagination* (Cambridge, MA, 1992); Monika Renneberg and Mark Walker (eds.), *Science, Technology and National Socialism* (Cambridge, 1994); Frank Trommler, "The Creation of a Culture of *Sachlichkeit*," in Geoff Eley (ed.), *Society, Culture and the State in Germany, 1870–1930* (Ann Arbor, 1996), 465–85; Helmut Lethen, *Cool Conduct: The Culture of Distance in Weimar Germany* (Berkeley, 2001). Some recent work still retains a fairly narrow focus on intellectual history. Thomas Rohkrämer, *Eine andere Moderne? Zivilisationskritik, Natur und Technik in Deutschland 1880–1933* (Paderborn, 1999); Mikael Hård, "German Regulation: The Integration of Modern Technology into National Culture," in Mikael Hård and Andrew Jamison (eds.), *The Intellectual Appropriation of Technology: Discourses on Modernity, 1900–1939* (Cambridge, MA, 1998), 33–67.

an "alternative modernity" that blended a disdain for rationalism, individualism, parliamentarianism, and feminism with a fascination for mass politics, social engineering, pronatalism, eugenics, productivism, and technology.[13] These studies have greatly enhanced our understanding of how the Nazi movement appropriated and reshaped enthusiasm for technology so as to cast itself as visibly "modern." Still, it remains unclear why a passionate fascination for technology existed in Germany in the first place, despite deep anxieties about the "modern age" in general and prominent critiques of technological innovation in particular since the late nineteenth century.

The distinct research agendas in British and German history have produced investigations that focus on different aspects of national debates about technological change. Although this contrast renders it necessary to launch a comparison of the public meanings of technology in Britain and Germany from two distinct platforms, studies of both countries have shared an interest in the complexities and ambiguities of modernity. As scholarship on modernity provides a crucial theoretical framework for this study, it is essential to define how this inquiry comprehends this concept to avoid confusion that can arise from several sources. To begin with, theories of "modernity" have been used in many analytical contexts, a practice that has frequently endowed the term with contradictory meanings. Furthermore, affirming "the 'modernity' of this or that historical phenomenon," as Fredric Jameson has recently observed, often awakens "a feeling of intensity and energy that is greatly in excess of the attention we generally bring to interesting events or monuments in the past." Considerations of modernity are prone to fuel contentious debate because they tend to touch on highly politicized assumptions about the character of the present.[14] Although figuring simultaneously as a multifaceted analytical category and as a polemical term, modernity does not have to prove a treacherous concept on which to build a comparison – quite the contrary.

For a start, modernity must be distinguished from artistic modernism, a term that describes the anti-academic, innovative, initially iconoclastic,

---

[13] For some recent scholarship along these lines, see Kees Gispen, *Poems in Steel: National Socialism and the Politics of Inventing from Weimar to Bonn* (New York, 2002); Margit Szöllösi-Janze (ed.), *Science in the Third Reich* (Oxford, 2001); Paul Weindling, *Health, Race and German Politics Between National Unification and National Socialism* (Cambridge, 1989); Erhard Schütz and Eckhard Gruber, *Mythos Reichsautobahn: Bau und Inszenierung der 'Straßen des Führers', 1933–1941* (Berlin, 1996); Ulrich Herbert, *Best: Biographische Studien über Radikalismus, Weltanschauung und Vernunft* (Bonn, 1996); Gabriele Czarnowski, *Das kontrollierte Paar: Ehe- und Sexualpolitik im Nationalsozialismus* (Weinheim, 1991).

[14] Fredric Jameson, *A Singular Modernity: Essay on the Ontology of the Present* (London, 2002), 35.

and stylistically heterogeneous aesthetic movements that sprang up with increasing frequency towards the end of the nineteenth century.[15] In our context, artistic modernism presents a concern only to the extent that it left an imprint on wider public debates about film, aviation, and passenger shipping. Furthermore, this investigation, while focusing on public attitudes to technological modernity, does not claim to contribute to longstanding scholarly exchanges in British and German historiography about the impact of cultural trends on long-term modernization, be it in the form of economic performance, social change, or democratization.[16] In German history, studies of modernization have not only sought to assess to what extent the National Socialist regime succeeded in systematically transforming the country's economy and society after 1933; they have also given rise to heated exchanges over the question whether these structural changes accidentally laid the economic foundations for the success of democracy in the Federal Republic after 1945.[17] While scholarship on Britain has been more hesitant to embrace modernization theory, it has highlighted a plethora of dynamic socio-economic developments to call into question assertions that the country headed for inevitable national decline from the 1890s.[18] Although these studies have invalidated assertions that Germany and Britain suffered from vastly dissimilar forms of traditionalism in the first half of the twentieth century, their claims tend to rest on problematic foundations. To begin with, it has often proved difficult to pin down causal links between cultural attitudes and the patterns of social, political,

---

[15] For a general evaluation of recent scholarship on modernism from a historian's perspective, see Robert Wohl, "Heart of Darkness: Modernism and Its Historians," *Journal of Modern History* 74 (2002), 573–621. On modernism in England, see Stella K. Tillyard, *The Impact of Modernism, 1900–1920: Early Modernism and the Arts and Crafts Movement in Edwardian England* (London, 1988); Peter Stansky, *On or About December 1910: Early Bloomsbury and its Intimate World* (Cambridge, MA, 1996); J. B. Bullen (ed.), *Post-Impressionists in England* (London, 1988). Classic German studies include Thomas Nipperdey, *Wie das Bürgertum die Moderne fand* (Berlin, 1988); Peter Paret, *The Berlin Secession: Modernism and its Enemies in Imperial Germany* (Cambridge, MA, 1980).
[16] Classics are Walt W. Rostow, *The Stages of Economic Growth* (Cambridge, 1960); Talcott Parsons, *The System of Modern Societies* (Englewood Cliffs, 1971).
[17] In German history, Ralf Dahrendorf and David Schoenbaum initiated a re-evaluation of the years between 1933 and 1945 by arguing that the Nazis unintentionally effected a modernization. See Ralf Dahrendorf, *Society and Democracy in Germany* (Garden City, 1969), 381–96; David Schoenbaum, *Hitler's Social Revolution: Class and Status in Nazi Germany, 1933–1939* (Garden City, 1966). For a measured intervention, see Axel Schildt, "NS-Regime, Modernisierung und Moderne: Anmerkungen zur Hochkunjunktur einer andauernden Diskussion," *Tel Aviver Jahrbuch für Geschichte* 23 (1994), 3–22.
[18] On Britain, see the literature in footnote 5 (p. 6) and Barry Supple's classic lecture: Barry Supple, "Fear of Failing: Economic History and the Decline of Britain," in Peter Clarke and Clive Trebilcock (eds.), *Understanding Decline: Perceptions and Realities of British Economic Performance* (Cambridge, 1997), 9–29.

and economic behavior that serve as indicators of modernization. More importantly in our context, research on modernization frequently operates with normative concepts of modernity that prioritize one model over another and thereby risk obscuring alternative interpretations of the modern that coexist at a given point in time.[19]

Since a wide range of phenomena struck Britons and Germans from many political backgrounds as "modern," normative concepts cannot capture the rich inconsistencies that shaped public debate about modernity. In fact, since the late nineteenth century it was semantic breadth rather than analytical precision that characterized the use of the term "modern" in the public sphere from the political Right to the Left. Virtually omnipresent in the public arena, the epithet "modern" provided an elastic category to celebrate and denounce a plethora of transformations and transitions in British and German society. Consequently, a vast number of interpretations of modernity were in public circulation at any given moment. While semantic flexibility contributed to the category's public ubiquity, and while everyday use often rendered the term vague, it still possessed a solid core meaning. Most importantly, the word "modern" captured the widespread conviction that the historical present was first and foremost an era of profound, irreversible, and man-made changes. In industrialized turn-of-the-century Europe, public calls for a straightforward return to the past had an increasingly unrealistic and anachronistic ring, because fewer and fewer people believed that the past provided recipes for present and future problems. As a consequence, history, as Jose Harris has written, became a "lost domain," and commentators across the political and cultural spectrum proclaimed that Europe had entered a new, unique historical era: "modern times."[20]

Of course, this sense of living in a novel age was not ahistorical. A host of historical tales threw into sharp relief the exceptional character of the modern age, no matter whether they stipulated a fundamental rupture between the present and the past, explained the outstanding features of the present as the culmination of continuous, incremental change, or construed heroic

---

[19] Critiques of normative concepts of modernity include Peter Fritzsche, "Nazi Modern," *Modernism/Modernity* 3 (1996), 1–22; Michael T. Allen, "Modernity, the Holocaust, and Machines Without History," in Michael T. Allen and Gabrielle Hecht (eds.), *Technologies of Power: Essays in Honor of Thomas Parke Hughes and Agatha Chipley Hughes* (Cambridge, MA, 2001), 175–214. It should be noted in this context that normative concepts of modernity underpin the arguments of both those who deny and those who emphasize the modernizing effect of the National Socialist regime. Influential examples are Rainer Zitelmann, "Die totalitäre Seite der Moderne," in Michael Prinz and Rainer Zitelmann (eds.), *Nationalsozialismus und Modernisierung* (Darmstadt, 1991), 1–20; Hans Mommsen, "Nationalsozialismus als vorgetäuschte Modernisierung," in Walter H. Pehle (ed.), *Der historische Ort des Nationalsozialismus* (Frankfurt, 1990), 31–46.

[20] Jose Harris, *Private Lives, Public Spirit: Britain, 1870–1914* (Harmondsworth, 1994), 36.

or tragic analogies between ancient mythology and the historical present.[21] The proliferation of historical narratives indicates that many contemporaries considered modernity as highly problematical because, as a time of transformation, it appeared unstable. The widespread sense of the instability of modern times hinged on conceptualizations of change. While many commentators emphasized the creative and positive dimensions of innovation, they were equally conscious of its disruptive aspects. In order to create the new, the old or traditional had to be displaced and often destroyed. From this perspective, public debate about technological change can be read as emblematic of contemporary ambivalence about the modern that resulted from perceptions of change as both creative and destructive.[22] After all, technological innovation added to the environment novel artifacts that signaled an end to previous eras. Technological change appeared literally to substantiate change and to lend physical expression to enunciations of "progress" and "decay." As was the case with architecture, technological artifacts objectified modernity in many contemporaries' eyes. As public debate about technology revolved around the key premise that radical change constituted modernity, there existed what at first sight may be viewed as a discrepancy between, on the one hand, the fast pace of technological developments and, on the other, a considerable degree of continuity in discussions that reflected upon these developments. Although they did not engender a sense of temporal stability among contemporaries, public debates displayed important structural continuities that played a crucial role in the embrace of technological innovation: they furnished widely applicable and persistent patterns of perception, thereby easing the burden of reflection in the face of frequently unsettling transformations. Despite the longevity of certain ideas, debates about technology were by no means static, and we shall have to pay particular attention to the turning points in assessments of the three technologies in question.

Given their prominence in Britain and Germany between the 1890s and World War II, public interpretations of modernity provide the conceptual framework for this comparison. While scholars have primarily emphasized the contrasts between these nations during this period, because

---

[21] Recent work has taken up Reinhart Koselleck's inquiries into the notions of temporality that texture attitudes towards modernity. Rudy Koshar, *From Monuments to Traces: German Artifacts of Memory, 1870–1990* (Berkeley, 2000); idem, *Germany's Transient Pasts: Preservation and National Memory in the Twentieth Century* (Chapel Hill, 1998); Reinhart Koselleck, *Futures Past: On the Semantics of Historical Time* (Cambridge, MA, 1985); idem, *Zeitschichten: Studien zur Historik* (Frankfurt, 2000); Matt K. Matsuda, *The Memory of the Modern* (New York, 1996).

[22] The origins of ambivalence in the simultaneously destructive and creative sides of change are discussed in David Harvey, *The Condition of Postmodernity: An Enquiry in the Origins of Cultural Change* (Oxford, 1989), 2–65; Marshall Berman, *All That Is Solid Melts Into Air: The Experience of Modernity* (London, 1983), 37–85.

of their diverging political histories, this book adopts a perspective that identifies both transnational similarities and national differences.[23] Britain and Germany are particularly suited to a comparative study of attitudes towards technology since each country regarded the other as a technological competitor and was consequently involved in a contemporary process of perception of the other. The British and German publics often reacted towards new technologies in similar ways, which highlight transnational *cultural* patterns that promoted innovation in a politically heterogeneous Europe. These shared cultural traits existed alongside divergent, primarily *political* evaluations of technology's significance for each nation that, while becoming most pronounced after 1933, pre-dated National Socialism's ascent to power. The comparative approach reveals how, in conjunction with wider Western European cultural aspects of public debate, liberal and illiberal political ideologies advanced technological change in Britain and Germany.[24] By extending its scope beyond 1933, this project complements the insightful comparisons of anti-democratic regimes during the interwar years as it locates National Socialist attitudes towards technological innovation within wider European contexts.[25] Rather than questioning the obvious political contrasts between interwar dictatorships and democracies, this comparison reveals how differing political surroundings could give rise to British and German celebrations of technological modernity after 1933 that can appear deceptively alike at first sight. This comparison, then, aims to identify transnational Western European and nationally specific dimensions in British and German public attitudes to technological modernity from the late nineteenth century to the Second World War.

---

[23] For considerations of how to combine contrasting and transnational patterns in studies of European history, see Johannes Paulmann, "Internationaler Vergleich und interkultureller Transfer: Zwei Forschungsansätze zur europäischen Geschichte des 18. bis 20. Jahrhunderts," *Historische Zeitschrift* 267 (1998), 649–85. More conventional are the thoughts expressed in Heinz-Gerhard Haupt and Jürgen Kocka, "Historischer Vergleich: Methoden, Aufgaben, Probleme: Eine Einleitung," in *idem* (eds.), *Geschichte und Vergleich: Ansätze und Ergebnisse international vergleichender Geschichtsschreibung* (Frankfurt, 1996), 9–45. An incisive exploration of differences is John Breuilly, "National Peculiarities?" in *Labour and Liberalism in Nineteenth-Century Europe: Essays in Comparative History* (Manchester, 1992), 273–295. Similarities in labor history are emphasized in Stefan Berger, *The British Labour Party and the German Social Democrats, 1900–1931* (Oxford, 1994).

[24] David Blackbourn and Geoff Eley established the argument that illiberal political cultures could generate modernity. See David Blackbourn and Geoff Eley, *The Peculiarities of German History: Bourgeois Society and Politics in Nineteenth-Century Germany* (Oxford, 1984).

[25] See Stanley G. Payne, *A History of Fascism* (Madison, 1995); Ian Kershaw and Moshe Levin (eds.), *Stalinism and Nazism: Dictatorships in Comparison* (Cambridge, 1997); Richard Bessel (ed.), *Fascist Italy and Nazi Germany: Comparisons and Contrasts* (Cambridge, 1996).

Public debates about technological innovation occurred in a wide range of media. The national daily press, parliamentary records, bestselling autobiographical books and articles, advertising material, periodicals, radio broadcasts, postcards, films, and photographs: all were sites for the production and dispersal of popular knowledge about new technologies. Despite their heterogeneous nature, these sources share the crucial characteristic of primarily addressing technological laypersons, who lacked the detailed expertise that familiarized scientists and engineers with new technological trends. The texts under consideration ascribed public meanings to spectacular innovations and thereby rendered the material transformation of the external world intelligible. Since these writings primarily addressed technological laypersons, they only rarely considered the complexities of technologies as systems. Rather, they focused on new objects, specific events, and individual heroes – discrete subjects that lent themselves readily to breathtaking dramatizations with artifacts in starring roles.

British and German societies grappled with technological change in public spheres that underwent dynamic transformation between the 1890s and World War II.[26] This period witnessed a reshaping of both countries' press landscape because of an explosion of book publications as well as a significant expansion and differentiation of print media as new political dailies and tabloids with circulations sometimes exceeding one million copies came into existence.[27] In the interwar years, the radio shed its status as a cultural oddity and became a popular medium that reached listeners from all walks of life.[28] The ascent of film had the most profound effect on the European

[26] Habermas's original treatment has been significantly modified by now. See Jürgen Habermas, *The Structural Transformation of the Public Sphere: An Inquiry into a Category of Civil Society* (London, 1989). Important modifications include Geoff Eley, "Nations, Publics, and Political Cultures: Placing Habermas in the Nineteenth Century," in Nicholas B. Dirks, Geoff Eley, and Sherry B. Ortner (eds.), *Culture/Power/History: A Reader in Contemporary Social Theory* (Princeton, 1994), 297–335; Margaret R. Somers, "What's Political or Cultural about Political Culture and the Public Sphere? Toward an Historical Sociology of Concept Formation," *Sociological Theory* 13 (1995), 113–44; Nancy Fraser, "Politics, Culture, and the Public Sphere: Toward a Postmodern Conception," in Linda Nicholson and Steven Seidman (eds.), *Social Postmodernism: Beyond Identity Politics* (Cambridge, 1995), 287–302.

[27] Systematic research on the transformation of the German press landscape is still rare. See Peter Fritzsche, *Reading Berlin 1900* (Cambridge, MA, 1996), 12–86; Karl Christian Führer, Knut Hickethier, and Axel Schildt, "Öffentlichkeit-Medien-Geschichte: Konzepte der modernen Öffentlichkeit und Zugänge zu ihrer Erforschung," *Archiv für Sozialgeschichte* 41 (2001), 1–38, esp. 32–4. Studies of British developments include LeMahieu, *A Culture for Democracy*, 17–55; James Curran and Jean Seaton, *Power Without Responsibility: The Press and Broadcasting in Britain* (London, 1981).

[28] Karl Christian Führer, "A Medium of Modernity? Broadcasting in Weimar Germany, 1923–1932," *Journal of Modern History* 69 (1997), 722–53, esp. 732–6; Inge Marssolek, "Radio in Deutschland 1923–1960: Zur Sozialgeschichte eines Mediums," *Geschichte und Gesellschaft* 27 (2001), 207–39; Asa Briggs, *The History of Broadcasting in the United Kingdom*, 5 vols. (Oxford, 1961–1995).

media scene. Within twenty years of its invention in the mid-1890s, the movie had given rise to an international industry with multinational production companies and a devoted clientele numbering tens of millions.[29] Film contributed to the proliferation of images in everyday life, a trend that was reinforced by the spread of photography, photojournalism, and a variety of novel printing techniques. In addition to serving as illustrations of texts, pictorial representations developed visual languages for notions of technology that were difficult to verbalize. The unprecedented saturation of the British and German public spheres with pictorial images renders it necessary to evaluate how pictorial material interacted with textual narratives in debates about technological change.[30]

The media allowed a broad range of participants to shape British and German public discussions about innovations. Often locked in fierce competition with one another, newspapers vied for readers and scoops through sensationalistic coverage of technological trends. In a media landscape brimming with exciting stories about innovation, publicity campaigns launched by the various industries achieved uneven results. While shipping firms were able to draw on large resources to mount effective public-relations campaigns, financial constraints hampered many initiatives taken by the early aviation industries. As the rapidly expanding film sector invested large sums in advertising, it battled with entrenched criticism that stressed the medium's moral and political dangers throughout the period. Interventions from speakers unaligned with industries supported or challenged the representations that emerged from public-relations departments. Furthermore, prominent political fault lines left their mark on public debate about technological change as adherents of varying political creeds struggled to lay exclusive ideological claim to individual artifacts in parliament and press. In short, debates about innovations often reflected and fueled political, social, and cultural conflicts in both countries, thereby casting novel artifacts as catalysts of public contestation. Of course, power relations between

---

[29] Miriam Hansen, "Early Cinema: Whose Public Sphere?" in Richard Abel (ed.), *Silent Film* (London, 1996), 228–44; Martin Loiperdinger, "The Kaiser's Cinema: An Archaeology of Attitudes and Audiences," in *A Second Life: German Cinema's First Decades* (Amsterdam, 1996), 41–50; Rachel Low, *The History of the British Film, 1906–1914*, 2nd edition (London, 1973), 25–6.

[30] On photojournalism, see Bernd Weise, "Pressefotografie I: Die Anfänge in Deutschland," *Fotogeschichte* 31 (1989), 15–40; idem, "Pressefotografie II: Fortschritte der Fotografie- und Drucktechnik und Veränderungen des Pressemarktes im Deutschen Kaiserreich," *Fotogeschichte* 33 (1989), 27–61; idem, "Pressefotografie III: Das Geschäft mit dem aktuellen Foto," *Fotogeschichte* 37 (1990), 13–36. Art historians have called for an integration of visual and textual sources in historical work. See Horst Bredekamp, *Antikensehnsucht und Maschinenglauben: Die Geschichte der Kunstkammer und die Zukunft der Kunstgeschichte* (Berlin, 1993), 16–17; Barbara Maria Stafford, "Introduction: Visual Pragmatism for a Virtual World," in *Good Looking: Essays on the Virtue of Images* (Cambridge, MA, 1997), 2–18; James Elkins, *The Object Stares Back: On the Nature of Seeing* (San Diego, 1997).

participants in discussions about technological transformations were never balanced, and it is necessary to pay attention to the mechanisms that ensured the dominance or marginalization of specific contributions. The aim of the book is to determine what could, and what could not, be said in public about technology in Britain and Germany between the 1890s and World War II, and how this interplay of contestations fostered a climate that pushed technological developments forward.

As much as contestation played an integral part in public debate about innovation, British and German negotiations of technology's meanings were anything but unstructured. Public debate shaped "complex field[s] of discourse" that, despite their irregularities, were patterned in several respects.[31] For a start, even when opponents were locked in argumentative exchanges, they had to take each other's views into account. As a consequence, various contributions to public debate "infected" one another, thereby establishing connections between conflicting positions.[32] Public controversies revolved around several main themes that provided agendas for debates about aviation, passenger shipping, and film. Even if discussants fervently disagreed with each other while addressing a topic, they often agreed on the topic's relevance in the first place. More than anything else, the dominant motifs of public debate about the technologies under consideration reveal a set of contemporary preoccupations that transcended the profound ideological divides responsible for the fervent civil and military strife in a host of contexts in the first half of the twentieth century. A cacophonous chorus, therefore, created the "basic social consensus" about the desirability of technological change that historians of technology have noted and that needs to be considered in relation to the increasing rather than decreasing socio-political contestation especially in interwar Britain and Germany.[33] What, then, were the main themes that framed British and German public debate about technological change?

---

[31] Michel Foucault, *The Archaeology of Knowledge* (New York, 1972), 23. Historians have theorized discourse analysis extensively. See Philipp Sarasin, "Subjekte, Diskurse, Körper: Überlegungen zu einer diskursanalytischen Kulturgeschichte," in Wolfgang Hardtwig and Hans-Ulrich Wehler (eds.), *Kulturgeschichte Heute* (Göttingen, 1996), 131–64; Peter Schöttler, "Mentalities, Ideologies, Discourses: On the 'Third Level' as a Theme in Social-Historical Analysis," in *The History of Everyday Life: Reconstituting Experiences and Ways of Life* (Princeton, 1995), 72–115; Gareth Stedman Jones, "The Determinist Fix: Some Obstacles to the Further Development of the Linguistic Approach to History in the 1990s," *History Workshop Journal* 42 (1996), 19–35.

[32] On "infection" between discourse and counter-discourse, see Richard Terdiman, *Discourse/Counter-Discourse: The Theory and Practice of Symbolic Resistance in Nineteenth-Century France* (Ithaca, 1985), 13–14.

[33] Wolfgang König and Wolfhard Weber, *Netzwerke, Stahl und Strom: 1840 bis 1914* (Berlin, 1997), 539. Joachim Radkau, *Technik in Deutschland: Vom 18. Jahrhundert bis zur Gegenwart* (Frankfurt, 1989), 199–221.

As already mentioned, Britons and Germans confronted flight, transatlantic passenger liners, and film with ambivalence because they lacked the expertise to understand the working of these mechanisms. The formula describing innovations as "modern wonders" formed the core of debates about all three technologies and lent expression to a fascination with, and fears about, the incomprehensibility of technology itself as well as the unpredictable changes implicit in technological advancement. While debate about technology oscillated between admiration and anxiety, it generated a specific form of ambivalence that gave voice to unease without necessarily endangering the acceptance of further change. Rather than as evidence for prominent German and British anti-modern sentiments, insecurity and disorientation in the face of change should be viewed as integral elements of debates about new technologies in both nations.

Important reasons why insecurity about new mechanisms did not develop into full-blown opposition emerge from the discussions, in both Britain and Germany, of the risks of technology. Chilling reports about airship explosions, shipwrecks, and aeroplane crashes confirmed latent fears of the physical dangers aerial and maritime technology posed. Contemporaries found it particularly disconcerting that the risks of flight and shipping technologies often escaped definitive explanations when inquiries failed to identify the causes of disasters that cost numerous lives. Cinematic technology, too, seemed risky, posing cultural if not physical dangers. Confusion about film's perils sprang from an inability to account for the seductiveness of the "illusions" it exhibited on the screen. As camera and projector appeared to obliterate the distinction between "fact" and "fiction," a host of critics became concerned about the effects of films on the moral and political behavior of the "mass" audiences flocking to the cinemas. Public debates about the physical, moral, and cultural perils of technology often generated strategies for containing and transforming anxieties about the consequences of innovation. Instead of misreading debates about risk as expressions of hostility to technology, this book accounts for strategies of risk containment and robust levels of risk acceptance in both countries at the time.

If ambivalence and risk awareness may well have signaled mixed feelings about new technologies, the fervent ardor with which Britons and Germans welcomed many innovative devices illustrates better than anything how strongly the public admired technology in this period. Powerful social fantasies, or dreams revealing desires that were shared across political and social divides, generated enthusiasm that rallied broad sections of German and British society behind technological progress. Observers were aware of the importance of amorphous desires that were present in this enthusiasm for technology. After all, the *Manchester Guardian* reporter who admired the *Queen Mary* so eloquently at the outset of this chapter called the ship a "realised dream." In the context of all three technologies under consideration, artifacts or their operators brought to the surface intangible yet potent wishes that often remained submerged in everyday life.[34] Crucially, since they straddled political and social camps, many social fantasies channeled support behind innovations from sections of the public that opposed one another on a host of contemporary issues. In part, social fantasies mobilized broad support because they were only rarely articulated in a clear, authoritative manner. Far from being apolitical, social fantasies were open to competing political interpretations, and innovative technologies often served as projection surfaces for similar fantasies with different political inflections. In the 1920s and 1930s, heroic pilots – male and female – rose to prominence because solo fliers faced and overcame lethal perils on their perilous long-distance flights. At a time when both countries still mourned the millions of victims of World War I, the icon of the aviator who voluntarily risked his life engaged an obsession with violent death in Britain and Germany. It was a widespread desire for transcendence that helped pilots to soar to unprecedented fame, thereby creating a fascination with dangerous aeroplanes that promoted aviation in general. Actively encouraged by the German and British shipping industries, the fantasy of the "floating palace" established a glamorizing and sanitized vision as the dominant representation of ships as sites of luxury consumption and leisure, thereby tapping into widespread yearnings for a carefree, comfortable life. Film technology, for its part, drew on idealizing notions of domesticity as producers of amateur equipment claimed that home-movie making strengthened the family as a core institution. In the last resort, many movie amateurs acted out fantasies of individual autonomy. Innovative artifacts as embodiments of social fantasies bring into view the manifold roles gender notions played in the public promotion of technology.[35] While innovations corresponded

---

[34] I employ the concept of "social fantasy" without the Lacanian references so prominent in Slavoj Žižek's work, from which I borrow the term. See Slavoj Žižek, *The Plague of Fantasies* (London, 1997), 3–44; idem, *The Sublime Object of Ideology* (London, 1989), 30–3, 87–9.

[35] Historians of the United States are leading the study of gender and technology. Nina E. Lerman, Ruth Oldenziel, and Arwen Palmer Mohun, "The Shoulders We Stand on and the View From Here: Historiography and Directions for Research," *Technology and Culture* 38 (1997), 9–30; Roger Horowitz (ed.), *Boys and Their Toys? Masculinity, Technology, and Class in America* (New York, 2001); Roger Horowitz and Arwen P. Mohun (eds.), *His and Hers: Gender, Consumption, and Technology* (Charlottesville, 1998); Virginia Scharff, *Taking the Wheel: Women and the Coming of the Motor Age* (New York, 1991).

with existing frameworks of masculinity and femininity in some contexts, they could also stretch and explode conventional gender models, giving rise to confusion as public depictions showed men and women grappling with the demands and opportunities brought about by new artifacts.

Finally, the conviction that national self-assertion depended on technological leadership engendered a multitude of arguments in favor of technological change in both countries. As acknowledgments of technology's national significance marginalized unease about mechanical artifacts and, indeed, highlighted their promise, they were also responsible for the stark political contrasts in German and British debates about technology that became particularly pronounced during the interwar years. At that time, British observers frequently considered technologies as instruments to stabilize an international *status quo* favorable to their nation, while Germans viewed products of engineering as tools to transform the international environment that stifled their political ambitions. Moreover, highly dissimilar accounts about technology and military violence accompanied the emergence of aerial warfare in Britain and Germany after 1914. Of course, these contrasts became particularly crass after the Nazis' rise to power in 1933. The marked divergence between German and British nationally minded pride in technology after 1918 brings into focus the impact of the First World War on technology's political meanings and allows us to evaluate the ways in which the Great War represented a turning point in public debate about technology.[36]

This book is organized thematically rather than chronologically, and moves from an examination of insecurity and risk awareness to the cultural and political sources of public enthusiasm to explain how public debate created cultural environments conducive to technological innovation in Britain and Germany between the late nineteenth century and World War II. Ambivalence, fears of risk, enthusiasm fueled by social fantasies, and nationally minded fervor – in short, highly emotional responses – textured public debate about technological innovation. Encompassing different levels of intensity ranging from moods via strong feelings to vehement passions, the public affects that underpinned the proliferation of technological artifacts ranged from warm-hearted admiration during festive celebrations

---

[36] In this context, my book touches on wider debates on how the First World War represented a watershed in European history. Two recent influential works that emphasize the profound political and cultural impact of the war are Modris Eksteins, *The Rites of Spring: The Great War and the Birth of the Modern Age* (Boston, 1989); and Mark Mazower, *Dark Continent: Europe's Twentieth Century* (New York, 1998). By contrast, Jay Winter has been influential in drawing attention to continuities between the prewar and interwar eras. See Jay Winter, *Sites of Memory, Sites of Mourning: The Great War in Cultural History* (Cambridge, 1995).

to violent hatred directed at enemies in times of war. Public discourse lent expression to and amplified an emotional landscape that on many occasions was conducive to the acceptance of new products of engineering, but that also signaled limits when contemporaries perceived risks and dangers as overwhelming or uncontrollable. While these responses ascribed cultural and political significance to products of engineering, they could not redress the knowledge problem that arose from the incomprehensibility of many artifacts. In fact, dynamic innovation processes constantly added to the fundamental knowledge problem that underlay public debates about technology by bringing into existence new, esoteric contraptions. While, of course, it would be erroneous to assume that Britons and Germans were completely devoid of any idea why an ocean liner remained afloat, how bulkheads worked, how airships and aeroplanes achieved lift, and that a film camera took a series of still images, technological laypersons' comprehension was bound to encounter limits in the face of increasingly sophisticated mechanisms that struck them as "modern wonders."

CHAPTER 2

# *"Modern wonders": technological innovation and public ambivalence*

Between the 1890s and World War II, new technologies occupied a prominent place in British and German public life. As the introduction of steel into shipbuilding triggered fundamental design changes to give birth to luxurious "floating palaces" in the late nineteenth century, launches of new vessels regularly attracted crowds in excess of 100,000 people, whose applause testified to the march of technological progress.[1] First flights of aeroplanes and airships also drew six-figure numbers of spectators, as did flight shows and air races from the first decade of the twentieth century.[2] Audiences numbering in millions left no doubt about the attraction of the novel technologies that gave rise to the cinema industry and, later, to radio broadcasting. Stoked by intensive industrial public-relations initiatives, a wide range of media amplified the frenzy that accompanied the public arrival of new technological objects. Time and again, these media storms hailed new technologies as stupefying "modern wonders."

The idea that technological innovations amounted to "modern wonders" was by no means new in the late nineteenth century; it had commanded prominence throughout the 1800s, and continued to do so well beyond. Novel engineering feats provoked profound surprise and literally made people wonder what kind of times they were living in throughout the nineteenth and twentieth centuries. "Wonderment," as Philip Fisher has recently emphasized, demarcates the boundary between the known and the

---

[1] On changes in shipbuilding, see Sidney Pollard and Paul Robertson, *The British Shipbuilding Industry, 1870–1914* (Cambridge, MA, 1979), 9–24; Arnold Kludas, *Die Geschichte der deutschen Passagierschiffahrt*, vol. 2: *Expansion auf allen Meeren 1890 bis 1900* (Hamburg, 1987), 181–3. On the public reception of new ships, see *The Times*, 9 September 1909, 10; *B. Z. am Mittag*, 23 May 1912, 1; *Berliner Tageblatt*, 16 August 1928 (morning edition), 4.
[2] Robert Wohl, *A Passion for Wings: Aviation and the Popular Imagination* (New Haven, 1994); Peter Fritzsche, *A Nation of Fliers: German Aviation and the Popular Imagination* (Cambridge, MA, 1992). In Weimar Germany, fascination with flight gave rise to dreams of space flight. See Michael Neufeld, "Weimar Culture and Futuristic Technology: The Rocketry and Spaceflight Fad, 1923–1933," *Technology and Culture* 31 (1990), 725–52.

unknown, and technological innovation dislodged established coordinates of knowledge by launching on to the historical stage previously unknown mechanisms.[3] Proclaiming new technologies as "modern wonders" testified to both the enthusiasm and the anxiety that the appearance of innovations generated: contemporaries, although they admired recent inventions, simultaneously found them beyond their comprehension and thus a worrying source of uncertainty. Historians of technology, while they have frequently noted that innovations triggered strong public reactions, have concentrated on the popular elation that welcomed technology in the late nineteenth and early twentieth centuries, but have stopped short of directing equal attention to the broad range of responses that registered unease and anxiety.[4] Examining the notion of the "modern wonder" allows us to study enthusiasm and insecurity about technology in conjunction, because this formula expressed ambivalent appreciations that were rooted in knowledge problems. Furthermore, classifying technological innovations as "modern miracles" linked technology to wider discussions about the "modern age," which had gained momentum during the nineteenth century. Assertions that technological developments furnished "wonders" contributed to the sense of living in a time of unprecedented modernity as well as a widespread, profound sense of ambivalence.

If, in the late nineteenth and early twentieth centuries, legions of European observers interpreted technology as an indicator of modernity in terms of uncertainty, we need to understand how ambivalence affected public support for new phenomena that struck people as modern at the time. While it is well known that anxieties could lead to a turn against the "modern," especially during the *fin de siècle* and the early twentieth century, public ambivalence could also have a different effect.[5] Between

---

[3] Philip Fisher, *The Vehement Passions* (Princeton, 2002), 1.
[4] This is true even of otherwise excellent inquiries into the public reception of technology. David Nye, for instance, does not reflect on the presence of terror in American understandings of technology as sublime. See David E. Nye, *American Technological Sublime* (Cambridge, MA, 1994). Other influential works that treat contemporary fears in a marginal fashion include Carolyn Marvin, *When Old Technologies Were New: Thinking About Electric Communication in the Late Nineteenth Century* (New York, 1990), 119–22; Cecilia Tichi, *Shifting Gears: Technology, Literature, Culture in Modernist America* (Chapel Hill, 1987), 53–4. An exception to this rule is Patrick Wright, *Tank: The Progress of a Monstrous War Machine* (London, 2000).
[5] The most influential statement of deep-seated anti-modernity in Germany in the late nineteenth century is Fritz Stern, *The Politics of Cultural Despair: A Study in the Rise of Germanic Ideology* (Berkeley, 1961). Recent work has emphasized the reforming impulses in German public debate about modernity at the time. See, as one example, Kevin Repp, *Reformers, Critics, and the Paths of German Modernity: Anti-Politics and the Search for Alternatives, 1890–1914* (Cambridge, MA, 2000). Scholarship about attitudes to technology has frequently identified both pro- and anti-modern sentiments at the time but stopped short of considering ambivalence in a comprehensive manner. In addition to the works cited

the 1890s and the late 1930s British and German discussions about technology as a "modern wonder" shaped a variant of ambivalence that supported, rather than inhibited, public acceptance of further change, thereby contributing to cultural environments conducive to innovation. The intersecting debates about aviation, passenger shipping, and film show how enthusiasm for, and unease about, innovative technology interacted in Britain and Germany between the 1890s and World War II. These discussions bring into view how the British and German publics judged both a changing world of objects and the most controversial media technology at the time. As we shall see, British and German debates before 1933 shared many evaluations of the artifacts that transformed the external world and the novel means of representing these ever-changing environments. Albeit modified by the Nazis in keeping with their ideological concerns, the rhetoric of the "modern wonder" emerges at the center of public debates about all three technologies that transcended national boundaries and lent expression to a pervasive sense of ambivalence throughout the period.

### "MIRACLES" AND "MONSTERS": ADMIRATION AND ANXIETY IN PUBLIC REACTIONS TO INNOVATIONS

Invocations of technological innovations as "modern wonders" were virtually omnipresent in public accounts during the late nineteenth and early twentieth centuries. The Hamburg-American Line and the Cunard Line promoted their liners as "gigantic wonder[s] of modern technology" on countless occasions.[6] Similarly, early British pilots extolled "these amazing modern days of aeroplane progress" and were proud to witness "all their wonder."[7] In the context of film, the formula of the modern miracle was employed to delineate equipment as well as individual movies even decades

---

in chapter 1 above, see works on anti-modern sentiment such as Frank Trentmann, "Civilization and Its Discontents: English Neo-Romanticism and the Transformation of Anti-Modernism in Twentieth-Century Western Culture," *Journal of Contemporary History* 29 (1994), 583–625; Correlli Barnett, *The Collapse of British Power* (Gloucester, 1984), 19–71.

[6] *Imperator auf See: Gedenkblätter an die erste Ausfahrt des Dampfers Imperator am 11. Juni 1913* (Hamburg, 1913), 11. See also Deutsches Museum, Munich, archive, brochure collection (hereafter DMBC), *Dampfer Kaiserin Auguste Victoria* (Hamburg, 1906), 5. For a British example, see University of Liverpool, University Archives (hereafter Cunard archives), D42/PR4/21; E. Keble Chatterton, *Aquitania: The Making of a Mammoth Liner* (1914), 5.

[7] Claude Grahame-White and Harry Harper, *The Aeroplane: Past, Present and Future* (London, 1911), 1. The leading National Socialist party paper described the airship *Hindenburg* in a similar vein as late as 1936. See *Völkischer Beobachter*, 5 March 1936, 1, 8.

after they had burst on to the scene in the 1890s.[8] E. A. Dupont's movie *Atlantic*, a feature from 1929 based on the wreck of the *Titanic*, struck critics as outstanding because it had overcome "formidable technical difficulties" to recreate the atmosphere on board a contemporary ocean steamer.[9] Just like the infamous ship on which a technological disaster plot unfolded, the film represented a "wonder of modern technology."[10] The rhetoric of the "modern wonder" thus not only linked technological change with wider debates about the nature of modernity; its semantic breadth also provided a flexible formula around which commentators could organize interpretations about a vast range of technologies. Which images, then, did the trope of the modern wonder of technology generate?

According to a host of accounts, technology left observers spellbound with pleasure and excitement.[11] In order to capture technologies' enchanting effects, factual reports poeticized and aestheticized innovative objects. On these occasions, writing suspended the prosaic logic of common sense and rationality, thereby elevating technology beyond the seemingly mundane realm of everyday life. Reports referred to flights as "romances" and "fairy tales" with aeroplanes and airships among their heroes.[12] Similarly, British and German ships reaped benefits from press reports that invested boats with an aura of the fairy tale. One journalist for the *Manchester Guardian* was struck by the sight of the *Lusitania*, a ship nearly 800 feet in length that was to recapture the Blue Riband from Germany, as she began her maiden voyage on a September night in 1907. He had found the boat neither "graceful" nor "beautiful" in broad daylight, but nightfall initiated a dazzling metamorphosis. "The hull itself was lost [in the darkness and] the parterres of lights from portholes and windows, every one illuminated, grew more and more brilliant," and the *Lusitania* began to resemble "a fireworks picture in steady light." Shining "red and white" through a "thin mist," the boat appeared to defy the laws of gravity as it embarked on its first voyage:

---

[8] For descriptions of cinematic equipment as a "modern wonder," see Rene Fülöp-Miller, *Die Fantasiemaschine: Eine Saga der Gewinnsucht* (Berlin, 1931), 31; Robert Gaupp and Konrad Lange, *Der Kinematograph als Volksunterhaltungsmittel* (Munich, 1912), 1.

[9] *The Times*, 29 October 1929, 14.

[10] Stiftung Deutsche Kinemathek Berlin, Schriftgutarchiv (hereafter SDKB archive), file 2853 "Atlantic", *Atlantic* (no place, no date), 6.

[11] On the relationships between a sense of wonder and pleasure, see Philip Fisher, *Wonder, the Rainbow, and the Aesthetics of Rare Experiences* (Cambridge, MA, 1998), 2, 20.

[12] Aviation is described as a "romance" in Alan Cobham, *Skyways* (London, 1925), v; B. Wherry Anderson, *The Romance of Air-Fighting* (London, 1917). On aviation as a "fairy tale," see *Luftfahrt voran! Das deutsche Fliegerbuch*, ed. J. B. Malina (Berlin, 1932), 107; *Vorwärts*, 15 May 1930 (morning edition), 9.

"the spectacle took wing and silently floated out to sea."[13] The description of a wondrous liner "in flight" not only blurred the technological differences between ships and aeroplanes, thereby creating the impression that the appearance of distinct new technologies stemmed from one process of innovation; it went so far as to attribute quasi-magical qualities to the vessel.

Film also appeared to possess miraculous properties. Cameramen were referred to as "modern magicians"; contemporaries were struck by films' ability to "change [audiences' moods] as if by a magic wand" and to leave patrons "vibrating with emotions" as they entered the fictional world of feature presentations.[14] One article in the *Daily Herald* even credited a production with the ability to work a wonder in the most literal sense. Under the headline "Talkie Cure for Deafness," the report told of a seventy-three-year-old woman who, attending a screening of the 1929 sound film *Singing Fool*, "heard two loud cracks in [her] head, and suddenly . . . found [she] could hear again" after decades spent in deafness.[15] Film came close to sorcery.

Thus reports went beyond emphasizing the enchanting and perplexing aspects of technological change. They also asserted that innovative objects resembled quasi-magical devices with supernatural properties. Of course, no one took this rhetoric of the "modern wonder" at face value. Yet the hyperbole in these public statements indicates how strong a surprise new technologies triggered. Despite vocal fears, especially in intellectual circles, that, in Max Weber's famous phrase, technological and scientific progress would transform contemporary society into an "iron cage" of cold, orderly rationality, a host of public assessments celebrated technology in exactly opposite terms, emphasizing its spectacular and miraculous qualities.[16] As technological change appeared to propel Britain and Germany into a modern age of unpredictable excitement, the public welcomed many "modern wonders" with open arms. The fascination that technological marvels cast over Britons and Germans, however, did not establish a stable mode for the public assessment of innovative objects. It was the language of the modern wonder itself that encouraged ambivalent appraisals of

---

[13] *Manchester Guardian*, 9 September 1907, 7.
[14] For the cameraman as a "modern magician," see *Ufa-Magazin*, 28 August 1926, 12. The quotation on the audience is from *Pathé Cine Journal*, 23 October 1913, 6.
[15] *Daily Herald*, 18 November 1929, 5.
[16] Max Weber, "Science as a Vocation," in H. H. Gerth and C. Wright Mills (eds. and trans.), *Max Weber: Essays in Sociology* (Oxford, 1959), 129–56, esp. 139. Interpretations of modernity following a reductionist Weberian reading have recently been effectively challenged in Simon During, *Modern Enchantments: The Cultural Power of Secular Wonder* (Cambridge, MA, 2002).

technological change. Public-relations campaigns and press accounts that glorified technologies as stunning miracles ran the risk of falling victim to their own sensationalism. While admiration for new artifacts sprang from surprise and amazement, this very sense of stupefaction could also leave contemporaries intimidated as they felt emotionally and intellectually overpowered. Hovering between elation and anxiety, ambivalent evaluations of technological change threatened to undermine enthusiasm as the rhetoric of the "modern wonder" brought into view the uncanny and even fearsome aspects of technological innovations.

As ships, airships, and aeroplanes signaled human mastery over nature in spectacular modes, their sheer size often led to uncertainty. Descriptions of these transport technologies as "giants" displayed the combination of astonishment, admiration, reverence, and fear with which Edmund Burke characterized the sublime.[17] After the launch of its largest ship the *Imperator* in 1913, the Hamburg-American Line published a glossy pamphlet claiming that the vessel's "gigantic appearance" had "enthralled (*ergriffen*) thousands and thousands" of onlookers.[18] While the texts radiated pride in the innovative artifact, this instance of "monumental seduction" also betrayed fear.[19] The promotional pamphlet contained passages noting that bystanders had been not only positively "enthralled" but also "shaken" (*erschüttert*) – an adjective suggesting intimidation – as they watched the huge ship glide into the water.[20] In this instance, ambivalence crept into the very advertising material that promoted innovative technologies as seemingly unproblematic. In fact, a host of promotional texts and press reports characterized liners and airships as threatening "monsters" or "colossuses" whose sight overwhelmed and frightened observers.[21] As sensationalistic

---

[17] Edmund Burke, *A Philosophical Enquiry into the Sublime and Beautiful and Other Pre-Revolutionary Writings*, ed. David Womersley (Harmondsworth, 1998), 101. On American descriptions of technology as sublime, see Nye, *American Technological Sublime*. On giantism, see Susan Stewart, *On Longing: Narratives of the Miniature, the Gigantic, the Souvenir, the Collection* (Durham, NC, 1993), 70–103.

[18] *Imperator auf See*, 15.

[19] This phrase is taken from Andreas Huyssen's discussion of large-scale monuments in contemporary German memory culture. See Andreas Huyssen, "Monumental Seduction," *New German Critique* 69 (1996), 181–200. Historical monuments in early twentieth-century Germany were often designed to "overwhelm" observers by their size. See Rudy Koshar, *From Monuments to Traces: Artifacts of German Memory, 1870–1990* (Berkeley, 2000), 34.

[20] *Imperator auf See*, 15.

[21] *The Times*, 9 September 1907, 10; 11 May 1922, 10; Cunard archives, D42/PR3/9/11a, press release, 17 April 1913, "Aquitania"; *Berliner Tageblatt*, 23 June 1912 (evening edition), 5. Technologies of flight are described in this vein in *Frankfurter Zeitung*, 4 August 1908 (evening edition), 2; Heinz Bongartz, *Luftmacht Deutschland: Luftwaffe – Industrie – Luftfahrt* (Essen, 1939), xii; Paul Bewsher, *Green Balls: The Adventures of a Night-Bomber* (Edinburgh, 1919), 20, 56.

descriptions extolled their outsize proportions, new technologies such as liners and flight machines often triggered a blend of emotions ranging from enthusiasm to uncertainty and fear.

Furthermore, locating these "giants" within established frameworks of perception posed problems. A journalist who watched Cunard's *Mauretania*, the *Lusitania*'s sister vessel, prepare for her maiden voyage in 1907, wrote that from close up "it was impossible to see enough of her at one view to make out the curves and lines which give a ship her appearance." Unless the observer literally created a distance between himself and the object, the ship's size rendered the very act of seeing difficult: "It was only by getting well out into the river that it was possible to realize that this great bulk was in the form of a ship." As "there was nothing to compare her with for the purpose of measurement," the boat's proportions interfered with existing visual habits.[22] Advertising materials also testify to the difficulties of representing liners in coherent or complete images. Some layouts, for instance, only partially accommodated the outlines of hulls or masts as the sizes of ships stretched established frames (see illustration 2.1).[23] Technological monumentalism, therefore, not only led to anxiety about the unprecedented dimensions of artifacts, but also contributed to representational dilemmas as the very proportions of innovative technological objects challenged existing modes of perception.[24]

The viewing of large objects was not the sole perceptual problem that encounters with novel technologies produced. As flight opened up new visual angles, pilots frequently struggled to lend expression to the sensation of being overwhelmed by airborne ways of perceiving the natural environment.[25] Nevertheless, it was film that mounted the most perplexing challenge to visual habits. The rise of the movies accentuated a range of questions about the nature of seeing that had surfaced during the nineteenth century, as a host of devices entertained audiences through visual surprises. From the early 1800s, magic-lantern shows entranced audiences with ghost-like "phantasmagorias" in darkened rooms all over Europe.

---

[22] *Manchester Guardian*, 18 November 1907, 8.
[23] The *Bremen* was not the only ship to create such representational dilemmas. See also the advertisement that only partially depicted the liner *Imperator* in *B. Z. am Mittag*, 24 May 1912, 9; Deutsches Schiffahrtsmuseum Bremerhaven, Library (hereafter DSB library), Norddeutscher Lloyd Bremen, *Bremen, Europa, Columbus* (Berlin, no date), 10.
[24] Similar examples can be given for aviation. Junkers promoted its aeroplane *G38* with the help of photos in which the wing tips were cut off by the borders. See Deutsches Museum, Munich, Junkers Archive (hereafter: Junkers archive), file 0303/T10/1, *The Junkers 'G38': Commercial Aeroplane*, 15.
[25] See Hellmuth Hirth, *20,000 Kilometer im Luftmeer* (Berlin, 1913), 173; Elly Beinhorn, *Ein Mädchen fliegt um die Welt* (Berlin, 1932), 111; André Beaumont, *My Three Big Flights* (London, 1912), 141.

"Modern wonders" 27

Illustration 2.1  Postcard of the liner *Bremen*. Giantism rendered visible as the *Bremen* dwarfs the surrounding tugs while its mast protrudes beyond the image frame.

Stereoscopic photographs, which were consumed at home as well as in commercial panoramas, owed their attraction to the display of three-dimensional camera images whose mimetic qualities stunned patrons from the 1840s onwards. In the second half of the nineteenth century, devices such as the zootrope, the praxinoscope, and the mutoscope impressed viewers by their ability to display discrete images in motion.[26] The proliferation of these new visual technologies launched inquiries into the foundations of human perception.[27] In particular, they intensified debates about the truth status of mechanically produced images and fueled epistemological discussions about the modes in which individuals related to the external world. Despite their obvious character as technological artifices,

---

[26] On visual entertainment in the nineteenth century, see Richard Altick, *The Shows of London* (Cambridge, MA, 1978); Terry Castle, "Phantasmagoria: Spectral Technology and the Metaphorics of Modern Reverie," *Critical Inquiry* 15 (1988), 26–61; Dolf Sternberger, *Panorama oder Ansichten vom 19. Jahrhundert* (Hamburg, 1946), 11–23; Helmut Gernsheim, *The History of Photography: From the Earliest Uses of the Camera Obscura in the Eleventh Century to 1914* (London, 1969); Deac Rossell, *Living Pictures: The Origins of the Movies* (Albany, 1998); Gerhard Kemner and Gelia Eisert, *Lebende Bilder: Eine Technikgeschichte des Films* (Berlin, 2000).

[27] For a survey of physiological research on vision during the nineteenth and twentieth centuries, see Charles G. Gross, *Brain, Vision, Memory: Tales in the History of Neuroscience* (Cambridge, MA, 1999), esp. 52–83.

these pictorial representations managed to override viewers' rational detachment and skepticism by arresting their attention and capturing their imagination. Time and again contemporaries pondered the implications of the fact that artificially produced images succeeded in producing cunning "reality effects."

Film's capacity to manipulate spectators' perceptual faculties counted as its defining technological characteristic. As cinema influenced viewers' inner lives, its universal appeal testified to the strong power this new technology possessed. One British critic from the prewar period found that the cinema offered an "immediate escape" from the onerous aspects of city life. Watching the flickering images from the comfort of a "well-wadded chair," a cold and windy street became "nothing or a dream" although it was "only a few yards away." He was stunned at how film was "strong enough to defeat circumstances . . . without any effort of the imagination." Film baffled audiences through its power to override their rational capacities by creating irresistible "illusions." With its ability to lure audiences into alternative and often overtly fictional worlds, the moving image obliterated the distinction between "truth and fiction."[28]

According to contemporary opinion, several features accounted for the extraordinary appeal of the new medium. Obviously, photographic realism was partly responsible for the power that flickering images exerted over audiences. From the very beginning, writers singled out "lifelike fidelity" as a crucial aspect of cinematography.[29] Above all, the camera captured – and the projector displayed – motion with "realistic" photographic precision. Several commentators attributed the ability of film to arrest and direct the audience's gaze to the fact that the photographic images displayed on the screen moved. When contemporaries described early films as "living pictures," they meant this term to be understood in a literal sense to underline the active nature of the objects perceived on the screen. The "modern wonder" of film technology created images with a "life" of their own that completely captivated viewers' attention.

The emphasis on the power of cinema's "exciting" and "cunning illusions" invested film technology with hypnotic qualities.[30] One writer felt that, by "bewitching" audiences, this medium challenged the fundamental

---

[28] *The Times*, 9 April 1913, 11.   [29] *The Times*, 22 February 1896, 15; 19 March 1929, 6.
[30] Frederick A. Talbot, *Moving Pictures: How They Are Made and Worked* (London, 1912), 3. Critics as late as the 1920s and 1930s continued to comment on the ability to create cinematic illusion. See *Neue Preußische Zeitung*, 13 January 1929 (edition B), 10; *The Times*, 27 March 1928, 14. On the hypnotic effects of objects, see James Elkins, *The Object Stares Back: On the Nature of Seeing* (San Diego, 1997), 20.

epistemological distinction between passive objects of perception and active subjects engaged in the process of perceiving them.[31] As they acknowledged the illusions the movies created, concerned observers found it impossible to evaluate the consequences of "the fury of seeing" initiated by what was widely considered to be a deceptive technology.[32] Given film's tremendous popular success, which gave rise to an international entertainment industry with millions of clients within less than two decades of its invention, a whole army of distressed cultural critics clamored against the psychological damage the movie craze allegedly wreaked. They expressed the fear that film would unleash "latent irrationalism" in individuals by bringing to life the ugly ghosts that lay dormant in the unconscious.[33] In conjunction with contemporary panic about the development of a commercial, so-called "mass culture," the visual properties of film generated a barrage of worries.

As film contributed to the visual turmoil that arose in Britain and Germany between the 1890s and the 1930s, a knowledge problem of a different nature further intensified public concern about technology. New technological objects overwhelmed people because their workings remained beyond the comprehension of most technological laypersons. To be sure, the popular accounts of science and technology that flooded the British and German public spheres had the potential to defuse anxiety by providing information on the ways rational and experimentally verifiable forms of knowledge laid the foundation for technological development. In addition, a number of reports detailing how mechanisms functioned sought to familiarize technological laypersons with novel objects.[34] Nonetheless, these narratives often achieved contradictory results because they demonstrated the limited intelligibility of innovative artifacts to those without expert knowledge.

With respect to the cinema, it was hard to find scientific models that would explain how audiences fell under the spell of individual films. To begin with, the technology of film posed the fundamental problem of

---

[31] For the quotation, see Harry Furniss, *Our Lady Cinema* (Bristol, 1914), ix–x. On film as a creator of illusions, see Urban Gad, *Der Film: Seine Mittel – Seine Ziele* (Berlin, 1921), 274; Thomas Langlands, *Popular Cinematography: A Book for the Camera* (London, 1926), 1; Ali Hubert, *Hollywood: Legende und Wirklichkeit* (Leipzig, 1930), 80. This was a constant theme in theories of vision during the late eighteenth and the nineteenth centuries. See Jonathan Crary, *Techniques of the Observer: On Vision and Modernity in the Nineteenth Century* (Cambridge, MA, 1992), 67–137.

[32] Furniss, *Our Lady Cinema*, 25. See also *The Times*, 19 March 1929, 6; 9 April 1913, 11.

[33] Castle, "Phantasmagoria," 29. Chapter 4 below deals with this issue in detail.

[34] W. H. Brock, *Science for All: Studies in the History of Victorian Science and Education* (Aldershot, 1996); Andreas W. Daum, *Wissenschaftspopularisierung im 19. Jahrhundert: Bürgerliche Kultur, naturwissenschaftliche Bildung und deutsche Öffentlichkeit* (Munich, 1998).

rendering it difficult to *imagine* how the camera and the projector created the impression of motion on the screen despite a generally accepted scientific explanation for this phenomenon. It was widely acknowledged that the movement to which the cinematic picture purportedly owed its suggestiveness was itself but an "illusion" or a "trick."[35] After all, to give the impression of motion on the screen, the projector displayed a succession of still images so rapidly as to render the human eye unable to distinguish individual frames. While technical accounts introduced the theory of "the persistence of vision in the eye" to describe how the human body translated a series of individual frames into visual motion, it remained impossible to explain the display of movement on the screen in positive terms. Instead, commentators argued that the very principle of film hinged on playing tricks on the imagination by "fooling" human perceptive faculties.[36] According to the theory of the "persistence of vision," the movie camera and the projector positively misled the observer by taking advantage of the physiological limits of the human perceptive apparatus. This theory, however, accounted only partially for the intense attraction the screen exerted on audiences during feature presentations. The forces drawing patrons into fictional worlds frustrated attempts at explanation, with contemporary assessments only reiterating that film relied on a "psychological deception that was difficult to explain," producing "an intentionally tolerated, enjoyable, gross, straight-forward deception of the consciousness."[37] The psychological workings of film technology continued to be described in the negatively charged rhetoric of pretense and simulation while it was unclear how the make-belief itself occurred. Since the scientific theory of "persistence of vision" merely provided an incomplete account for the attractiveness of the cinema, film retained its magical attributes throughout the period. Some considered the film camera "a modern mechanical miracle" as late as 1930.[38]

In the context of aeroplanes and airships, descriptions of technical equipment created seemingly paradoxical effects: they laid bare the limited ability of non-experts to comprehend the functioning of aerial technology. While

---

[35] Talbot, *Moving Pictures*, 3.
[36] For two early examples, see *The Times*, 22 February 1896, 15; F. Paul Liesegang, *Das lebende Lichtbild: Entwicklung, Wesen und Bedeutung des Kinematographen* (Düsseldorf, 1910), 10–38. The history of the theory of "persistence of vision" is explored in Joseph and Barbara Anderson, "Motion Perception in Motion Pictures," in Teresa de Lauretis (ed.), *The Cinematic Apparatus* (London, 1980), 76–95.
[37] Andor Kraszna Krausz (ed.), *Kurble! Ein Lehrbuch des Filmsports* (Halle, 1929), 2. See also Talbot, *Moving Pictures*, 3.
[38] Edward Hobbs, *Cinematography for Amateurs: A Simple Guide to Motion Picture Taking, Making and Showing* (London, 1930), 19.

some pilots penned texts that set out to explain "the large number of instruments" in the cockpit, many popular narratives written by technological laypersons capitulated at the sight of dials, control lamps, switches, and more.[39] They saw the cockpit as an "inner sanctum" (*Heiligtum*) that the layman entered in a spirit of humility. On the German flying boat *Do X*, a visitor in the year 1931 noted how the board engineer's activities instilled awe as "he wandered restlessly from one manometer to the next, always conscientiously registering on big charts the data from the pointers. I tried to count the many measuring instruments which he had to guard; there were sixty-four, but there may have been even more."[40] Descriptions such as this invested apparatuses on the flight deck with esoteric qualities that rendered technology inaccessible to non-experts. Accounts of airship travel portrayed the atmosphere in the control room in a similar vein. When crew members laid out the principles of airship flight to travelers, such expositions affirmed aviation's seemingly mysterious nature at least as much as they provided popular "enlightenment." According to one German journalist in 1929, the captain resembled "an alchemist pointing out in a learned and matter-of-fact way how to create an artificial human being. I understood nothing at all."[41] Complicated mechanisms required complex explanations that, however, exceeded the layperson's analytical faculties. Despite detailed descriptions of the technological environments, innovative objects preserved an opaque character as their workings remained unintelligible to technological laypersons. At best, non-experts could aspire to a state of "informed awe."[42]

The fact that companies strictly controlled information about the production processes to which prominent artifacts owed their existence also added to the mystique surrounding technology. The reluctance to describe assembly procedures openly did not stem from worries that such accounts would make tedious reading. Rather, individual firms cultivated an aura of secrecy to keep specific procedures and organizational arrangements from competitors' probing eyes. Furthermore, silence about production processes helped innovative artifacts to retain a sanitized public image. After all, extensive descriptions of construction routines risked dragging into view the labor conditions under which technologies were built, thereby potentially highlighting the social tensions in industrial environments in

---

[39] S. H. Long, *In the Blue* (London, 1920), 21; W. A. Bishop, *Winged Warfare: Hunting the Huns in the Sky* (London, 1918), 30–2.
[40] E. Tilgenkamp (ed.), *Do X* (Zürich, 1931), 39.
[41] Max Geisenheyner, *Mit "Graf Zeppelin" um die Welt: Ein Bild-Buch* (Frankfurt, 1929), 19.
[42] Tichi, *Shifting Gears*, 26.

the late nineteenth and early twentieth centuries.[43] As a result of this silence, technological laypersons found it difficult to comprehend how "giants" such as airships and passenger vessels proceeded from drawing board to finished product. Most texts written for technological laypersons concentrated on the final objects in all their supposed glory, reifying innovative artifacts by casting them as products of anonymous and autonomous acts of creation. And even when, on rare occasions, promotional material mentioned construction sites, they chose to portray production processes as bewildering.

In one German pamphlet, shipyards appeared as inscrutable sites where huge quantities of materials and the presence of a multitude of workers created chaotic impressions. The "tangle of beams and scaffolding" among which workmen moved confused observers.[44] A British promotional journalist, for his part, invited readers in the mid-1920s to lose themselves "in the maze of girders [and] regard with marvel and amazement the long avenues of steel" in a builders' yard. Asking his audience to consider that it required only a brief time to transform the "lifeless mass of metal" into a transatlantic liner, this writer conceived the construction process in strict conformity to the narrative conventions of the technological miracle: "certainly, you never saw anything made by man which is half as wonderful as a modern ship."[45] Although the text went on to document the various construction stages, it affirmed the notion that technology remained a mind-boggling "wonder," since shipyards overawed people because the processes involved in assembling the vessels eluded them. Material about German airships directed at a wider readership tended to achieve similar effects.[46] Non-experts, therefore, received only limited information about the ways in which some of the most prominent symbols of technological modernity came into existence.

---

[43] Workers' representatives in both Britain and Germany repeatedly criticized shipyards and aviation companies as harsh employers. In 1910, a Social Democratic parliamentarian castigated shipping companies for their disregard of workers' security in the face of frequent deaths and mutilations. See *Verhandlungen des Reichstags: Stenographische Berichte*, vol. 260 (Berlin, 1910), 1656–7. The Social Democratic press emphasized workers' hard labor conditions in the construction of airships. See *Vorwärts*, 15 October 1924 (evening edition), 1. The topic also featured in the Communist press. See *Rote Fahne*, 14 April 1928, 1; 16 August 1928, 8. In 1924, the Labour *Daily Herald* sided with Imperial Airways employees who resisted an extension of working hours; see *Daily Herald*, 1 April 1924, 5.

[44] DSB library, Norddeutscher Lloyd Bremen, "*Bremen*", "*Europa*": *Die kommenden Großbauten des Norddeutschen Lloyd Bremen* (Zwickau, Hamburg, no date), 15.

[45] Chatterton, *Aquitania*, 2–5.

[46] For some examples see *Graf Zeppelin und sein Luftschiff: Luxusausgabe* (Nuremberg, 1908), 8; Luftschiffbau Zeppelin GmbH, *LZ 126: 20 Originalphotographien vom Amerikaluftschiff* (Friedrichshafen, 1924), 3; Hugo Eckener, "Das Schiff," in Rolf Brandt (ed.), *Die Amerikafahrt des "Graf Zeppelin"* (Berlin, 1928), 20–5.

As people oscillated between admiration and insecurity, their surprise at the continual appearance of "modern" technologies contributed to a widespread sentiment that Britain and Germany had entered a new historical era: "modern times." As commentators included technology in the material inventory of modernity, they interpreted innovations as symptomatic of the times they were living through, while only rarely spelling out the exact reasons behind such assertions. Public debate about technological change, therefore, usually operated with vague notions of modernity. Of course, descriptions of technology as a modern miracle were nothing new after 1890. Railways, factories, canals, bridges, electricity and printing presses had attracted the same epithet throughout the entire nineteenth century.[47] The circumstance that Britons and Germans living between the 1890s and 1930s were not the first ones to conceive of technology as a "modern marvel," however, did not interfere with their belief in the exceptionality of *their* historical present. Despite its long existence, the rhetoric of the modern wonder of technology failed to translate into a coherent tradition because it blocked the establishment of a sense of continuity that could have arisen from its repeated use. While *continually* announcing new eras, it exclusively cast the past as a foil and conceived of the present as a time of inception devoid of historical precursors. Each renewed proclamation that hailed the arrival of a "modern miracle" thus overshadowed preceding statements to the same effect, thereby pushing them into oblivion. If the formula describing technology as a "modern wonder" was itself fairly old, its rhetoric of exceptionality helped to perpetuate the conviction that the respective moment of the historical present was radically distinct from previous points in time. This sense of temporal rupture with the past contributed to insecurities about the present because contemporaries felt unable to turn to history for guidance about the future when past and present seemed to bear such little resemblance.[48] The proliferation of technological innovations thus intensified the conviction that the historical present was a time of extraordinary modernity.

Before the First World War, passenger liners transformed maritime travel to such an extent that it proved impossible to compare the present with

---

[47] Wolfgang Schivelbusch, *The Railway Journey: The Industrialization of Time and Space in the Nineteenth Century* (Berkeley, 1986); Marvin, *When Old Technologies Were New*; Merrit Roe Smith, "Technological Determinism in American Culture," in Merrit Roe Smith and Leo Marx (eds.), *Does Technology Drive History? The Dilemma of Technological Determinism* (Cambridge, MA, 1994), 1–35.

[48] Some inroads into these issues are provided by Reinhart Koselleck, "'Space of Experience' and 'Horizon of Expectation': Two Historical Categories," in *Futures Past: On the Semantics of Historical Time* (Cambridge, MA, 1985), 267–88; Matt K. Matsuda, *The Memory of the Modern* (New York, 1996).

the past. In 1907, a British newspaper quoted an eighty-four-year-old man who had first crossed the ocean in a sailing ship half a century earlier. Personifying "history," the elderly traveler "could not give adequate utterance to the comforts of the present compared with the discomforts of the first voyage."[49] The speaker's statement illustrates how difficult contemporaries found it to create coherent links between past and present. As spectacular engineering breakthroughs occurred at such high rates that the phenomenon of technological change defied sober, dispassionate analysis, the relationships between past, present, and future appeared enigmatic. In a similar manner, early chroniclers of film claimed that the fast pace of innovation prevented them from giving coherent interpretations of the new medium's history.[50] In the context of aviation, even experts lacked the confidence to formulate authoritative prognoses about the future as late as the interwar period. When manufacturers publicly presented prototypes of a helicopter and a flying wing at the RAF air pageant at Hendon in 1926, the specialist journal *The Aeroplane* asked whether these flying machines should be considered "practical jokes," "epoch-making inventions," or "interesting intellectual exercises."[51] Innovative technologies, then, enhanced the temporal confusion that contemporaries considered characteristic of their "modern" present. The silence surrounding the processes governing the production of several high-profile objects only reinforced this dilemma. Since most observers lacked detailed knowledge of how passenger liners and airships came into existence, these markers of modernity appeared devoid of any past. With their individual histories hidden, they seemed to burst into the present from nowhere.

Tales that cast innovative technologies as reified objects, therefore, contributed to a "genesis amnesia" that complicated attempts to locate the modern present firmly within a coherent temporal order or a continuous chronology.[52] The difficulty of grounding processes of technological innovation in history amplified ambivalent evaluations of change. Even

---

[49] *Daily Mail*, 12 October 1907, 7.
[50] See Gerard Ford Buckle, *The Mind and the Film: A Treatise on the Psychological Factors in the Film* (London, 1926), xi; Colden Lore, *The Modern Photoplay and its Construction* (London, 1923), 9; *The Times*, 19 March 1929, 6.
[51] *The Aeroplane*, 7 July 1926, 10.
[52] Richard Terdiman employs the term "genesis amnesia" to draw attention to the forms of forgetting that underlie concepts of modernity as a crisis of temporality. See Richard Terdiman, *Present Past: Modernity and the Memory Crisis* (Ithaca, 1993), 12–13. The term was coined by Pierre Bourdieu, *Outline of a Theory of Practice* (Cambridge, 1984), 79. As Terdiman points out, interpretations of reification as a means of covering up an object's production history date back to Marx's treatment of commodity fetishism. See Karl Marx and Friedrich Engels, *Werke*, vol. XXIII: *Das Kapital*, vol. 1 (Berlin, 1974), 85–98.

Hitler, whose heroic posturing usually excluded public admissions of anxiety, could not help but feel "almost frightened" at the "speed" of technological "progress" when he addressed the annual party rally in Nuremberg in 1934.[53] The surprising and inscrutable "modern wonders" of technology repeatedly seemed to change the path of history in unpredictable ways. Indeed, from the turn of the century an intellectual current gained strength that viewed developments in engineering as threatening because technology's dynamism appeared to obey autonomous laws beyond human control. As a result, public fears intensified that artifacts had wrested history from man. In 1931 Oswald Spengler, who was Germany's "most successful political writer in terms of book sales during the twenties" and a leading exponent of the "Conservative Revolution," gave typical expression to this staple argument: "The creation rises against its creator. The ruler of the world becomes the slave of the machine. Without exception, it forces each and everyone of us to follow its course." According to Spengler, the knowledge problem that we encountered earlier crucially contributed to this state of affairs: "It is part of the tragedy of our time that, once unbound, human thought can no longer grasp its own consequences . . . Technology has become esoteric."[54]

## THE CULTURAL FOUNDATIONS OF PUBLIC ENTHUSIASM FOR "MODERN WONDERS"

Although German and British reactions to new technologies oscillated between admiration and anxiety, public ambivalence did not undermine a cultural climate conducive to innovation, because anxiety about new artifacts only rarely turned into public fear or terror. Spectacular accidents such as aeroplane crashes, airship explosions, and shipwrecks could generate passionate fears about new machines, which could lead to vociferous resistance to further experiments in specific technological disciplines. In

---

[53] Adolf Hitler, "Kulturrede," in Julius Streicher (ed.), *Reichstagung in Nürnberg 1934* (Berlin, 1934), 140–74, here 142.
[54] Oswald Spengler, *Der Mensch und die Technik: Beitrag zu einer Philosophie des Lebens* (Munich, 1931), 52, 54. This work received a warm review in the leading journal of the "Conservative Revolution." See *Die Tat* 23:1 (1931–2), 405–8. On Spengler, see Rolf Peter Sieferle, *Die Konservative Revolution: Fünf biographische Skizzen* (Frankfurt, 1995), 106–31; Jeffrey Herf, *Reactionary Modernism: Technology, Culture and Politics in Weimar and the Third Reich* (Cambridge, 1984), 49–69. Spengler's ideas on technology are explored in the wider context of the "Conservative Revolution" in Stefan Breuer, *Anatomie der Konservativen Revolution* (Darmstadt, 1993), 70–8; Mikael Hård, "German Regulation: The Integration of Modern Technology into National Culture," in Mikael Hård and Andrew Jamison (eds.), *The Intellectual Appropriation of Technology: Discourses on Modernity, 1900–1939* (Cambridge, MA, 1996), 33–67.

1930, for instance, the explosion of the British airship *R101*, which killed fifty-one crew members and passengers in a fireball, led to the termination of the airship construction program in the United Kingdom. This decision, which resulted from public shock at yet another hair-raising airship accident, nevertheless remained exceptional. Usually, commentators managed to convince the public that mishaps represented temporary setbacks that led to design modification, thereby fueling technological "progress."[55] In general the ambivalence that surrounded technological innovation did not block further change.

One reason was that highly emotional responses of revulsion such as terror or horror remained exceptional in British and German reports about Europeans' encounters with new artifacts. Britons and Germans took great pains to show that their uncertain stance towards technological innovation did not lead to panic or loss of self-control, claiming that Europeans possessed the mental strength to subdue technophobic urges. In fact, displays of unbridled fear and terror were widely associated with non-European reactions to technology, based on "savagery" and "superstition." Commentators vilified the supposedly irrational and uncivilized manifestations of horror that they considered typical of colonized non-Europeans' responses to technological objects such as aeroplanes. Depictions of the fear that the sight of advanced technologies allegedly instilled into "natives" overseas not only served to emphasize Western dominance and superiority; they also differentiated between "primitive" and "advanced" modes of responding to innovative objects. Africans appeared to exhibit particularly "primitive" fears of all things technological.[56] When German aviatrix Elly Beinhorn landed on the fringes of the Sahara desert, for instance, she described spotting "a horrified black man running away with raised arms."[57] Alan Cobham, a pilot who made three widely publicized flights through the British empire during the mid-1920s, attributed the cause of such reactions to the fact that "the native has a very stupid temperament."[58]

By contrast, Europeans were advised to conquer their anxieties about technological innovations through a "manly" demeanor that signaled self-discipline and emotional self-control, which compared favorably with "effeminate" attitudes towards technology in Africa and Asia.[59] Spengler,

---

[55] See the following chapter.
[56] See Michael Adas, *Machines as the Measure of Man: Science, Technology, and Ideologies of Western Dominance* (Ithaca, 1990), esp. 153–65.
[57] Beinhorn, *Ein Mädchen fliegt um die Welt*, 32.
[58] Alan Cobham, *Australia and Back* (London, 1926), 41.
[59] Gender stereotypes underpinning distinctions between Europeans and non-Europeans have been explored by Mrinalini Sinha, *Colonial Masculinity: The "Manly Englishman" and the "Effeminate Bengali" in the Late Nineteenth Century* (Manchester, 1995).

for his part, argued that, given the overwhelming qualities he ascribed to technology in early twentieth-century society, the individual should face new developments calmly, although, in his fatalistic view, the "decline of the West" was inevitable. Despite his sizeable following in German intellectual circles during the Weimar Republic, Spengler's "courageous pessimism," which exhausted itself in calls for "gestures of heroic dignity," remained, on the whole, a minority position.[60] Others adopted considerably more optimistic stances, like the German commentator who advised his contemporaries "not to be overpowered or perplexed" despite the presence of "relatively mystical apparitions [*relativ mystischer Erscheinungen*] in everyday life about which we know little more than that they are based on iron, wheels, pistons and similar things."[61] In 1933, a British journalist adopted a similarly defiant tone as he admonished his readers not to "be machine coward[s]," because "the brain that devised the machine can surely control the machine of its devising."[62] Thus Britons and Germans were credited with the ability to face up to the psychological pressures of modern life that technological change created. It was argued that Europeans, rather than succumbing to states of horror, should and could confront the technological wonders of the present with calm self-confidence. Of course, calls for emotional restraint did little to remove the causes that gave rise to anxieties about technological innovation. Still, they pointed towards a habitus that enabled Britons and Germans to live in a world of disquieting change generated by artifacts that eluded most observers' imaginative and intellectual capacities.

Drawing attention to the close links between "science" and "technology" provided an effective argument that helped laypersons to achieve self-control and self-restraint *vis-à-vis* innovative artifacts. Technological "wonders," commentators argued, were firmly based on secular, rational, and verifiable forms of knowledge and therefore obeyed human commands. Public-relations departments had a strong interest in impressing this argument on to the British and German publics because it promised to generate acceptance for innovations. Debates about passenger shipping and aviation construed the relationship between science and technology in two ways. First, commentators pointed out that technological improvements either furthered scientific knowledge directly or laid the foundations for new research projects.[63] Contributions to scientific research, therefore, justified

---

[60] Sieferle, *Konservative Revolution*, 130.  [61] Kraszna Krausz (ed.), *Kurble!*, 1.
[62] *Daily Mail*, 11 February 1933, 10.
[63] This was a standard argument in the context of aviation. For an early example, see *Daily Mail*, 13 July 1910, 6. German promoters of the airship – including those on the radical Left – repeatedly played up the scientific potential of aerial travel. See Internationale Studiengesellschaft zur Erforschung der Arktis mit dem Luftschiff (ed.), *Das Luftschiff als Forschungsmittel in der Arktis: Eine Denkschrift*

innovative technological experiments. Second, it was frequently emphasized that scientific insights laid the theoretical foundations for technological change. At times conceptualizing technology as "applied science," the public-relations departments of aviation and shipping firms took particular pains to point out that recent scientific findings guided the construction of new ships and flying machines. Thus advertising material sought to convince contemporaries and potential customers that technological progress followed the logic of science. Both shipping and aviation firms also boasted about their commitment to research. During the 1920s, the Junkers works consistently claimed that a unique culture of "industrial research" organized all aspects of production processes, calling their laboratories the "soul" of their enterprise.[64] Junkers's style of "industrial research" integrated the analysis of raw materials, aerodynamic research, production methods, and market surveys with the aim of delivering goods of the "highest use value" to customers.[65] Photographs depicted clean aeroplane engines and implied that they had been assembled in laboratory environments.[66] The company branded itself as a firm whose organization and products, such as the "gigantic" *G38* aeroplane, embodied pure rationality, regularly garnishing prospectuses with charts, tables, and drawings that documented devotion to "scientific" research.[67]

The strategy of emphasizing technology's scientific dimensions achieved a variety of effects. First of all, highlighting intimate links between "science" and "technology" invested engineering with prestige by aligning innovative objects with quests for unalterable and objective truths.[68] Moreover, crediting "science" with a prominent role in technological innovation inspired trust in new technologies. Even if bystanders such as the visitor on the control deck of a German airship, whom we encountered earlier, did

---

(no place, 1924); *Rote Fahne*, 29 July 1931, 11; 1 August 1931, 17; *Vorwärts*, 30 July 1931 (evening edition), 4. Popular science on the Left in Weimar Germany is explored in Nick Hopwood, "Producing a Socialist Popular Science in the Weimar Republic," *History Workshop Journal* 41 (1996), 117–53.

[64] Junkers archive, file Propaganda, Firmenschriften, *Die Junkers-Werke* (no place, 1921), 23. Junkers public relations efforts are analyzed in detail in Michael Geyersbach, *Wie verkauf' ich meine Tante? Corporate Design bei Junkers, 1892 bis 1933* (Dessau, 1996). Examples from the shipping sector include *The Cunard Passenger Logbook* (Glasgow, 1897), 29; *Dampfer Kaiserin Auguste Victoria*, 5.

[65] Junkers archive, file Firmenschriften, Junkers und sein Werk, *Die Junkers-Werke* (Dessau, 1926), 5.

[66] *Die deutsche Luftfahrzeugindustrie auf der Deutschen Verkehrsausstellung München 1925* (Munich, 1925), 35.

[67] Junkers archive, file 0303/T.10/3, press information *Entwicklung der Junkers-G38*, 4.

[68] See Lorraine Daston and Peter Galison, "The Image of Objectivity," *Representations* 40 (1992), 81–128. For an examination of the ways American engineers publicly boosted their status claims by representing technology as "applied science," see Ronald Kline, "Construing 'Technology' as 'Applied Science': Public Rhetoric of Scientists and Engineers in the United States, 1880–1945," *Isis* 86 (1995), 194–221.

not understand the science underlying technology, he still found listening to experts "nice. Since every now and then, some connections became obvious, and I got a notion of the precision of technology."[69] Despite a limited grasp of mechanical intricacies, this aerial traveler gained faith in technology because of the scientific exactness of technology. Playing up technology's scientific dimension also deflected public attention away from the profit motives that necessarily influenced the adoption of specific types by individual companies. Instead, focusing on the relationship between "science" and "technology" as symbiotic allowed companies to emphasize that they adopted individual designs for purportedly objective reasons rather than for their own financial gain. Claims that new technological objects owed their existence to scientific findings also diminished the need for descriptions of production processes that, as we saw earlier, risked drawing unwanted attention to social inequalities in shipyards and aeroplane factories. If innovative artifacts were built in accordance with the findings of incontrovertible research, it became superfluous to detail the varying stages of assembly processes as well as the workers involved in them. Viewed from this perspective, the public image of a close nexus between "science" and "technology" enhanced reified understandings of technology.

Nonetheless, emphasizing that processes of technological innovation conformed to the logic of scientific research only partially soothed public anxieties about innovation, since many Britons and Germans lacked the expertise to evaluate such claims. No matter how intensively public-relations departments highlighted the links between "science" and "technology," public unease about innovation lingered on. At the same time, promotional campaigns setting out the scientific foundations of technological change powerfully established the argument in Germany and Britain that the emergence of new technologies – surprising as they might appear – rested on rationally intelligible procedures and could, therefore, be trusted. Thus constant reassurance that the technological innovations of flight and passenger shipping represented man-made marvels subject to the authority of science allowed observers to abandon themselves to a sense of wonderment.[70]

---

[69] Geisenheyner, Mit "Graf Zeppelin" um die Welt, 19.
[70] This sense of wonderment is, therefore, fundamentally different from that of the early modern period. Before 1750, surprise at "wondrous" phenomena acted as an impulse for scientific inquiry, while in the nineteenth century the astonishment at industrial technologies was, in contemporary understanding, the result of scientific advances. On "wonder" in Europe before 1800, see Lorraine Daston and Katharine Parks, Wonders and the Order of Nature, 1150–1750 (New York, 1998).

In contrast to aviation and shipping, the film industry could not resort to the authority of science to ease public fears about its products. After all, there existed (and exists) no coherent and convincing scientific explanation of why and how moving images had such a strong appeal to audiences. Furthermore, the long-term effects on patrons who watched movies that obliterated the distinction between "fact" and "fiction" remained initially unclear. As a consequence, heated debates about the political and moral dangers of the cinema engulfed the political and cultural elites in Germany and Britain once American, French and German companies had established an international entertainment industry after 1910. Of course, the socially diverse clientele that flocked to the movies paid little attention to the Cassandras who denounced film. Still, strict censorship regimes testified to the suspicions harbored in politically and culturally influential circles about the new medium, which, as recurring conflicts about individual feature films indicate, persisted in both countries until the 1930s. To be sure, many public attacks on film were part of a wider backlash against a developing commercial "mass" culture. Yet the fact that the workings and effects of film technology resisted scientific explanation amplified fears about the new medium.

While public statements emphasized the scientific foundations of technological change in both Britain and Germany, this argument was given less prominence in Nazi Germany. Although National Socialist publications acknowledged the links between science and technology, they did not exclusively conceive of engineering as a form of applied science. The anti-intellectual and anti-rational poses that leading National Socialists struck regularly stood in the way of an unqualified affirmation of technology as an expression of systematic scientific research.[71] Like a wide range of critics of civilization, Hitler and his followers subscribed to the view that the proliferation of rational and, by logical extension, scientific explanatory models since the nineteenth century had begun to exert a stifling effect in contemporary society. In the words of chief ideologue Alfred Rosenberg, excessive "worship of the goddess of reason" sidelined the "instincts" and the "will" as driving forces of historical development, which, according to Nazi doctrine, amounted to a permanent "struggle" with the natural world and between individual "races."[72] Consequently, the National Socialists considered technology an important tool to energize the various struggles

---

[71] The anti-scientific slant of National Socialist ideology has been emphasized in Helmuth Trischler, "Self-Mobilization or Resistance? Aeronautical Research and National Socialism," in Monika Renneberg and Mark Walker (eds.), *Science, Technology and National Socialism* (Cambridge, 1994), 72–87, esp. 73.
[72] Alfred Rosenberg, "Rede," in *Reichstagung in Nürnberg 1934*, 129–39, here 129.

in which they saw themselves engaged, including the "conquest of nature" and the reinvigoration of a German nation supposedly beleaguered by internal and external enemies. In 1934 Otto Dietrich, the party press secretary, published an unabashedly self-congratulatory bestseller about the "Faustian efforts, the phantastically modern strategies, the genuinely heroic struggle" of the early 1930s, that culminated in the "national revolution." According to Dietrich, technology contributed crucially to Hitler's "triumph," not only because the Führer availed himself of the "most modern means of transport" such as the aeroplane during campaigns, but also because "the exploitation of the newest technological instruments" enabled the Nazis to withstand the "physical, intellectual, and spiritual demands" of a "thoroughly modern life style," which, in turn, purportedly enhanced campaign initiatives.[73]

Accounts of domestic trials and tribulations were not the only writings in which the National Socialist press employed highly stylized narratives to cast technology as a means to strengthen vital energies in a conflict-ridden environment that required constant, forceful self-assertion. Subject to substantial restrictions laid down by the international community in the Versailles Treaty during the decade after World War I, the expansion of civilian and – especially – military aviation in the 1930s came to be celebrated by Hermann Göring as a manifestation of the "energy of the new National Socialist Germany."[74] A volume published in 1938 to honor the fiftieth birthday of aerial engineer Ernst Heinkel linked the rise of German air power directly to the "human need to satisfy sporting and fighting desires."[75] Unlike many German intellectuals affiliated to the "Conservative Revolution," the National Socialists did not view technological progress as an overbearing and stifling factor in contemporary life. Instead, they interpreted technology as a material manifestation of energetic assertiveness. Viewing technology as the material expression of the will to struggle rather than solely as the outcome of scientific advances enabled the Nazis to reconcile their devotion to the irrational with their love of technology. As they went beyond stressing technology's scientific foundations, the Nazis emphatically attributed innovations in engineering to vitalistic drives.[76] As a result, National Socialist writings affirmed

---

[73] Otto Dietrich, *Mit Hitler an die Macht: Persönliche Erlebnisse mit meinem Führer* (Munich, 1934), 11, 66, 72.
[74] Hermann Göring, "Einleitung," in Heinz Bongartz, *Luftmacht Deutschland: Luftwaffe – Industrie – Luftfahrt* (Essen, 1939), vii–xiv, here xii.
[75] *Kameradschaft der Luft* (Berlin, 1938), 7. Lufthansa's promotional literature placed similar emphasis on the notion of struggle. See *Die Luftreise*, March 1936, 72–5.
[76] The links between vitalism, will, and anti-rationalism were, of course, staples of the "philosophy of life" (*Lebensphilosophie*) that the Nazis had appropriated. For a concise introduction, see Herbert Schnädelbach, *Philosophy in Germany, 1831–1933* (Cambridge, 1984), 139–49.

technological developments as inscrutable wonders that derived their origins from motivations located in inaccessible recesses of human nature and that ultimately eluded rational understanding. In the last instance, the National Socialists celebrated and mythologized technology as a mystery. It was thus much more than a desire for enhanced productivism and economic efficiency that prompted the Nazis to embrace technology.[77] Of course, because they maintained that technology sprang directly from instinctual motivations, the National Socialists did little to remove the knowledge problems that fueled ambivalence about technology in the first place.

In addition to the insecurity that stemmed from a limited understanding of the technical aspects of innovations, confusion arose about the temporal disorder that the appearance of novel artifacts caused. On countless occasions commentators stressed that spectacular breakthroughs in film, aviation, and passenger shipping signaled the advent of a new, uniquely "modern" era that bore little resemblance to the past. Accounts of the disorienting temporal ruptures effected by technological change, however, existed alongside a plethora of public tales locating technological objects in coherent histories of dynamic yet continuous development. These "inventions of traditions" forged links between the past and the "modern" present to demonstrate that the succession of dazzling engineering feats, confusing as they may have been, did follow identifiable historical patterns of progress.[78] In other words, commentators repeatedly counteracted the widespread "genesis amnesia" regarding innovative artifacts by pointing out that technological change amounted to a consistent, positive tradition of modernity in Britain and Germany.

The "propaganda" departments of the flight, shipping and, to a more limited extent, film industries recognized and exploited the image benefits to be reaped from playing up the historical qualities of their products. At times, mythology provided the cultural context for innovative artifacts. While Neptune featured as the defeated sea god in promotional material for new ships, the legend of Daedalus and Icarus rather predictably served to contrast current aerial accomplishments with past failures.[79] Associating aerial and naval technologies with ancient myths conferred cultural

---

[77] This motive has received considerable attention in recent times. See Michael T. Allen, "The Puzzle of Nazi Modernism: Modern Technology and Ideological Consensus in an SS-Factory at Auschwitz," *Technology and Culture* 37 (1996), 527–71; idem, "Modernity, the Holocaust, and Machines Without History," in Michael T. Allen and Gabrielle Hecht (eds.), *Technologies of Power: Essays in Honor of Thomas Parke Hughes and Agatha Chipley Hughes* (Cambridge, MA, 2001), 175–214.

[78] *The Invention of Tradition*, ed. Eric Hobsbawm and Terence Ranger (Cambridge, 1983).

[79] See *Imperator auf See*, 23; SDKB archive, file SD 1244, *Wings: Ein Paramount Film der Parufamet* (no place, 1929), 6.

"Modern wonders" 43

prestige on the achievements of technological change in ways that highlighted "progress." Alternatively, historical interpretations construed continuities between past and present developments. Aviation and shipping easily lent themselves to the invention of long traditions. After Alan Cobham had completed his flight between Britain and Australia in 1926, Australian Prime Minister S. M. Bruce rejoiced in his feat because it illustrated that "the spirit of Drake and Raleigh . . . still lives in the modern world."[80] While references to Elizabethan explorers remained an exclusively British and imperial affair, the press as well as advertising departments invoked Columbus as a venerable precursor to contemporary maritime and aerial intercontinental travel in both Britain and Germany.[81] The White Star Line, for its part, cast historical continuity between past and present engineering in pictorial terms. A promotional postcard from 1912 depicted the *Titanic* flanked by several North American architectural monuments, including skyscrapers on the left and Cologne Cathedral, St. Peter's in Rome, and a pyramid on the right (see illustration 2.2).[82] The image encouraged multiple readings. First, it stressed the boat's monumental nature by placing it next to various large buildings. Second, it invested the *Titanic* with cultural prestige by ranking it among several world-famous landmarks. Third, it introduced a historical and geographical narrative: not only did the ship's central position between architectural examples from the "old" and the "new" worlds suggest that liners physically maintained ties between America and the locations where "western civilization" had originated; the card also placed the vessel prominently among revered artifacts of both the past *and* the "modern" present. The drawing, therefore, positioned the passenger ship in a long line of monumental artifacts that stretched from the distant past to the immediate present, thus grounding the ship in a history.

Furthermore, the British and German public spheres abounded with reports that measured "progress" in practical terms throughout our entire time period. In the context of film, lineages highlighted improvements to cameras and projectors as well as increasingly lavish sets.[83] The expanding size of aeroplanes, airships, and passenger liners demonstrated tangible technological improvements, as did the growing speeds at which new models

---

[80] Cobham, *Australia and Back*, vi.
[81] *The Times*, 12 December 1910, 6; *B. Z. am Mittag*, 15 October 1924, 1; *Berliner Tageblatt*, 15 October 1924 (morning edition), 1.
[82] *White Star Line R. M. S. "Titanic"*, postcard, author's collection.
[83] *The Times*, 29 March 1929, 6, 16; Valentia Steer, *The Secrets of the Cinema* (London, 1920); Henny Porten, *Vom "Kintopp" zum Tonfilm: Ein Stück miterlebter Filmgeschichte* (Dresden, 1932).

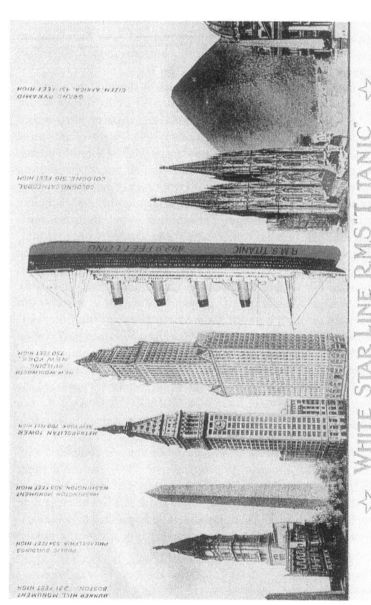

Illustration 2.2 Postcard of the *Titanic*. Placing *Titanic* between architectural wonders of the "Old" and the "New" world underlined the ship's historical stature.

traversed the oceans and the air.[84] Greater comfort and security for aerial and naval travelers also indicated positive advance.[85] Enhanced human control of the environment, or, to put it in contemporary terms, the ongoing "conquest of nature," defined technological "progress." Public-relations departments and the national press employed the image-enhancing qualities of ancient mythology and history to counteract a widespread sense of historical dislocation. In addition to bestowing cultural prestige, these historical references embedded aviation, passenger shipping, and, to a lesser degree, film in historical continuities and demonstrated that, despite fast rates of change, technological developments followed coherent trajectories. Dynamic rather than static notions of historical continuity informed these interpretations. No matter how fast-paced, technological developments, the argument ran, adhered to a teleology of open-ended "progress" whose logic structured historical change.

To be sure, the First World War had the potential to deal a blow to contemporary beliefs in a European march towards a bright future. After the military had unleashed unprecedented violence with the help of novel technologies, the slaughter on European battlefields drew attention to a dark side of technological innovation. For instance, German and British peace movements repeatedly voiced their concern about the prospect of a future war fought with new weapons of mass annihilation. Yet notions of progress continued to saturate retrospective accounts of technological change during the interwar period, despite widespread acknowledgment of technology's destructive capacities in Britain and Germany. Rather than condemning novel technological developments in principle, observers on the Left tended to conceive of the First World War as a period of protracted abuse at the hands of rampant "militarism" and aggressive nationalism.[86] Firm opposition to aggressive nationalism and imperialism allowed the British and German Left to support innovative technologies energetically after the war. Detailed and enthusiastic reports in the Labour newspapers covered spectacular long-distance flights by airships and aeroplanes.[87]

---

[84] *Cunard Passenger Logbook: A Short History of the Cunard Steamship Company and a Description of the Royal Mail Steamers Campania and Lucania* (Glasgow, 1904), 26.

[85] *Dampfer Kaiserin Auguste Victoria*, 35; *Resolute, Reliance*, 13; *Royal Mail Triple-Screw Steamers "Olympic" and "Titanic"* (Liverpool, 1911), 16; *Die Luftreise*, August 1938, 188–9; *Deutsche Lufthansa Summer Time Table 1938* (no place, no date).

[86] For some examples of this argument in the Social Democratic press, see *Vorwärts*, 14 October 1924 (morning edition), 2; 21 June 1928 (morning edition), 6; 28 August 1931 (morning edition), 5. For Communist critiques of militarism in the context of new technological objects, see *Rote Fahne*, 12 October 1928, 7; 14 October 1924, 3. For critical reports about militarism in the British press, see *Daily Herald*, 2 October 1926, 1; 2 August 1930, 1; 5 August 1930, 1.

[87] See *Daily Herald*, 2 August 1930, 1; 2 October 1926, 1; *Vorwärts*, 12 April 1928 (morning edition), 1; 24 May 1932 (morning edition), 3.

During the 1920s, moderate and radical German Socialist opinion formers considered airships to be agents of progress despite their involvement in raids on British and French cities during the First World War. In 1924, the Social Democratic daily *Vorwärts* hailed the successful transatlantic maiden flight by a German aerial "giant" as exemplifying the "progress of human civilization" (*Kulturfortschritt*) because it extended human control of nature.[88] No left-wing publication called for an end to experiments in the air and on the sea on the basis of the argument that it was technological change that laid the foundation for the misuse of technology in the first place. Skepticism about the military implications of technological change did not undermine support for innovation on the Left, since an implicit differentiation between civilian and military uses of technology bolstered the Left's faith in technological progress.

Industrial promoters, Liberals, Conservatives, and National Socialists maintained a belief in the progressive properties of technological change without the distinction between civilian and military uses of technology. Although they denied neither the violent nature nor the horrors of the First World War, they argued that its battles had confirmed the need for innovative technologies as an indispensable prerequisite for successful national defense.[89] The Cunard Line, for instance, drove this point home in brochures and company histories, detailing the contributions of merchant vessels to supplying the British war effort. A host of German publications not only praised the roles of aeroplanes and airships in military campaigns between 1914 and 1918; they also described how the war had accelerated technological developments.[90] These British and German accounts thus integrated the experience of war directly into histories of technological progress. Rather than witnessing a full-scale assault on the concept of technological progress itself, the 1920s threw into sharper relief political conflicts about definitions of technological progress that had been established before 1914. While the Left denounced technologies' military potential as regressive and affirmed their civilian uses as progressive, Liberal, Conservative, and National Socialist circles considered

---

[88] *Vorwärts*, 14 October 1924 (morning edition), 1. Similar praise could be found in the Communist press. See *Rote Fahne*, 16 October 1924, 6.

[89] David Edgerton has detailed these arguments with respect to aviation in David Edgerton, *England and the Aeroplane: An Essay on a Militant and Technological Nation* (Basingstoke, 1991), 38–49. See also chapter 8, on technology and the nation, below.

[90] F. Lawrence Babcock, *Spanning the Atlantic* (New York, 1931), 188–98; Cunard archives, file D42/PR3/8/8, *R.M.S. Carmania* (no place, 1919); file D42/PR3/9/12b, *Cunard Line Peace Day: July 19th, 1919* (no place, 1919); *On War Service* (London, 1919); v. Hoeppner, *Deutschlands Krieg in der Luft: Ein Rückblick auf die Entwicklung und die Leistungen unserer Heeres-Luftstreitkräfte im Weltkriege* (Leipzig, 1921); *Kameradschaft der Luft* (Berlin, 1938), 26–32.

both the military and civilian uses of new artifacts as factors that symbolized and furthered progress. Albeit in differing degrees, technological artifacts retained their status as positively defined symbols of modernity across the political spectrum in interwar Germany and Britain before and after World War I.

Finally, addressing new and surprising objects as "wonders" proved to be an elastic device of perception and knowledge production about technological change. The formula of the "modern wonder" assumed the characteristics of a fundamental category around which public debate could be organized, thereby easing the classification of a wide range of artifacts and phenomena. Of course, from an engineering perspective the language of the technological wonder produced unspecific public knowledge about individual mechanisms. Movie cameras and projectors, aeroplanes, airships, ships, as well as bridges, electrical generators, cars, canals, and guns, could all be viewed as creating a world filled with "wonders." The vagueness of the rubric "modern wonder" enhanced its flexibility and proved to be of crucial importance for British and German debates about technological change. The trope of the "modern wonder" accommodated a plethora of diverse mechanisms in existing frameworks of public knowledge and included individual innovations in one universal maelstrom of material transformation. Thus depictions of technological achievements as the wonders of the present tended to blur distinctions between individual artifacts and contributed to the construction of "technology" as a new totalizing category. Furthermore, as a result of its vagueness, the language of the "modern wonder" remained politically multivalent, thereby favoring the ideological appropriation of innovative artifacts by opposing political movements. Consequently, novel technologies attracted support from across the political spectrum. The flexibility of this trope, therefore, eased the acceptance of innovative technologies and helped speakers from a broad range of political contexts to classify a variety of novel artifacts.

To summarize, then: in late nineteenth- and early twentieth-century Britain and Germany the public shared the conviction that technological "wonders" contributed to an era of unprecedented, dynamic, and exciting man-made changes. Although commentators had identified a broad range of engineering feats as "modern wonders" throughout the nineteenth century, encounters with spectacular innovations in a plethora of social contexts perpetuated and intensified the belief that the historical present between 1890 and World War II was exceptional. Because the idea that engineering created miracles persisted as a ready-made formula to classify a host of transformations over a long period of time, it rarely generated a sense of continuity. After all, the formula of the "modern miracle" repeatedly highlighted the breathtaking aspects of the present to establish contrasts with a past that was quickly becoming obsolete. As both societies underwent fundamental transformations, the trope of the "modern wonder" provided a persistent device to describe mind-boggling changes by maintaining a complex of interpretations that fluctuated between admiration and anxiety. Novel mechanisms, while they commanded passionate enthusiasm, simultaneously challenged fundamental categories that structured interpretational and perceptual frameworks and thereby created problems of knowledge on several levels. Spectacular modes of advertising involuntarily contributed to public insecurity about technological change, as promotional material played up the breathtaking dimensions of innovations. These mounted challenges to maps of the contemporary imagination on several levels. The monumental size of passenger ships and airships exploded established observers' visual frames and left them wondering how to look at these new artifacts. By creating hypnotic images that arrested the gaze of the observer, film subverted conventional distinctions between passive objects of perception and subjects engaged in active perception. Given the epistemic confusion surrounding the appearance of novel artifacts, contemporaries found it impossible to ascribe stable meanings to new technologies, and some prominent intellectuals began to feel intellectually overpowered by new technologies that seemed to have turned into autonomous historical agents beyond human control.

At the same time, the rhetoric of the modern wonder possessed elements that prevented anxieties about technology from developing into outright technophobia. Rational demeanor, it was argued, enabled Europeans to face novel objects with composure. Moreover, public debate not only emphasized the links between science and technology to demonstrate technology's rational foundations, thereby building public trust; it also alerted contemporaries to the idea that technological development followed a history of persistent European progress – and this despite the slaughter of the First World War. Finally, the formula of the "modern wonder" eased the classification of a wide array of new technological objects and rallied support for innovation across the fractured political spectrum. While the characteristics of the rhetoric of the "modern wonder" mentioned so far commanded prominence in both countries in our time period, a notably new and unique element appeared in Germany after 1933. Given its prominent anti-rationalism, the National Socialist movement advanced the claim that technological wonders originated from an inscrutable will to power that energized the historical present.

Still, public ambivalence about aviation, passenger shipping, and film did not vanish in both countries throughout the entire period because the knowledge problems that accompanied technological change persisted. Since no guidance emerged on how to look at the new giants of the air and sea without feeling overpowered, problems of perception continued to haunt debate as the sight of novel artifacts threw contemporaries off balance before and after the Great War. Concern about the epistemic consequences of technological innovation was particularly virulent with respect to the "illusions" created by film technology, whose power eluded attempts at scientific explanation. In conjunction with widespread fears about the detrimental moral effects that supposedly resulted from the rise of a commercial "mass" culture, film technology appeared to have given birth to a mysterious industry with the ability to spread powerful visual fictions among a clientele numbering millions. The existence of stringent censorship regimes in both countries bears out the deep suspicions with which the British and the German states regarded film technology at the time. Thus the impact of new forms of visual information underpinned public insecurity about technological change.

Furthermore, technological devices retained an opaqueness because their functioning eluded the grasp of technological laypersons. Thus assurances that contemporaries could confront novel artifacts with calm composure demanded a deeply irrational leap of faith. After all, most Britons and Germans could not develop trust in innovative technologies on the basis of a detailed understanding of the intricacies of new mechanisms. A tension, therefore, existed between demands for rational conduct in the face of innovations and the the fact that many contemporaries could only partially base this conduct on an informed, or scientifically grounded, knowledge of new technologies.

On the whole, welcoming sentiments outweighed skepticism about technological innovations as commentators underlined technology's scientific foundations as well as its historical role in promoting progress. Moreover, press reports frequently described the enthusiastic excitement that the introduction of novel artifacts generated. While these arguments sought to dispel fears, they never completely alleviated public anxiety about technological change. Nonetheless, insecurity did not lead to the appearance of forceful opposition to technological change in Britain and Germany. The rhetoric of the "modern wonder" formulated ambivalence in public debate while simultaneously launching a variety of arguments meant to counteract uncertain evaluations of innovation. Thus the trope of the "modern wonder" allowed contemporary discussions to embrace technological change in the process of registering anxiety, and therefore crucially contributed to a cultural climate conducive to change.

At the same time, the public insecurities that resulted from the multiple knowledge problems lingering beneath the surface of debates about technological innovation possessed the potential to rear their heads and to grow into full-blown fears and phobias in times of crisis. Spectacular accidents, such as shipwrecks, aeroplane crashes, and airship explosions, provided such occasions, which threatened to disrupt enthusiasm for technological innovation. Mishaps brought to light the physical perils of the mechanisms that most contemporaries understood only imperfectly. At times, the public rejected devices that proved too dangerous to operate safely. On many other occasions, however, attitudes to technology's physical risks moderated fears that could have amplified the anxieties that sprang from a limited grasp of technological intricacies, thus complementing the strategies that contained public ambivalence, which we just encountered. Public assessments of the physical perils that revealed themselves when mechanisms violently malfunctioned provide crucial clues to why British and German societies supported the development of technological innovations whose dangers nobody denied.

CHAPTER 3

*Accidents: the physical risks of technology*

No matter how enthusiastically the British and German public welcomed innovations, they were also aware of their potential and actual dangers. While, as we have seen in the preceding chapter, German and British assessments of new technologies ambivalently vacillated between fascination and uncertainty, accidents, often resulting in mutilation and death, dramatically demonstrated technologies' perils and confirmed latent fears of new objects. Airship explosions, plane crashes, and shipwrecks regularly featured in the main news pages as well as in cinema newsreels.[1] Lengthy reports used eye-witness evidence to describe in detail both how individual accidents unfolded and how victims died in flaming infernos or icy waters. In fact, some disasters, such as the sinking of the *Titanic* in 1912 or the explosion of the airship *Hindenburg* in 1937, have left deep traces in the collective memory to the present day, in part because they can retrospectively be viewed as symbolic harbingers of the devastating wars that erupted soon afterwards. Despite occasional claims that these technical catastrophes undermined exuberant optimism about technological progress born out of a collective sense of hubris, these events did not constitute turning points in the popular understanding of technology that might have led to a rejection of innovation at the time.[2] Indeed, within a year of the *Titanic* disaster, British and German shipyards launched even bigger liners to great public acclaim, and immediately after the *Hindenburg* had been destroyed, the National Socialist regime ordered the completion of a new airship that took to the skies only sixteen months later. Generating immense public attention, these accidents brought to the surface widespread fears that usually

---

[1] The explosion of the German airship *L2* in 1913 was international cinema news. See *Pathé Cine Journal*, 23 October 1913, 35–6.
[2] This is argued by several scholars. See Hermann Glaser, *Bildungsbürgertum und Nationalismus: Politik und Kultur im Wilhelminischen Deutschland* (Munich, 1993), 98–9; Donald Read, *England, 1868–1914: The Age of Urban Democracy* (London, 1988), 419. For an interpretation stressing the wide range of contemporary reactions to the *Titanic*, see Stephen Kern, *The Culture of Time and Space* (Cambridge, MA, 1983), 107–8.

remained submerged. Steven Biel's assessment that the loss of the *Titanic* "expose[d] and [came] to represent anxieties about modernity" holds true of many other technological mishaps as well.³

Though presently still rare, historical studies of risk have produced a heterogeneous body of knowledge. Elaine Freedgood's recent book emphasizes how popular Victorian writings that celebrated the dangerous exploits of mountaineers and colonial explorers sought to banish danger "from the domestic scene" and relocate it "in the world outside British borders." Freedgood credits nineteenth-century British culture with a considerable aversion to risk, as do studies of traffic legislation that aimed to prevent car accidents in the wake of the proliferation of the automobile, which maimed and killed increasing numbers of innocent bystanders.⁴ Rather than drawing attention to the limits of risk tolerance, investigations into labor conditions in the nineteenth and early twentieth centuries have tended to come to different conclusions, pointing out that persistent high levels of danger in factories, mines, and other workplaces indicate a widespread acceptance of physical risk in industrial environments.⁵ The spread of amusement parks between 1880 and 1930, with their fast roundabouts and roller-coasters, alerts us to yet another phenomenon: urban crowds thoroughly enjoyed the sensation of risk in "thrill rides" as long as it remained just that – a fleeting sensation.⁶ These diverse examples illustrate that the uneven contours of risk societies result from complex social, cultural, and political processes that reject certain dangers while accepting or even perhaps reveling in others.⁷

---

³ Steven Biel, *Down With the Old Canoe: A Cultural History of the Titanic Disaster* (New York, 1996), 8.
⁴ Elaine Freedgood, *Victorian Writings about Risk: Imagining a Safe England in a Dangerous World* (Cambridge, 2000), 1; Sean O'Connell, *The Motor Car in British Popular Culture* (Manchester, 1998), 112–49; Clive Elmsley, "'Mother, What Did Policemen Do When There Weren't Any Motors?' The Law, the Police and the Regulation of Motor Traffic in England, 1900–1939," *Historical Journal* 36 (1993), 357–81.
⁵ Arne Andersen and René Ott, "Risikoperzeption im Industriezeitalter am Beispiel des Hüttenwesens," *Archiv für Sozialgeschichte* 28 (1988), 75–109; Helmut Trischler, "Arbeitsunfälle und Berufskrankheiten im Bergbau 1851–1945," *Archiv für Sozialgeschichte* 28 (1988), 111–51; Anson Rabinbach, "Social Knowledge, Social Risk, and the Politics of Industrial Accidents in Germany and France," in Dietrich Rueschemeyer and Theda Skocpol (eds.), *States, Social Knowledge, and the Origins of Modern Social Policies* (Princeton, 1996), 48–89. Mark Aldrich provides a variation on this theme by arguing that American business interests demanded an improvement of labor security to ensure steady operation that brought predictable profits. See Mark Aldrich, *Safety First: Technology, Labor, and Business in the Building of American Work Safety* (Baltimore, 1997).
⁶ Arwen P. Mohun, "Designed for Thrills and Safety: Amusement Parks and the Commodification of Risk, 1880–1929," *Journal of Design History* 14 (2001), 291–306.
⁷ On the selection and rejection of risk, see Mary Douglas and Aaron Wildavsky, *Risk and Culture: An Essay on the Selection of Technological and Environmental Dangers* (Berkeley, 1983), esp. 29–48. The term "risk society" is, of course, drawn from Ulrich Beck, *Risikogesellschaft. Auf dem Weg in eine andere Moderne* (Frankfurt, 1986).

## Accidents 53

This chapter examines how British and German public debates evaluated the physical risks – or, put more bluntly, the dangers to life and limb – that aerial and naval technology created and that played a prominent part in defining Britain and Germany as risk societies. It does not explore contemporary discussions about the physical dangers of the cinema. To be sure, picture shows before 1910 exposed patrons to physical dangers because exhibitions used highly flammable celluloid film, which started cinema fires on more than one occasion. Towards the end of the first decade of the twentieth century, however, British and German legislation forced the film industries to distribute their products on fire-resistant celluloid, thereby effectively reducing the physical risk of "going to the pictures."[8] From the 1910s on, the cinema began to be considered as a cultural rather than a physical threat, and later we shall treat the specific dangers that Britons and Germans associated with the cinema between the 1900s and World War II.

In the context of shipping and flight, Germans and Britons primarily focused on three aspects of their physical dangers. First, reports often noted the destructive violence of accidents and determined their impact on the human body and psyche. Second, public inquiries pursued the causes of disasters. Third, the British and German public engaged in vociferous arguments about the best ways of improving the safety of ships, aeroplanes, and airships to reduce their risks in the future. Britons and Germans generally championed new flight technologies and passenger vessels *despite* their risks, although they reserved the right to speak out against individual artifacts that did not improve their safety records. The prominent and forceful debates that contributed to cultural environments that favored the acceptance of technologies' physical risks reveal a host of similarities between Britain and Germany before 1933. It was only after the rise of the Nazi regime that major contrasts began to emerge in the two countries' assessments of technological risk. Irrespective of the form of interwar political government in Britain and Germany, debates about the dangers of aerial and maritime travel during the interwar years differed in one fundamental respect from public exchanges before 1914: the focus of attention shifted from passenger shipping to aviation. While aeroplane

---

[8] One particularly spectacular cinema fire occurred in Paris in 1897. See H. Mark Gosser, "The *Bazar-de-la-Charité* Fire: The Reality, the Aftermath, the Telling," *Film History* 10 (1998), 70–89. On safety legislation in Britain, see Rachel Low, *The History of the British Film, 1906–1914*, 2nd edition (London, 1973), 58–62. On Germany, see Gary D. Stark, "Cinema, Society and the State: Policing the Film Industry in Imperial Germany," in Gary D. Stark and Bede Karl Lachner (eds.), *Essays on Culture and Society in Modern Germany* (Arlington, TX, 1982), 122–66.

crashes and airship explosions occurred with almost predictable regularity, transatlantic shipping companies were spared major accidents after 1918. Consequently, passenger liners were widely viewed as much safer than aerial modes of transport, which retained their public image as a high-risk technology until the outbreak of World War II. Nonetheless, debates about aviation's perils took up a host of motifs that had already shaped older considerations about transatlantic shipping, and established continuities between attitudes to technology's physical dangers across the watershed of the Great War. Discussions about safety and danger in aerial and maritime travel concentrated, by and large, on similar issues, and we can therefore treat them in conjunction.

### "IT IS IMPOSSIBLE TO DESCRIBE IT": THE HORRORS OF TECHNOLOGICAL ACCIDENTS

Entering the cockpit of an airship that was about to set out on its transatlantic maiden flight in 1924, a German journalist was struck by the objects he found on display. He noted, among other items, a canary, a teddy bear, a red wooden parrot, and a little carved swallow, all of them "tokens of good luck" (*Glückszeichen*) with which the crew had decorated their workspace.[9] While they playfully signaled other, more practical preparations for problems on the voyage, the talismans also tacitly betrayed the anxiety among the crew at the beginning of a technological feat that entailed placing their lives in significant danger. After all, the airship received its lift from 70,000 cubic meters of highly flammable hydrogen gas, which could be ignited by a tiny spark. Although the launch of this airship did not explicitly give rise to public considerations of the dangers of airship flight, fears of accident and death were indirectly apparent through the mascots.

Such diffuse manifestations of anxieties about technology were widespread and persisted throughout our entire time period, a fact that illustrates the remarkable degree of continuity in attitudes towards technology's physical dangers. Vague apprehensions about the ship's safety supposedly led a wide range of people to decide against traveling on board the *Titanic*.[10] Superstition provides another indicator of the unease with which many passengers embarked on voyages and aerial journeys. One passenger later maintained he "felt nervous" when he learnt that he had been assigned

---

[9] *Neue Preußische Zeitung*, 13 October 1924 (evening edition), 1.
[10] Such stories can be found in the quixotic volume by George Behe, *Titanic: Psychic Forewarnings of a Tragedy* (Wellingborough, 1988). For a discussion of this work, see Biel, *Down with the Old Canoe*, 186–8.

cabin number 13 on a voyage from South Africa to England in 1896. Somewhat ironically, it turned out to be his "lucky" number because he became one of only three survivors of a shipwreck that claimed 143 lives.[11] Three decades later, an aeroplane passenger concealed her anxiety with humor as she joked about her imminent departure to another life and insisted on repaying a friend before take-off "in case we do not meet again."[12] They didn't. After her aeroplane had crashed between Paris and London, the passenger's words lost their ironic quality and instead resembled an uncanny premonition. All these tales illustrate how deep-seated anxieties about technological devices pushed their way to the surface. Beneath a veneer of irony and playfulness, apprehension left visible traces. If the possibility of losing their lives because of technological failure unsettled people, actual accidents that could claim high death tolls confirmed their fears and led British and German observers to articulate their alarm at technology in drastic terms. Technological mishaps thus brought into the open the silent apprehensions that stalked processes of innovation.

Survivors of accidents, as well as reporters visiting the scenes where machines had come to grief, often emphasized how shocked they were at the sights they confronted. A journalist standing next to the burnt-out wreck of a Lufthansa passenger aeroplane in 1927 professed to be "deeply shaken by the terrible misfortune" he beheld.[13] Those fortunate enough to escape from shipwrecks or plane crashes were often traumatized by recurring nightmares that made them unable to eat or sleep for days.[14] In fact, shock affected survivors so strongly that many were left in a "state of collapse," which rendered it impossible to furnish "anything like a coherent account" of events.[15] Experiences and images of technological accidents thus often proved too powerful to be readily contained in narratives. One injured survivor of an airship explosion found that words failed him when he tried to convey the horror he had lived through: "the fire was awful – awful. It is impossible to describe it."[16] In 1935, a journalist on the staff of the *Daily Herald* – a tabloid that frequently entertained its readership with hair-raising tales – refrained from providing a detailed account of the victims who had perished in an aeroplane crash because "their injuries were so shocking" that they defied description.[17] The scale of destruction, suffering, and confusion caused by accidents rendered straightforward narration

---

[11] *The Times*, 22 June 1896, 12.    [12] *Daily Mail*, 4 October 1926, 9.
[13] *Berliner Tageblatt*, 23 September 1927 (evening edition), 1.
[14] See the account by a survivor of a shipwreck in 1896, *The Times*, 22 June 1896, 12.
[15] *Daily Herald*, 30 May 1914, 3.    [16] *The Times*, 6 October 1930, 14.
[17] *Daily Herald*, 10 December 1935, 1.

difficult.[18] The numerous "rumors" that sprang up in the wake of major misfortunes added to a general atmosphere of uncertainty and disorientation.[19] Thus technological mishaps took the established signifying order to its limits as the British and German public struggled to make sense of the disruption caused by malfunctioning machines.

Of course, initial speechlessness in the face of disaster never lasted long. On the contrary, accidents attracted wide public coverage exactly because of the representational challenges they mounted. Technological misfortunes invited intensive attempts to interpret them, since they signaled the spectacular breakdown of a man-made external order. Although they did not ascribe definitive meaning to mishaps, tales of the confusion that accidents left in their wake used several narrative devices to overcome speechlessness. Disaster scenes struck observers as disjointed sites of chaos. While reports of shipwrecks often featured outbreaks of panic among passengers, accounts of aerial accidents focused on the physical appearance of the wrecks themselves to emphasize the havoc that the accidents had wreaked on erstwhile stable structures.[20] A journalist inspecting the remnants of an airship found it difficult to orient himself among the "weird twisted shapes of metal through which quivers of smoke still edd[ied]."[21] Others confronted a "shapeless mass" of debris or a "heap of wildly tangled tubes" that bore no resemblance whatever to flying machines.[22] Photographs depicting twisted girders and other metal parts accompanied such dispatches from crash sites.[23]

In the eyes of many observers, the places at which malfunctioning aeroplanes and airships had hit the ground turned into death zones replete with traces of apocalyptic tragedy. Approaching the wreck of an aeroplane in 1926, one newspaper man considered himself face to face with a "steel skeleton" composed of a "scarred mass of snapped and twisted, fire-blackened fragments."[24] The remains of the British airship *R101*, which was destroyed in an explosion that killed fifty-one people in 1930, also offered a desolate

---

[18] Medical anthropology has investigated how confrontations with death affect people's ability to form coherent narratives about their life events. See Gay Becker, *Disrupted Lives: How People Create Meaning in a Chaotic World* (Berkeley, 1997), 37–79.

[19] The relationships between rumor and anxiety are investigated in Patricia A. Turner, *I Heard it Through the Grapevine: Rumor in African-American Culture* (Berkeley, 1993). Many stories later called "rumors" sprang up around the wreck of the *Titanic*. See *The Times*, 16 April 1912, 9; *Manchester Guardian*, 16 April 1912, 8; *Daily Mail*, 16 April 1912, 9. *Daily Herald*, 30 May 1914, 3.

[20] For a detailed investigation of stories about panic during the wreck of the *Titanic*, see Biel, *Down with the Old Canoe*, 46–53.

[21] *Daily Herald*, 6 October 1930, 1.

[22] *The Times*, 6 October 1930, 14; *Vorwärts*, 24 September 1927 (morning edition), 6.

[23] *Flight*, 10 October 1930, 1113.    [24] *Daily Mail*, 4 October 1926, 9.

sight. Among a pile of rubble, only a row of pillars had stayed intact, and they "seemed to grin garishly like false teeth in a skull."[25] Metaphors of the death of machines and humans merged, conveying the eerie transformations of the spaces in which accidents had occurred. According to his account of the night-time rescue efforts after an airliner had crashed into a wooded hillside in December 1935, a journalist who had picked his way through the dark suddenly found himself "within a foot of a mangled body. People moved about in a ghostly sort of way, and every now and then a shout announced that one of the party had found another victim."[26] This accident transformed a stretch of countryside into a somber deathscape in which the living uncannily began to resemble, and mingle with, the perished machine and its passengers. As the lethal violence of technological breakdowns destroyed the boundaries separating the world of the living and the realm of the dead, descriptions of disaster sites during the interwar period carried overtones of the horrors of No Man's Land and the battlefields of the First World War.[27]

Accounts of lethal accidents commonly noted that it was impossible to initiate effective counter-measures to reassert control over mechanisms once mishaps began to unfold. From the moment mechanisms started to malfunction, technology appeared to slip from human control altogether. Often, the sheer swiftness with which disasters occurred overwhelmed eyewitnesses and survivors. After hitting the rocks off the French Atlantic coast in 1896, virtually everyone on board the *Drummond Castle* died because she sank with such an "appalling . . . suddenness" as to leave no time for emergency measures.[28] Similarly, the *Empress of Ireland* foundered with "terrifying rapidity," taking over 1,000 people to the bottom of the sea in 1914.[29] Plane crashes and airship explosions occurred with even more disconcerting suddenness. A radio operator who escaped the "masses of flames" that engulfed the British airship *R101* in 1930 told reporters that "it was all over in a minute."

The abruptness so distinctive of many accidents rendered it very difficult to take life-saving measures. Steamships often sank too quickly for crews to initiate rescue operations or for other vessels in the vicinity to come to their aid. Airships and plane crashes instantaneously engulfed travelers in

---

[25] *The Times*, 6 October 1930, 14.  [26] *Daily Herald*, 11 December 1935, 2.
[27] For descriptions of battlefields, see Modris Eksteins, *The Rites of Spring: The Great War and the Birth of the Modern Age* (London, 1989), 139–69; Paul Fussel, *The Great War and Modern Memory* (Oxford, 1975).
[28] *Manchester Guardian*, 18 June 1896, 7. See also *The Times*, 19 June 1896, 10.
[29] *Daily Herald*, 30 May 1914, 3.

fireballs from which there was no escape. The radio operator of the *R101* "went along [the airship] to see if [he] could get anybody out, but nothing could be done."[30] Similarly, German farmers hurrying towards a wrecked plane in 1929 soon realized "that any help . . . came too late" after the machine caught fire upon hitting the ground. These bystanders felt "awful" because of their helplessness, for they "could do nothing but look on" as the passengers burnt to death.[31] Aerial and naval disasters thus appeared to obey a destructive logic which resisted human intervention and lent lethal results an aura of inevitability.

To highlight the destructive violence of such accidents, reports were keen to describe accident victims – or their remains. While shipwrecks left few traces of the deceased, many aerial disasters took place in full view of the public, who witnessed horrific deaths. Since airship and aeroplane crashes frequently started as, or resulted in, huge explosions, crew members and passengers burnt to death. Newspaper articles did not spare readers the grim details of such events, telling how victims' bodies visibly bore the signs of the violence accidents unleashed. Graphic coverage of aeroplane and airship crashes was not only the preserve of the penny press; it also featured in the respectable broadsheets and formed part of a long-standing tradition of sensationalism that reached back to the nineteenth century, when railway accidents, boiler explosions, and urban firestorms had received similar coverage.[32] One survivor of the *R101* explosion told journalists that he had heard the "shrieks and cries of men trapped in the airship who saw themselves entombed and engulfed in a prison of fire." On the same occasion, a witness at the scene "saw one man . . . trying to get out of a cabin but he was terribly burned and we saw him fall back helpless into the flames."[33] More than anything else, the agony of victims testified to the brutality with which technologies maimed humans. After bystanders had rescued three soldiers from the smoldering wreckage of the German airship *L2* in 1913, it was by no means clear whether they had done the victims a service. "Burnt all over [and] moaning dreadfully," one man

---

[30] *The Times*, 6 October 1930, 14.  [31] *Vorwärts*, 24 September 1929 (morning edition), 6.
[32] See Wolfgang Schivelbusch, *The Railway Journey: The Industrialization and Perception of Time and Space in the Nineteenth Century* (Berkeley, 1986), 129–33. Such coverage could be found on both sides of the Atlantic. See Michael Barton, "Journalistic Gore: Disaster Reporting and Emotional Discourse in the *New York Times*, 1852–1956," in Peter Stearns and Jan Lewis (eds.), *An Emotional History of the United States* (New York, 1998), 155–72; Scott Gabriel Knowles, "Lessons from the Rubble: The World Trade Center and the History of Disaster Investigation in the United States," *History and Technology* 19 (2003), 9–28. For a bloodthirsty American report of a railway accident from the 1870s, see Aldrich, *Safety First*, 14.
[33] *Daily Herald*, 6 October 1930, 1, 3.

begged onlookers: "Kill me, kill me, I'm suffering too much."[34] According to the British press, Captain Lehmann, the commander of the *Hindenburg*, had "staggered from the wreckage" with "his clothes ablaze" and died in a hospital "terribly burned . . . staring at the ceiling all day crying 'Water, water.'"[35] Even when witnesses arrived too late on the scene to see people actually die, their accounts still contained graphic portrayals of victims' bodies that bore the unmistakable marks of a violent death. Hardly any report from the site of an airship or an aeroplane crash held back from describing human physical disfigurement in great detail. After explosions, victims' bodies were often too charred to be identified with certainty.[36] Similarly, when a Lufthansa aircraft hit the ground killing all on board in 1927, newspaper articles depicted the accident site as having been littered with "pieces of human flesh." In fact, the victims were so mutilated that the German authorities decided not to grant the relatives permission to see the remains of their loved ones.[37] Reports of other accidents told readers that a dead pilot's body had been reduced to a "torn-up mass" (*zerfetzte Masse*), or that rescue workers had resorted to "picks and shovels" to extract two bodies that had been "found embedded deep in the sodden soil" at a crash site.[38]

Such graphic descriptions of human physical disfigurements not only underlined the disruptive violence of technological accidents; they also indicated the personal dimensions of aerial and naval disasters. Chilling reports rendered technological dangers comprehensible by illustrating how individuals came to grief through novel artifacts. In short, disaster reports translated the abstract concept of "risk" into descriptions of concrete events with tragic human implications. Yet powerful public associations between technological havoc and personal risk provided an evocative background for individual heroic feats. Public representations foregrounded those who strove to rise above the destructive chaos at the scene of an accident. Acts of personal heroism, however, often turned out to be futile, only underscoring the dangers of the aeroplane, the airship, and the passenger liner. Numerous people who rushed to crash sites engaged in rescue attempts that proved both perilous and futile. In one instance, two men who had hurried to a machine right after it had blown up in flames were unable to save any passengers, although they rushed "into the heat" to pull burning wreckage

---

[34] *Neue Preußische Zeitung*, 17 October 1913, 1; See also *B. Z. am Mittag*, 17 October 1913, 2.
[35] *Daily Mail*, 8 May 1937, 9.
[36] For some examples, see *The Times*, 6 October 1930, 14; *Manchester Guardian*, 7 October 1930, 9; *Vorwärts*, 18 October 1913, 2.
[37] *Berliner Tageblatt*, 24 September 1927 (morning edition), 1; *Vorwärts*, 24 September 1927 (evening edition), 3.
[38] *Berliner Tageblatt*, 19 September 1930 (morning edition), 4; *Daily Mail*, 11 December 1935, 13.

from the cabin. They suffered facial burns themselves, and their heroism was of no avail.[39] On another occasion, a man who arrived at the burning wreck of an aeroplane "saw a man's legs sticking out close to the . . . engine . . . His clothing was alight, but his face was quite untouched. I put the fire out with a fire extinguisher . . . As I and other men were pulling him away there was a big explosion." In this instance, rescuers failed to save the victim and were fortunate to escape alive.[40]

Accounts of shipwrecks also stressed the "tragic splendour" of self-sacrificial acts through which individuals transcended the chaos of each unfolding technological catastrophe.[41] Although stories about personal heroism began to circulate in the wake of many naval disasters, the best-known and most elaborate examples emerged after the *Titanic* had gone down with its illustrious passengers.[42] "Duty and self-sacrifice" allegedly ruled supreme during the ship's last hours.[43] In accordance with conventional gender prescriptions, reports by survivors celebrated the sacrificial heroism "of women refusing to leave their husbands, and of husbands forcing their wives into the boats before themselves quietly joining the ranks of doomed men." The press singled out prominent individuals such as John Jacob Astor, Benjamin Guggenheim, and Major Archibald Butt, who displayed particularly commendable feats of "unselfish bravery of which Anglo-Saxons on both sides of the Atlantic may well be proud."[44] Above all, "order" was said to prevail in the face of impending destructive chaos. Although most travelers on the *Titanic* faced certain death, their behavior confirmed "the absolute insignificance of tremendous danger in the presence of the calm courage of the white man."[45] Such individual heroic acts were claimed to have calmed passengers from "dubious" social and racial backgrounds who appeared on the verge of uncontrollable panic, thereby allowing rescue efforts to continue.[46] Reports of naval disasters thus not only pointed out that the havoc of technological accidents endangered human lives; shipwrecks also illustrated how individuals disregarded danger through exceptional acts of bravery to uphold a social order under threat.[47] As accounts of shipwrecks addressed both the personal and the

---

[39] *Daily Mail*, 27 December 1926, 9.
[40] *The Times*, 4 October 1926, 14–16. See also *Daily Mail*, 4 October 1926, 8.
[41] *Daily Mail*, 17 April 1912, 6.
[42] For an examination of British reactions to the sinking of the *Titanic*, which inexplicably does not draw on the daily press, see Richard Howells, *The Myth of the "Titanic"* (Basingstoke, 1999). On contemporary American reactions to the *Titanic*, see Biel, *Down with the Old Canoe*, 23–85.
[43] *Manchester Guardian*, 20 April 1912, 8.   [44] *The Times*, 20 April 1912, 9.
[45] *The Times*, 17 April 1912, 6. See also *Daily Mail*, 20 April 1912, 4.
[46] *The Times*, 20 April 1912, 10.
[47] Similar accounts emerged when the *Empress of Ireland* foundered. See *Daily Herald*, 1 June 1914, 3.

social dimensions of technological risk, they emphasized human agency in the face of certain perdition and simultaneously underlined the destructive nature of technological accidents. Even heroes could not avoid going down with the ship, nor could they save victims from crash sites.

Above all, the social dimension of technological disasters manifested itself in public rituals of mourning. Many aerial and naval accidents made an emotional impact on British and German societies that went far beyond the families who lost relatives. After the *Titanic* had foundered, the *Manchester Guardian* noted a "sense of national depression" that found expression in memorial services and contributions to a relief fund for the survivors.[48] Similarly, the explosion of the airship *R101* triggered ostentatious national grief in Britain in 1930. Two days after the airship had blown up in a fireball, thousands waited in the drizzling October rain past midnight outside Victoria Station to witness the return of the victims' remains to the capital.[49] Before the memorial service was held in St. Paul's two days later, 250,000 people queued to file past the coffins of the perished, which were on display in Westminster Hall. The *Daily Mail* considered this event to be proof of national unity as Britons from all walks of life paid their respects to the dead, while the *Daily Herald* chose to emphasize the mourners' extraordinary dedication: "Hundreds fainted. The pavements became temporary hospitals. Those who recovered from faintness asked to go back to the queue." In fact, entire streets had to be blocked off to keep the mourners out of harm's way.[50] On this occasion, the grieving community publicly acted out the very spirit of sacrifice that, as virtually every leader article stressed, had commended the crew of the ill-fated airship, thereby expressing reverence for the victims and simultaneously asserting that the nation in mourning shared and affirmed this quality of the perished "heroes."

Some technological disasters even led to international demonstrations of compassion that transcended divisions between nations. When the *Titanic* sank, official condolences poured in from European nations including Germany, France, Italy, and Belgium, as well as from Brazil, Chile, China, and Japan.[51] One British paper approvingly quoted a German parliamentarian who called the wreck of the *Titanic* a "cosmopolitan calamity" because "men [were] brought together in sympathy and mutual help and good will all over the world" by the "tragedy."[52] The explosion of the *R101* entailed a comparable international display of sympathy as condolences again arrived from all over the world. Hugo Eckener, the director of Germany's leading

---

[48] *Manchester Guardian*, 18 April 1912, 6; *The Times*, 18 April 1912, 9; *Daily Mail*, 19 April 1912, 6.
[49] *Daily Mail*, 8 October 1930, 11; *Daily Herald*, 8 October 1930, 1.
[50] *Daily Mail*, 11 October 1930, 11; *Daily Herald*, 11 October 1930, 1.
[51] *The Times*, 17 April 1912, 10; 19 April 1912, 10.   [52] *The Times*, 17 April 1912, 9.

airship company, traveled to Britain to attend the funeral of the airshipmen who had burnt to death.[53] Exhibitions of mourning on an international scale, however, occurred only after exceptionally spectacular accidents, and especially when prominent travelers had lost their lives. The *Titanic* went down with leading society and business figures on board, while the destruction of the *R101* wiped out nearly the entire top tier of Britain's aviation establishment, not to mention several titled politicians, including the Secretary of State for Air, Lord Thomson.

Despite the national and international repercussions of aerial and naval accidents, it remains difficult to determine how the British and German population perceived these technological disasters on a personal level. On more than one occasion, the spectators who flocked to disaster sites displayed a distinct lack of piety when they immediately initiated a blossoming trade in plundered wreckage. Although the authorities lost no time to post guards around the smoking hull of an airship that had exploded in Berlin in October 1913, they could not prevent "thousands of souvenir hunters" from scrambling for the parts that "fell from the vans" that were transporting the wreck back to its mooring station. "Knots of people threw themselves at individual pieces," dividing them up, and either taking their trophies home or selling them to the highest bidder "at quite steep prices." Fighting over scrap metal like dogs over a bone, onlookers of this variety were said to disgrace themselves through a potent absence of dignity.[54] Albeit in a less blatantly sacrilegious fashion, publishers also strove to turn accidents into profits. Within days of the crash of the German airship *L1* in a storm off Heligoland on 9 September 1913 that killed fifteen soldiers, a small firm from Kiel issued four postcards depicting the event in a series of photomontages. A set was purchased by a man holidaying in Wilhelmshaven, who, rather than using the cards to reflect upon this recent technological mishap that the press had declared a national disaster, covered them with platitudes about the weather and the quality of the "pretty" photos, eventually sending them "with a thousand kisses" to his fiancée (see illustration 3.1).[55] In this instance, the accident triggered a multi-layered private response. Although national news about the event induced this man

---

[53] *Daily Herald*, 6 October 1930, 3; *Manchester Guardian*, 13 October 1930, 10.
[54] *B. Z. am Mittag*, 18 October 1913, 1. "Souvenir hunters" also flocked to crash sites in Britain. In 1936, a volunteer complained that scavengers had disturbed rescue operations after an aeroplane crash outside Croydon. See *Manchester Guardian*, 17 January 1936, 5.
[55] Press reports evaluating the national significance of the explosion of *L1* can be found in *B. Z. am Mittag*, 10 September 1913, 1; *Neue Preußische Zeitung*, 10 September 1913, 1; *Frankfurter Zeitung*, 10 September 1913 (morning edition), 1. The postcards are kept by Deutsches Schiffahrtsmuseum Bremerhaven, archive, postcard collection, file 90/9.

Illustration 3.1  Photomontage of *L1*. The third postcard in a series of four, this photo shows the *L1* breaking up after falling into the sea. The text on the reverse reads: "My good, dear, little Berta, here you see how the airship crashes into the sea. The images are rather clear and good, aren't they? Tell me please if you got all four cards. How is the weather back home? It's excellent here. A thousand kisses from your tramp."

to buy the set of cards, in his writing he did not join the chorus that described the crash as a national misfortune. Despite its character as a memorable event, the accident remained of marginal personal significance for this holidaymaker.

This limited private impact of aerial and naval disasters may have been a consequence of the fact that many public accounts referred to their social *and* individual dimensions. While a host of press reports ascribed national or even international significance to accidents, the many individualized depictions of victims and heroes described above could also allow contemporaries to dissociate themselves from technologies' potential horrors. After all, they signaled that such misfortunes happened to distinct and exclusive groups of people – in these cases, soldiers and travelers. Unlike car accidents, which had harmed increasing numbers of passive bystanders such as pedestrians since the turn of the century and had led to vociferous calls for safety legislation, the development of aerial and naval technologies entailed only limited "risk impositions" on society.[56] Thus, personal

[56] For this term, see Andreas Teubner, "Justifying Risk," *Daedalus* 119 (1990), 235–54, here 236.

stories about victims and heroes in accidents not only rendered the abstract category of "risk" imaginable; they also emphasized that only select people were exposed to the dangers of passenger shipping and aviation. Individualizing narratives, therefore, restrained the cultural impact of descriptions of the violence that aerial and naval technologies could unleash as they left nothing but chaos, destruction, and maimed bodies in their wake. An awareness of the unequal social distribution of technological danger therefore helped Britons and Germans both to live in and to affirm the "risk societies" of their day. Nonetheless, given the spectacular and shocking effects ascribed to technological accidents, British and German societies felt a need to demonstrate their grief over mishaps through rituals of mourning on both a national and an international scale. These public demonstrations were, however, only one dimension of the attempt to come to terms with disasters; Britons and Germans also demanded information about their "causes."

## "STILL MANY PROBLEMS ARE IMPERFECTLY UNDERSTOOD": PURSUING THE ELUSIVE CAUSES OF ACCIDENTS

Although a wide range of investigations strove to explain why people lost their lives as a result of accidents, they often contributed little towards a greater public understanding of technology. After all, many analyses involved a degree of technical and scientific complexity that rendered the subject matter at hand beyond the grasp of non-experts. Furthermore, inquiries into accidents also revealed the concept of "cause" to be a problematic category as specialists sometimes admitted that they themselves were at a loss to explain why technologies broke down. Disasters frequently pushed experts and technological laypersons alike towards and beyond the limits of their analytical faculties. Investigators of air crashes and shipwrecks made no secret of their inability to determine the specific factors that had led to particular accidents or what had rendered them so deadly. Rather helplessly, they called the origins of misfortunes a "mystery" or impressed upon observers that the search for evidence was "exceptionally difficult."[57] At times, conflicting eye-witness accounts created more confusion than clarification about what had happened and why.[58] Up to a point, investigators could blame a lack of evidence for their problems. Shipwrecks took all clues to the bottom of the sea, while explosions of airships and aeroplanes

[57] *Daily Herald*, 27 December 1924, 9; *Vorwärts*, 23 September 1927 (morning edition), 1.
[58] *The Times*, 27 December 1927, 10.

destroyed most pieces of helpful information. As one commentator pointed out to readers, "the crashing of an aeroplane usually involves a destruction of the machine and passengers so thoroughgoing that there is little evidence of the causes of the mishap."[59] Often, the most fundamental and obvious reasons for deadly accidents puzzled experts. Engineers expressed surprise after the ship *Drummond Castle* had sunk within three minutes of hitting a reef in 1896, since it was equipped with bulkheads that should have given it at least enough time to launch the lifeboats.[60] Similarly, when a German aeroplane lost its left wing in mid-air in 1927, the German aviation authorities, the Lufthansa representatives, and the machine's manufacturers all failed to come up with a convincing explanation in public.[61] Ten years later, all observers agreed that the highly inflammable hydrogen gas that had provided buoyancy for the airship *Hindenburg* was responsible for the giant's explosion at Lakehurst. Still, since none of the survivors and eyewitnesses, or an examination of the wreck, managed to provide conclusive evidence, the immediate cause for the ignition of the gas eluded empirical detection, and subsequent explanations consequently remained on the level of theoretical speculation.[62]

In this atmosphere of uncertainty, contemporaries pinned their hopes on formal, institutionalized inquiries, which they expected to probe the reasons of technological failure "impartially." Formalized procedures, which had developed during the second half of the nineteenth century to investigate industrial accidents, had a comforting effect on the public because they required authorities to confront the disturbing aspects of technological development and promised to uncover useful information that would serve to prevent their recurrence. Nonetheless, even concentrated scientific and technological expertise provided no guarantee that the causes of accidents would be found. A formal investigation into an aircrash on the outskirts of London on Christmas Eve 1924 was unable to determine the reason for the misfortune.[63] Philip Sassoon, the Under Secretary of State for Air at the time, was not astonished at this outcome, since there were "still many air

---

[59] *Manchester Guardian*, 27 December 1924, 8.    [60] *The Times*, 19 June 1896, 10.
[61] *Neue Preußische Zeitung*, 23 September 1927 (evening edition), 1; *Berliner Tageblatt*, 24 September 1924 (morning edition), 2; *Vorwärts*, 24 September 1924 (morning edition), 3. British examinations could also come up empty-handed. In 1936, an inquest into the crash of a flying boat that killed ten people concluded that "it was almost impossible to state the cause of the accident with complete certainty." The main body of evidence elaborates on various scenarios that could have led to the mishap. See *The Aeroplane*, 25 March 1936, 393.
[62] See the articles published in the British press. *Manchester Guardian*, 8 May 1937, 12; *Daily Mail*, 8 May 1937, 8.
[63] The report can be found in *Flight*, 12 February 1925, 83–5.

problems imperfectly understood."⁶⁴ The press found the inquiry report "reassuring . . . in its thoroughness," but did not hide its disappointment at its inconclusive findings.⁶⁵

Of course, inquiries did produce results in other instances. The investigation into the reasons for the shipwreck of the *Titanic* ruled that "the loss . . . was due to collision with an iceberg, brought about by the excessive speed at which the ship was being navigated."⁶⁶ Yet Britain's exoneration of White Star Line manager Bruce Ismay, who had survived the trip, contrasted sharply with accusations against him during an American investigation, which had held Ismay responsible for ordering the ship to travel too quickly through an icefield.⁶⁷ The question of who had caused the accident was to remain a contested issue that was evaluated differently on both sides of the Atlantic. Even when inquiries appeared to draw unequivocal conclusions, they typically failed to answer all the questions. The official inquiry into the explosion of the *R101* ruled that the vessel had leaked hydrogen in high winds. Subsequently losing lift, the airship had been caught in a downward air current; when the airship hit the ground, sparks from the propeller engines ignited the hydrogen in the gas cells and the vessel exploded.⁶⁸ Although this analysis identified a sequence of causally related events, a crucial technical issue remained unresolved, as one journalist pointed out: "How the vessel began to lose gas can never be definitely ascertained."⁶⁹ Thus finding a "cause" for a technological accident proved elusive when expert analyses opened as many questions as they answered. Although inquiries into disaster causes started as formalized exercises with the purpose of generating authoritative knowledge about technology, they commonly brought new, unresolved issues to the fore, which, in turn, required further investigation. Rather than settling technological problems once and for all, expert investigations thus initiated an open-ended process of scientific and technological inquiry.

If contemporaries considered it difficult to determine the causes of accidents, they found it equally problematic to decide which decision makers and institutions bore responsibility for technological misfortunes. Heated

---

⁶⁴ *Parliamentary Debates: House of Commons* (hereafter *Parliamentary Debates*), fifth series, vol. 180 (London, 1925), 2324.
⁶⁵ *Manchester Guardian*, 11 February 1925, 8; *The Times*, 11 February 1925, 13, 17.
⁶⁶ *Report on the Loss of the "Titanic"*, Cmd. 8194 (London, 1912), 1.
⁶⁷ John P. Eaton and Charles A. Haas, *Titanic: Triumph and Tragedy* (Wellingborough, 1986), 248.
⁶⁸ *Daily Mail*, 1 April 1931, 7; *Daily Herald*, 1 April 1931, 1; *The Times*, 1 April 1931, 15.
⁶⁹ *Manchester Guardian*, 1 April 1931, 4. This passage is a direct quotation from the report presented by the official inquiry. See the reprint of Cmd. 3825 (1931) *R 101: The Airship Disaster, 1930* (London, 1999), 153. The *Daily Herald* followed a similar line of reasoning; see *Daily Herald*, 1 April 1931, 1.

debates about who was to blame for disasters rarely produced agreement as participants exchanged accusations and rebuttals. In particular, British and German elected representatives fought over the issue of political accountability for accidents in their respective parliaments. After the *Titanic* had sunk, the British Board of Trade became the target of severe criticism because it had not forced the White Star Line to provide a sufficient number of lifeboats. In fact, when it transpired that the *Titanic*'s supply of lifeboat accommodation had exceeded the official regulations that had been in force since 1894, but were inadequate for the unprecedented giants launched since the turn of the century, MPs did not hide their indignation. One speaker demanded that the Board of Trade should "really be censured seriously for allowing obsolete rules to remain in force."[70] The President of the Board of Trade, Noel Buxton, defended himself by arguing that the regulations were premised on the assumption that the *Titanic*'s bulkheads represented "a factor in its safety, and that [consequently] such a ship . . . would really require a smaller proportion of lifeboats." Buxton hastened to add that his consideration had met with the endorsement of experts "throughout the country."[71] A fellow parliamentarian came to his aid, pointing out that "everybody was under the impression that the water-tight compartments would make the ship unsinkable."[72] The Board of Trade thus denied political responsibility for the tragedy because it had followed expert advice.

Similarly, the inquiry into the explosion of the airship *R101* ruled that the authorities had not acted irresponsibly when they pressed ahead with the plan for a long-distance flight, although the vessel had not undergone proper trials. Rather, it was pointed out, the Secretary of State for Air, who had insisted on a tight project schedule, had "relied on his experts and it must never be forgotten that he was entitled to do so." The question of who was actually to blame for the accident went unanswered.[73] In Germany, attempts to identify agents accountable for accidents also often came to nothing. After two German naval airships had come down within a month in 1913, German government officials denied allegations that military requirements had necessitated design modifications that had rendered the airship hazardous.[74] Alfred Tirpitz himself, Chief Admiral of the German navy, stepped forward to deny that he had pursued any

---

[70] *Parliamentary Debates*, fifth series, vol. 37, 646.   [71] *Ibid.*, 651.   [72] *Ibid.*, 652.
[73] *Manchester Guardian*, 1 April 1931, 4. Again, this is a direct quotation. For the original, see *R 101: The Airship Disaster*, 157.
[74] This point was put forward by the Social Democrat Gustav Noske. See *Verhandlungen des Deutschen Reichtages* (hereafter *Verhandlungen*), vol. 293 (Berlin, 1914), 7492.

initiatives in opposition to expert advice.[75] As in Britain, political responsibility for airship accidents remained undetermined in Germany as decision makers hid behind the argument that they obeyed the recommendations of prevailing engineering opinion.

Thus Britons and Germans not only encountered seemingly insurmountable obstacles when they searched for the causes of accidents; they also found it difficult to single out the agents responsible for technological disasters. Given these problems of determining the origins of accidents, many contemporaries asserted that technological mishaps were inevitable. No event appeared to confirm this view more than the sinking of the *Titanic*. One editorial laconically commented that the shipwreck showed "that no ship is unsinkable."[76] The *Titanic*'s fate demonstrated the extent of human fallibility and revealed its hubris:

> We think we have mastered Nature, we build these monsters of steel, fit them up with a luxury beyond the imagination of a Roman Emperor . . . and forget there is any danger at all. And then, Nature, the silent, the inscrutable . . . just puts out her finger, and at her touch man is reduced to a helpless pigmy and all his works are swallowed up by nothingness.[77]

"Danger" had therefore been present all the time, despite significant technological advances, and reminders that "risks" could not be altogether eliminated came from many quarters. Naval politician Charles Beresford did not mince words when he declared in a parliamentary debate on the *Titanic* that "you cannot make regulations to guard against everything at sea," and the German liberal press would have agreed with this line.[78] Other commentators took this reasoning further and pointed out that not even a larger number of lifeboats on board would necessarily have reduced the number of victims; more lifeboats might have clogged up deck space, so that their "close proximity" might have prevented their safe launch.[79] While rescue mechanisms might indeed lessen the scale of the disaster, they could not be expected to prevent deaths completely.

With respect to air travel, there also existed widespread agreement that risk could not be avoided altogether. Potential for human error was one factor whose impact engineers and constructors could not exclude. "Improvements on the scientific side both in knowledge and material will not always prevent a mistake being made on the ground or in the air," claimed a

---

[75] *Ibid.*, 7498.  [76] *The Times*, 17 April 1912, 9.  [77] *Ibid.*
[78] *Parliamentary Debates*, fifth series, vol. 37, 1767; *Berliner Tageblatt*, 16 April 1912 (evening edition), 1.
[79] *The Times*, 18 April 1912, 10; *Parliamentary Debates*, fifth series, vol. 37, 519, 1766, 1762.

parliamentarian in 1925.[80] In addition, observers argued that some risks, such as bad weather, simply evaded human influence. Airships in particular appeared to be "at the mercy of the elements."[81] In 1936, an inquiry into an aeroplane accident conducted by the Air Ministry found that bad weather "had undoubtedly contributed" to the disorientation of the pilot, who misjudged his landing approach and crashed short of Croydon, killing all eleven people on board. An earlier coroner's inquest had already freed the pilot of all "blame" because of the confusing weather conditions, which included gales and fog.[82]

Above all, participants in public debates stressed that unpredictable risks were an inevitable facet of novel technological designs. In fact, these dangers appeared to be part of the process of innovation. Questioned about the frequent explosions of airships in 1913, Admiral Tirpitz emphasized to parliamentarians "that dangers in connection with new achievements such as airships develop which no one expects on the drawing board." Ernst Bassermann, a leading National Liberal, agreed with Tirpitz, and considered "construction errors," regrettable as they undoubtedly were, to be unavoidable.[83] To counter charges that the Reichstag was ignoring the dangers faced by soldiers who volunteered to serve in aviation units, it passed an insurance law in 1912, awarding victims of military air crashes and their families the same compensations as war victims.[84] By putting the dangers of human flight on a level with the hazards of war, the German legislature firmly acknowledged the dangers of aviation. Like work safety legislation, which, in the late 1880s, began to compensate laborers injured in highly mechanized workplaces because, so the legal argument ran, industrial accidents were unavoidable, the law for airship crews rested on the premise that innovative machinery inevitably exposed its operators to potentially debilitating, if not lethal, risks.[85] Britons were also convinced that technological innovations generally contained an "element of luck, the unforeseeable, the incalculable," whose threatening side was revealed in mishaps.[86] One journalist for the Labour newspaper the *Daily Herald* employed the idea of ubiquitous risk to relativize the dangers of air travel.

---

[80] *Parliamentary Debates*, fifth series, vol. 180, 2324.
[81] *Manchester Guardian*, 6 October 1930, 8; *B. Z. am Mittag*, 10 September 1913, 1; *Vorwärts*, 10 September 1913, 1; *Neue Preußische Zeitung*, 10 September 1913, 1.
[82] For the report by the Air Ministry, see *The Aeroplane*, 25 November 1936, 698. For the reports of the inquest, see *Manchester Guardian*, 17 January 1936, 5; *Daily Herald*, 17 January 1936, 7.
[83] *Verhandlungen*, vol. 293, 7498, 7514.
[84] The bill can be found in *Verhandlungen des Reichstages: Anlagen zu den Stenographischen Berichten*, vol. 299 (Berlin, 1914), Nr. 485.
[85] Rabinbach, "Social Knowledge," 56–61.   [86] *Daily Herald*, 1 April 1931, 1.

Commenting on an aeroplane crash in 1924, he argued that "risk is common to every form of travel." Enlarging this claim by asserting that "everywhere is risk, everywhere is danger," he found contemporary fears of flying exaggerated. After all, travelers by car or train also faced dangers.[87] British Conservatives and Liberals shared this conviction, maintaining in 1926 that "a certain risk in air travel must continue to exist."[88] Both British and German societies, then, exhibited robust levels of risk acceptance before and after the Great War, considering danger an essential element of innovation processes.

To some extent, moreover, Britons and Germans embraced the dangers of technological innovation because they feared the "risks of inaction."[89] As we shall see in chapter 8, on the national importance of technology, commentators worried about falling behind in technological competition – fears very visibly played out in the direct economic, political, and military rivalries between Britain and Germany. Defending positions at the cutting edge of international technological developments was often seen as a national necessity. Nonetheless, the risks of inaction could be formulated in more general terms, as when the German aviation firm Dornier issued a press release as it sent its flying boat *Do X* on a transatlantic trip in the early 1930s. In its public relations material the company readily conceded that "each ocean crossing represent[ed] a risk in the present state of technology," yet it simultaneously stressed that it confronted potential dangers because advances required hazardous forays into unknown spheres. To demonstrate this point, the press release resorted to a historical analogy, stating that "if Columbus's caravels had been subjected to the regulations which apply to present-day steamships, America would never have been discovered."[90] Put differently, the Dornier managers claimed to have made up their minds in favor of the experimental flight because risk-taking represented an integral element of a quest for exploration that, in their eyes, provided a justification in itself.

Thus British and German debates associated risk with epistemic uncertainty in several ways. Public hearings tended to illustrate the complexities faced by investigators whose formalized inquiries, conducted in the wake of accidents, either were either impaired by a lack of crucial evidence, or, alternatively, uncovered sequences of events whose root cause eluded

---

[87] *Daily Herald*, 27 December 1924, 4.
[88] *The Times*, 4 October 1926, 15; *Manchester Guardian*, 4 October 1926, 8.
[89] Douglas and Wildavsky, *Risk and Culture*, 27.
[90] Deutsches Museum, Munich, archive, Luft- und Raumfahrtdokumentation, LRD 02244–03 (hereafter LRD), "Kritik am Do.X," 3.

detection. As both a shortage and an abundance of empirical information could frustrate explanatory efforts, determining which actors ought to assume political responsibility for technological disasters also proved problematic. After all, British and German government officials sought to exonerate themselves after accidents by drawing public attention to the fact that their policies had complied with expert advice. If even expert knowledge provided no reliable safeguard against mishaps, risk-taking figured as an integral element of the processes of innovation, with accidents the inevitable price society had to pay for advances. At the same time, however, acknowledging the ubiquitous presence of risk posed a potential threat to the public acceptance of technological innovations. If the causes of mishaps remained in the dark, critics could accuse experts and politicians of lacking authority over technologies whose autonomous dynamism culminated in injury and loss of life. While this argument rose to prominence after World War II, in our time period a host of interventions focused on the measures that improved technologies' safety records.[91]

## "INFANT DISEASES?" CONTESTING THE LIMITS OF TECHNOLOGICAL SAFETY

The issue of accident prevention fueled public controversies between, on the one hand, a loose coalition of promoters from industry, Liberals, and Conservatives, who praised the manifold arrangements to minimize the risks of innovative artifacts, and, on the other, the Left, which challenged existing safety regimes. Fierce public debates described the quest for technological safety as a matter of life and death, a trend that newspaper coverage supported by concentrating on fatal accidents. Since the media tended to remain silent when safety systems proved effective and saved potential victims from perdition, mishaps came to be equated with loss of life. In 1907, for instance, after the White Star Line's 12,500-ton ship the *Suevic* had run aground off the Cornish coast in thick fog, a rescue operation managed to save all passengers because bulkheads protected large parts of the vessel against flooding. When it proved impossible to tug the ship off the reef it had struck, engineers decided to sever the hull by a series of controlled explosions, and transported its aft section to Southampton, where it had a

---

[91] Langdon Winner, *Autonomous Technology: Technics-out-of-Control as a Theme in Political Thought* (Cambridge, MA, 1977); Willibald Steinmetz, "Anbetung und Dämonisierung des Sachzwangs: Zur Archäologie einer deutschen Redeform," in Michael Jeismann (ed.), *Obsessionen: Beherrschende Gedanken im wissenschaftlichen Zeitalter* (Frankfurt, 1995), 293–333.

new bow fitted.⁹² Although the accident had threatened over 600 lives and generated considerable local interest, neither the successful rescue operation nor the salvage attempts received wide attention in the national press.⁹³ The White Star managers were unlikely to encourage widespread coverage that might raise uncomfortable questions about the competence of the captain who had navigated the boat on to the rocks, and as a result they lost an opportunity to praise a safety regime that averted considerable loss of life.

The *Suevic* episode illustrates the fact that sensationalistic instincts among the national press alone were not responsible for the restricted coverage of accidents involving few or no deaths. Rather, industrial public-relations departments devoted considerable energy to concealing technological problems from the public eye as long as possible. When Junkers presented its "giant" aeroplane *G38* to an audience of journalists in 1929, its staff were told not to discuss the "engine question." With a weight of over 20 tons and driven by four engines with a combined 3,000 horsepower, the new machine was severely underpowered, a fact that reporters might have interpreted as a potentially life-threatening hazard.⁹⁴ To avoid premature stigmatization, Junkers remained silent about the aeroplane's motors. Dornier also refused to acknowledge similar technical problems relating to its flying boat, *Do X*, labeling engine troubles as an "infant disease."⁹⁵ Thus companies strove to hide or belittle those risks that were not lethal – at least not yet.

The reluctance of the industry to disclose non-fatal operational difficulties stemmed partly from apprehensions that the media would seize on such incidents to portray new technologies as life-threatening devices, thereby undermining public trust. Even when crews managed to maintain control of flight technologies in the face of technical problems, reporters penned dramatic accounts that presented these events as narrow escapes from death. It proved impossible for the Zeppelin company to regulate the way in which a journalist wrote about an unexpected occurrence when the airship *Graf Zeppelin* encountered turbulence that tore apart some of the fabric covering its stabilization fin on its maiden flight across the Atlantic in 1928. To convey how calmly and efficiently the crew handled even a potentially

---

⁹² For a detailed account, see John P. Eaton and Charles A. Haas, *Falling Star: Misadventures of White Star Line Ships* (Wellingborough, 1989), 84–98.
⁹³ The reports in *The Times* remained brief and did not follow all the stages of the event. See *The Times*, 19 March 1907, 8; 20 March 1907, 5; 3 April 1907, 13.
⁹⁴ Deutsches Museum, Munich, Junkers archive (hereafter Junkers archive), file 0303 T10/36, note, 10 October 1929, 1.
⁹⁵ LRD 02244–03, "Kritik am Do.X," 2.

"fatal" mishap, flight commander Hugo Eckener couched his account in sober terms as a primarily technical challenge mastered by successful crisis management. His version, however, was overshadowed by a highly melodramatic report of the incident by journalist Rolf Brandt, which cast the crew's efforts to fix the fin as a struggle for life and death, and described how Eckener worried about his son, who had volunteered to carry out repairs in mid-air "during the fateful moments."[96] Although the expert's account of the successful emergency operation was intended to dispel fears that technological accidents automatically resulted in deaths, the journalist interpreted the same event as evidence for the life-threatening character of technological risks.

In consequence, the industry resorted to other strategies to demonstrate the safety of innovative technologies. Rather than singling out examples of successful crisis management, companies drew attention to the various design features intended to prevent life-threatening situations from occurring in the first place. Promotional pamphlets and press releases abounded with praise for the technical security systems within new artifacts, to stress that innovation did not occur at the expense of safety. Shipping lines routinely listed a whole series of precautions that kept vessels and travelers out of harm's way. Double hulls, bulkheads, electrical fire-alert systems, multiple steering systems, internal communication by telegraph, valve systems, and countless other features – all guaranteed that passengers would arrive at their destinations safely.[97] Furthermore, aviation firms reassured the public that their products included many back-up provisions that safeguarded the proper functioning of flying machines even when part of their standard equipment failed. In the early 1930s, for instance, Dornier boasted that the *Do X* could still fly after four of its twelve engines had been shut off in a test, while the Junkers works similarly considered the *G38*'s ability to land on a flat tire an important safety feature.[98] Strict adherence to technological specifications necessary for safe operations was a conspicuous feature of airship and airline advertising, exemplified by the publicized weighing of passengers by Imperial Airways and the German Zeppelin Airship Company to avoid overloading their craft.[99]

---

[96] Hugo Eckener, *Die Amerikafahrt des "Graf Zeppelin"*, ed. Rolf Brandt (Berlin, 1928), 34–40, 82–6.
[97] These examples are taken from *Die Entwickelung des Norddeutschen Lloyd Bremen* (Bremen, 1912), 123–35; *Royal Mail Triple-Screw Steamers "Olympic" and "Titanic"* (Liverpool, 1911), 55–66.
[98] *Do X*, ed. E. Tilgenkamp (Zurich, 1931), 11; Junkers archive, 0303/11/41, press release, 16 December 1930, 4.
[99] Imperial Airways ran an advertisement headed "Why Passengers are Weighed!" in several newspapers. See *Daily Mail*, 29 July 1930, 2; *The Times*, 1 August 1930, 18. See also H. Stuart Menzies, *All Ways by Airways* (London, 1932), 4; Eckener, *Die Amerikafahrt*, 29, 75.

In addition to high-quality technical equipment, rigorous inspection regimes ensured that machinery remained in optimal working order. Cunard informed its customers in 1897 that "before every voyage, a thorough examination of the ship is made . . . and there is not one single thing, from stem to stern, from boiler to button, that is not in working condition."[100] Decades later, Lufthansa described its complicated inspection procedures, which required its staff to fill in multiple copies of reports, in similar terms. Although one journalist found some aspects exaggerated, he considered Lufthansa's arrangements "exemplary."[101] Moreover, the aviation industry stressed that strict maintenance standards extended to spare parts. The Vickers company, a producer of spares, knew that it could not provide "cheap, shoddy equipment . . . The risks are too big."[102] In an advertisement in the *Daily Express*, automobile manufacturer Morris assured potential customers in 1936 that its products underwent the same stringent tests as aeroplanes, thereby publicly acknowledging the "exacting" regulations prevalent in the aviation world.[103] The aviation and shipbuilding industries thus successfully represented themselves as guarantors of security because they insisted on levels of quality control that set model standards beyond their sectors.

Once in operation, competent staff also ensured that ships, airships, and aeroplanes were handled properly. Like Cunard, North German Lloyd congratulated itself upon its "well disciplined . . . crew . . . under the command of energetic, circumspect and responsible officers" who steered vessels clear of danger.[104] Aviation literature stressed that pilots had to undergo demanding training courses before they qualified to fly commercial aeroplanes. Commercial airmen passed through a competitive selection process and studied a wide range of topics including engine maintenance, "material science" (*Materialwissenschaft*), aerodynamics, navigation, and chemistry.[105] Just like naval captains, aviation pilots had to be highly skilled experts to be able to control the complicated mechanisms in their charge.[106] Taken together, high-quality equipment, rigorous inspection regimes, and competent personnel guaranteed "ample margins of safety," as one advertisement

[100] *The Cunard Passenger Logbook* (Glasgow, 1897), 27.   [101] *Vorwärts*, 25 September 1927, 6.
[102] Brooklands Museum, Weybridge, Technical Archives, brochure, *Vickers Accessories for Aircraft* (no place, 1929), 2.
[103] *Daily Express*, 15 May 1936, 9.
[104] *Cunard Passenger Logbook*, 27; *Entwickelung des Norddeutschen Lloyd*, 141.
[105] See *Aviaticus: Jahrbuch der deutschen Luftfahrt, 1931* (Berlin, 1931), 2; J. B. Malina (ed.), *Luftfahrt voran! Das deutsche Fliegerbuch* (Berlin, 1932), 112–22.
[106] Captains' and officers' training courses are outlined in Norddeutscher Lloyd Bremen, *Die Geschichte des Norddeutschen Lloyd* (Bremen, 1901), 116–32.

for Imperial Airways boasted.[107] A German shipping company pursued a similar line when arguing that it undertook all preparations "to structure the ship organism in such a way that any conceivable danger can be eliminated or coped with."[108]

In this manner, airlines as well as operators of airships and passenger liners insisted that they were anything but helpless against the risks faced by innovations. Moreover, companies foregrounded the constant improvements of their safety records. Airlines and aeroplane production companies alike published statistics of the numbers of passengers transported and the mileage covered without accident. The German aviation industry proudly stated that it had managed to improve its security statistics enormously. While an accident leading to injuries or deaths had occurred every 426,000 kilometers covered by plane in 1926, the ratio had increased to 1,497,779 kilometers per accident by 1929. To demonstrate the minimal risks of flying, the German aviation industry also reminded the public that only 6 out of over 56,000 aeroplane landings resulted in passenger injuries.[109] Because they gave an impression of impartial objectivity, statistics appeared to provide indisputable proof that flight was becoming an ever-safer means of transport. Shipping companies also employed figure-laden evidence to drive home claims about improving security records.[110] Whether Britons and Germans turned their eyes to aviation or shipping, then, they found that charts demonstrated the decrease of technological risks in traveling.

In the light of assertions that safety standards were improving, the industries and their supporters strove to convince the public that those accidents that did occur represented mere temporary setbacks. Liberal and Conservative publications as well as the trade press abounded with expectations that technological change would travel a thorny road, and claimed that any failure had to be understood as a constituent element of an ultimately progressive innovation process. They urged citizens in both countries to consider misfortunes as spurs to strive harder for advancement, because engineers and entrepreneurs drew lessons from the mistakes that had led to accidents. One German commentator adopted a defiant pose in the Left-liberal *Berliner Tageblatt* after the *Titanic* had sunk, asserting that "we will stand paralyzed many a time, watching how nature cruelly humiliates

---

[107] *Daily Mail*, 29 July 1930, 2.   [108] *Entwickelung des Norddeutschen Lloyd*, 122.
[109] *Aviaticus*, 46–47. The most thorough overview was presented by Lufthansa in 1929. See Erhard Milch, *Die Sicherheit im Luftverkehr auf Grund der Betriebsergebnisse der Deutschen Luft Hansa, 1926–1928* (Berlin, 1929). For similar statistical evidence, see *The Aeroplane*, 20 February 1924, 151; 6 October 1926, 483; *Imperial Airways Gazette*, September 1928, 2; February 1929, 2; September 1932, 3; June 1934, 6; December 1937, 3.
[110] For just one example, see *Norddeutscher Lloyd Bremen*, 55.

human pride in human works; despite this, our battlecry must be: forward!" As "martyrs (*Blutzeugen*) for the future," the victims who had perished on board the *Titanic* beckoned the living to work towards the prevention of disasters by eradicating their causes.[111] Thus reports conceived of technological progress as a never-ending dialectical process that placed society under a moral obligation to intensify its quests for technical perfection.

Many appeals drew on the familiar, religiously charged language of redemptive sacrifice that, of course, borrowed heavily from celebrations of military heroism. A leading journal serving the German aviation community assured its readers after a lethal crash that "engineers will learn very much from this accident so that this sacrifice for flight is not in vain."[112] British reports from crash sites also resorted to the language of redemption to place accidents in wider narratives of progress, whose promise pointed towards a better future. After the *R101* had exploded in 1930, the Labour press approvingly quoted a vicar who declared that the crew's "sacrifice was no waste ... They have died for the cause of knowledge and human advancement."[113] Some editorials foresaw no end to the quest for "advancement." With respect to airship technology, the *Manchester Guardian* declared that "we do not know how long it will go on before its final success, but we know that when this time comes, some other objective will have arisen which calls for more effort, more martyrs."[114] Given National Socialism's embrace of a spirit of combat, the loudest calls for a redemption of human sacrifice emerged in Germany after the *Hindenburg* disaster, when the party-led press considered it the country's duty to "give meaning to these deaths through a new project ... Go forward over the tombs!"[115]

Britons and Germans thus agreed that accidents should lead to systematic investigations to eradicate fatal engineering errors. Research too, however, could become a contentious issue, despite this consensus about the importance of systematic investigations in improving technological safety. British and German exchanges about the roles played by research in processes of innovation were not restricted to security issues; they also considered the impact of research strategies on the ability of both countries to maintain a position of leadership in international technological competitions. In debates about the military potential of innovations, especially in the context of aviation, parliamentarians argued that excessive caution led to over-conscientious and lengthy research programs whose cumbersome procedures delayed tangible results, thereby endangering a nation's chances to

---

[111] *Berliner Tageblatt*, 16 April 1912 (evening edition), 2.   [112] *Luftfahrt*, 29 (1925), 246.
[113] *Daily Herald*, 6 October 1930, 1.   [114] *Manchester Guardian*, 10 October 1930, 12.
[115] *Völkischer Beobachter*, 8 May 1937, 1; see also *ibid.*, 8 May 1937, 2; 9 May 1937, 1.

preserve its position at the international forefront of technology. On these occasions, speakers disagreed about the appropriate extent of the systematic inquiries that were needed to eliminate risks without jeopardizing the international technological status of their nations.[116] Thus assurances about the importance of research could only partially dispel fears about technological risks, because Britons and Germans knew that safety issues had to be balanced with other considerations that were deemed crucial for the nation.

Furthermore, even public proclamations of unconditional dedication to research into safety improvements on the part of industry and political institutions failed to placate Left-leaning critics of existing security provisions. British and German Socialists did not evaluate accidents as inevitable "fiasco[s] of technology" (*Fiasko der Technik*) whose origins rested in ignorance of technological intricacies. Rather, the Left singled out "greed for profit" and "capitalistic addiction to records" (*kapitalistische Rekordsucht*) as the main causes of many disasters.[117] According to Labour adherents and German Social Democrats as well as Communists, "capitalism" demonstrated that it ranked profit above safety and thereby increased technological risks. One incident in 1924 that provoked criticism from the British Left was a dispute over an Imperial Airways salary scheme that required pilots to raise their annual flying time from roughly 600 to 900 hours to maintain existing levels of annual income. The Labour newspaper the *Daily Herald* condemned this proposal because it intensified "the nervous strain upon pilots," which, in turn, would lead to more accidents; the paper criticized the airline for its willingness to jeopardize passengers' safety for financial reasons.[118] Although the German Communist press resorted to more radical and overtly ideological attacks than did Social Democratic or Labour organs, its critiques identified similar sensitive points. *Rote Fahne*, the Communist daily from Berlin, indicted "monocapitalistic greed" after the explosion of the *R101*, because the United States government had refused to export the helium that would have provided a much safer source of lift for the airship than highly inflammable hydrogen. According to this interpretation, American enterprises preserved their advantage over other nations' airship industries, which had no access to artificially produced helium, at the expense of airship security.[119]

According to the Left, however, it was the *Titanic* that demonstrated most glaringly how cynical profit motives overrode safety concerns in prevailing

---

[116] *Parliamentary Debates*, fifth series, vol. 8, 1565–8, 1574–5; vol. 180, 2207; *Verhandlungen*, vol. 293, 7487, 7492, 7503, 7514.
[117] The phrases are from *Vorwärts*, 17 April 1912, 1; 20 April 1912, 5.
[118] *Daily Herald*, 1 April 1924, 5.   [119] *Rote Fahne*, 7 October 1930, 13.

entrepreneurial practice. In addition to the fact that "prestige" considerations had induced the company to make excessive haste, the lack of lifeboats outraged the Labour press. As an article in the *Labour Leader* accused, "it has become the custom to excuse companies, if the letter of the law is obeyed, but the fact that providing lifeboat accommodation was not borne simply because such was not compulsory is an eloquent illustration of the inhumanity of privately controlled industry." It also came as no surprise to Labour journalists that a disproportionate number of first-class passengers secured places in lifeboats.[120] The *Titanic* thus proved "the truth of the Socialist maxim that so long as the machinery of industry is privately owned human life will be sacrificed to profit and to the luxury of the rich."[121]

Critiques of accidents by the German and British Left did not reject technological change as such, however. On the contrary, Socialists of different persuasions in both countries welcomed innovations when they promised to improve working conditions, drive up consumption, and increase national prestige, or, from more radical perspectives, when they held out the prospect of accelerating "concentration processes" that would bring capitalism closer to its supposedly inevitable collapse.[122] Nonetheless, as in debates about security in the workplace, the Left berated the business world when it believed it had detected instances in which technological innovations were brought about at the expense of human lives and health.[123] A sustained critique of technological risk in our period, therefore, emanated from the political Left – but with decidedly ambiguous and muted effects. Left-wing critiques of technological risk focused on the claim that capitalistic enterprises neglected security issues in their desire to enhance profit margins. While this core conviction ensured the coherence and stability of Left-leaning critiques of prevalent regimes of risk tolerance throughout the period, the anti-capitalistic overtones of these critiques tainted them with the stigma of political radicalism, thereby restricting their wider impact in public debate. After all, Liberal, Conservative, and pro-business circles

---

[120] *Labour Leader*, 19 April 1912, 241.
[121] *Ibid.*, 26 April 1912, 257. Similar arguments emerged after the *Empress of Ireland* had foundered. See *Daily Herald*, 1 June 1914, 8.
[122] Positive appraisals of technological innovation on the Left are covered in Charles Maier, "Between Taylorism and Technocracy: European Ideologies and the Vision of Industrial Productivity in the 1920s," *Journal of Contemporary History* 5 (1970), 27–61; Mary Nolan, *Visions of Modernity: American Business and the Modernization of Germany* (New York, 1994), 51–4, 79–82, 118–120; Hans Albert Wulf, *"Maschinenstürmer sind wir keine": Technischer Fortschritt und sozialdemokratische Arbeiterbewegung* (Frankfurt, 1987). See also Eva Cornelia Schöck, *Arbeitslosigkeit und Rationalisierung: Die Lage der Arbeiter und die kommunistische Gewerkschaftpolitik, 1920–1928* (Frankfurt, 1977).
[123] For case studies, see Trischler, "Arbeitsunfälle und Berufskrankheiten"; Rabinbach, "Social Knowledge."

could not but automatically reject charges that comprised a fundamental attack on the dominant economic order. Several examples bear this out. In the context of the wreck of the *Titanic*, the establishment expressly rejected accusations of both class-bias during rescue operations and the neglect of security for financial reasons. *The Times* countered the charges leveled by the Labour press as motivated by "malicious hatred . . . [which] is not the spirit of British workmen."[124] In the light of tense industrial relations in 1912 – not least in the shipping sector, where a strike had threatened the scheduled departure of the *Titanic* – the establishment's intransigent stance comes as no surprise.

Most revealingly, Left-wing criticism played no role in the decision to terminate Britain's airship construction program after the explosion of the *R101*. Although initial press reactions in the immediate aftermath of the disaster employed the conventional language of redemptive sacrifice, described above, to demand a continuation of airship development in Britain, the mood had changed by the time the report of an official inquiry was published six months after the crash. Citing the findings of the inquiry, the press pointed out that airships had failed to improve their security standards, despite assurances to the contrary. It appeared pointless to call for further potential "sacrifices," given that these had not led to reductions in risk. "The airship is so frail a thing that it is liable to be destroyed by every kind of accident . . . We presume that no more money will be wasted in building such craft . . . which are perilous to their crew and entirely useless for . . . commercial traffic," stated the *Daily Mail*.[125] In this instance, commentators argued that there was no realistic prospect of technical improvement in the foreseeable future. What dealt the fatal blow to British airship construction schemes, then, was not criticism from the Left, but rather the dialectical understanding of technological progress that demanded that the victims of accidents would be redeemed through advances. Because airship development, with its spectacular and recurring mishaps, did not result in demonstrable improvements, British public opinion opposed the continuation of an airship program in 1931. Crucially, the availability of aeroplane technology facilitated this decision, allowing the nation to pursue active air policies through other channels. Thus the public expectation that processes of innovation would show discernible patterns of improvement in safety defined the limits of risk acceptance even when

---

[124] *The Times*, 20 April 1912, 9. For another rebuttal of Left accusations, see *Report on the Loss of the "Titanic,"* 40.
[125] *Daily Mail*, 1 April 1930, 10. See also *The Times*, 1 April 1930, 15; *Manchester Guardian*, 1 April 1930, 8.

the immediate physical consequences of accidents remained confined to specific and clearly circumscribed social groups.

As we have seen, state authorities and industry found it difficult to control the intense and controversial public debates that arose in the wake of technological mishaps in the air and on the water. In order to prevent public exchanges that could have revealed technical shortcomings in German flight technologies, addressed negligence on the part of the authorities, or tarnished the German government in other ways after 1933, the National Socialist regime used its extensive power over the media to reshape the public treatment of accidents. Bringing the public portrayal of technological risks into line with Nazi ideology entailed a significant departure from established modes of coverage. The Nazi rulers advanced a distinctive vision of Germany as a risk society that diverged from the interpretations that had held sway in Germany before 1933 and remained dominant in Britain throughout the 1930s. It was National Socialist ideology, rather than new technological developments, that disrupted the continuities in German attitudes towards the physical risks of technology in the first half of the twentieth century.

The regime began by instituting a far-reaching ban on coverage of aerial accidents occurring over the Reich. The official German press agency *Deutsches Nachrichtenbüro* routinely issued circulars that either completely prohibited reports about individual accidents or restricted the publication of articles to newspapers in the immediate area in which a mishap had occurred. Directives to newspaper editors usually included draft notices in a factual tone that served as guidelines for short articles covering crashes, and unequivocally reminded journalists not to incorporate eye-witness accounts or to write unauthorized editorials, since both could easily have challenged the officially sanctioned version of events.[126]

While these arrangements kept aerial accidents out of the national news, they did not allay all public concerns about aerial safety in Germany. In fact, National Socialist press policy may have unintentionally fueled suspicions about aerial technologies, leaving some experts wondering whether German manufacturers possessed especially bad safety records that required a state-coordinated cover-up. The outline of a sales pitch devised by the Junkers company for in-house training purposes touched on this issue in 1937, when a fictitious customer questioned the claim that Junkers aeroplanes were more reliable than American models and alleged that "this can

---

[126] Such directives can be found in Karen Peter (ed.), *NS-Presseanweisungen der Vorkriegszeit: Edition und Dokumentation*, vol. V *1937* (Munich, 1998), 131, 184, 218, 325, 480, 626, 678.

be attributed to the fact that one never hears about accidents in German air travel while coverage is sometimes immensely frank abroad." The training manual instructed company representatives to correct this "error" by drawing attention to the low accident rates of Junkers machines operated by foreign airlines.[127] On the one hand, maintaining silence about air crashes averted undesirable public attention from the safety problems that beset the aviation sector. On the other hand, however, suppressing this information deprived industry and state of opportunities to celebrate improving safety records. After all, this promotional technique required knowledge of the frequency of previous accidents as a comparative unit of measure. It is thus unclear to what extent the culture of secrecy that shrouded air crashes in mystery during the Third Reich succeeded in reversing the prevalent image of the aeroplane as an accident-prone technology.

Of course, not even the tightly controlled media environment of Nazi Germany managed to ban reports of spectacular disasters from the main news pages. When the airship *Hindenburg* exploded in Lakehurst in May 1937, the propaganda apparatus focused on news-management strategies that simultaneously heroicized the crew and diverted responsibility for the accident from the German engineers and authorities. Initially prohibited from speculating on the causes of the explosion, the media quickly focused on the theory that lightning from a brewing thunderstorm had ignited the hull.[128] According to press coverage, it was a force of nature beyond human control, rather than a design failure or human error, that had triggered the disaster – an analysis that left intact the assertion that "the operational safety of German airships is a proven fact."[129] These efforts to shield designers and politicians from potential criticism regarding the explosion were combined with initiatives to hide the physical suffering of both crew and travelers. While the first reports from Lakehurst had included a few isolated passages that described the "persistent screams of a mechanic in pain," official guidelines immediately reminded journalists not to print the "horror stories that the American press tells in keeping with its prevalent methods."[130] As a result, the German public saw none of the harrowing and

---

[127] RLD 02718, typescript draft for a sales talk for JU 90, 19 August 1937, 6–7. Criticism about the secrecy surrounding aerial accidents also emerged in generally pro-German British publications such as *The Aeroplane*. See *The Aeroplane*, 6 May 1936, 544–5.

[128] *NS-Presseanweisungen*, vol. v, 359; *Völkischer Beobachter*, 9 May 1937, 1; 12 May 1937, 4. See also the article that reproduces the officially sanctioned press agency report issued by *Deutsches Nachrichtenbüro* on 12 October 1937 in RLD, file 01999-01.

[129] For the quotation, see *NS-Presseanweisungen*, vol. v, 359.

[130] For the passage describing the mechanic's suffering, see *Völkischer Beobachter*, 8 May 1937, 1. The directive is included in *NS-Presseanweisungen*, vol. v, 359.

sensationalistic photos that appeared in American broadsheets and tabloids, depicting naked corpses of burns victims and injured travelers and crew members in their scorched clothes being led from the flaming wreck by rescue workers.[131] The German press reports about Captain Lehmann, the airship's commander, who died within twenty-four hours of the explosion, also refrained from detailing his agony, a topic that featured prominently in a *Daily Mail* article about the "terribly burned" man.[132] Instead, *Völkischer Beobachter* portrayed Lehmann as the "shining example of a fighter" who safeguarded the survival of sixty-five out of ninety-seven people on board by holding out at his command post amid a flaming inferno until "his burning uniform fell off his body."[133] Not only did this eulogy place Lehmann in a gallery of heroic martyrs who had willingly sacrificed themselves for the National Socialist cause; the sanitized image of his stoic heroism also signaled that, despite raging chaos, discipline had prevailed during this technological disaster.[134] As the commander defiantly sought to assert his will over the exploding vessels, he set an example and rose above the catastrophe. His conduct in the face of death, therefore, provided a gesture of human superiority over technology that, according to the official press, bore no trace of fatalism.

As a result, articles appearing in Germany culminated in unanimous demands to continue the airship program. "Nevertheless!" (*Dennoch!*) cried the headline of the leading editorial in the *Völkischer Beobachter* under a photograph of the blazing airship.[135] Unflagging public support for this high-risk technology became a question of principle for a regime that prided itself in its commitment to struggles against seemingly insurmountable obstacles. The press reports about the *Hindenburg* reveal that the Nazis celebrated risk-taking as an end in itself that demonstrated the nation's vitality as well as its willingness to "fight."[136] Overshadowed by the international tensions generated by the Sudeten crisis, the next aerial giant took to the skies sixteen months later as a proof of Germany's fighting mindset.[137] National Socialism thus pursued a twofold and ultimately schizophrenic strategy in its promotion of dangerous technologies. While their combative

---

[131] Such photos can be viewed in the collection of press cuttings in RLD, file 02000. The *New York Times* also printed such images; see *New York Times*, 7 May 1937, 20; 9 May 1937, section 10.
[132] *Daily Mail*, 8 May 1937, 9.   [133] *Völkischer Beobachter*, 9 May 1937, 2.
[134] On the creation of martyrs in National Socialism, see Sabine Behrenbeck, *Der Kult um die toten Helden: Nationalsozialistische Mythen, Riten und Symbole, 1923 bis 1945* (Vierow, 1996), esp. 134–48; Rudy Koshar, *German Travel Cultures* (Oxford, 2000), 134–49.
[135] *Völkischer Beobachter*, 8 May 1937, 1.   [136] *Ibid.*
[137] *Völkischer Beobachter*, 22 September 1938, 4. Further press reports, especially from the *Deutsche Börsen Zeitung*, can be found in RLD 02002.

ethos predisposed the Nazis to embrace high-risk technologies such as the airship, the regime also suppressed information about many accidents that would have associated authorities and designers with failures. This dual strategy allowed the National Socialist government to support an airship program years after Britain had cancelled a similar project because of safety considerations.

Multifaceted media coverage about naval and aerial technologies shaped the public depictions of Britain and Germany as risk societies between the 1890s and the late 1930s. In the immediate aftermath of accidents, public accounts constituted symbolic risk management of the most basic variety: they overcame the speechlessness that could afflict eye-witnesses surveying disaster sites. Although often describing the effects of accidents in unsettling detail, dispatches from crash scenes initiated a process of articulation that made sense of raw events. Technological risks not only affected the individuals who were mutilated or killed in air crashes or shipwrecks; they also shocked communities on a national and even international level. Contemporaries found it particularly worrying that the physical risks of technological innovations escaped analysis when inquiries failed to identify the causes of accidents. It often proved impossible to establish with certainty who or what was responsible for individual disasters. Moreover, commentators pointed out that it was impossible to avoid risks altogether, since new designs revealed their dangers only once they were used. Prominent interventions, therefore, suggested that the physical risks of innovations were deadly, uncannily unintelligible, and, crucially, an inevitable, intrinsic element of processes of innovation. While, as we saw in the previous chapter, scores of public commentators marveled at the mechanisms that came into existence at unprecedented rates in our time period, they did not expect these contraptions to be altogether safe. As technology defined modernity in many eyes, the modern, viewed from this perspective, necessarily included danger because technology had limited perfectibility. Public debate did not push the notion of risk to the margins of British and German culture, nor did the concept possess the exclusively negative connotation that it acquired in the late twentieth century, according to Ulrich Beck and other sociologists.[138]

In part, the fact that only specific groups, such as military personnel, employees handling equipment, and travelers, faced direct exposure to the

---

[138] Beck, *Risikogesellschaft*, 28; Anthony Giddens, *Modernity and Self-Identity: Self and Society in the Late Modern Age* (Stanford, 1991), 109–143; Richard Sennett, *The Corrosion of Character: The Personal Consequences of Work in the New Capitalism* (New York, 1998), 76–97.

risks of aerial and maritime technologies facilitated the social acceptance of hazards. Despite the wide-ranging collective impact of technological catastrophes, which revealed itself in public mourning and sensational media coverage, most Britons and Germans could dissociate themselves from accidents on a personal level. With respect to passenger shipping and aviation the risk impositions appeared limited – quite in contrast to present-day perceptions of technology's dangers as ubiquitous and inextricably woven into the social fabric. If the Left attributed disasters to profit motives that overrode safety considerations and rendered innovations unnecessarily dangerous, industrial promoters insisted that engineers did not confront safety issues helplessly and pointed to a variety of security measures that testified to active efforts to prevent risk. In the end, accidents did not dampen enthusiasm as long as public debate conceived of technological progress in dialectical terms that saw accidents as opportunities to improve designs, ultimately resulting in safer mechanisms. At the time, public observers self-confidently insisted that it was possible to create highly reliable devices, a stance betraying none of the skepticism arising from the relatively recent realization that safety provisions add to the complexities of technological systems and thus often introduce new potential hazards.[139] In short, contemporaries accepted those risks that appeared to diminish over time and threatened only a limited number of people. Arguments against technological dangers, however, became effective when it proved impossible to reconcile recurring disasters with narratives of dialectical progress, as was the case when Great Britain abandoned airship construction in 1931.

Considerable continuities and transnational similarities marked British and German attitudes towards the physical risks of technology as the focus of public debate shifted from passenger liners to aviation before 1933. As was the case with the way in which public ambivalence towards technological innovation developed, it was only the rise of the Nazi government that added a new element to public risk assessments and thus became responsible for a major contrast between both countries. Its combative ethos predisposed National Socialism to continue support for airship technology even after the explosion of the *Hindenburg*. Despite the decision to back this high-risk technology and publicly to parade the party's dedication to the idea of struggle for its own sake, Nazi media policy also displayed clear limits of risk acceptance when official authorities sought to silence news

---

[139] This phenomenon has been examined in Charles Perrow, *Normal Accidents: Living with High-Risk Technologies* (Princeton, 1999), esp. 123–231. On the failure of safety procedures from a sociological perspective, see Diane Vaughan, *The Challenger Launch Decision: Risky Technology, Culture, and Deviance at NASA* (Chicago, 1996).

of aeroplane crashes that could have associated the government with failure on a regular basis. A fundamental contradiction thus lay at the heart of National Socialist attitudes to technological danger: fear of being publicly linked with the consequences of technological failure coexisted with clamorous celebrations of risk taking as an end in itself. On the whole, the British and German societies displayed high levels of tolerance for the physical risks of innovative technologies, a disposition that was a crucial dimension of a cultural climate conducive to technological change. Neither the passenger liner nor the aeroplane nor the airship could have developed had contemporaries consistently shied away from the dangers that accompanied them.

When we turn our attention to discussions that accompanied the rise of the movies, it becomes clear that attitudes to the dangers they posed to life and limb provide only part of the assessment of the risks generated by technological change. Altogether different dangers stood at the center of public debate about the transformation of the contemporary media environment. By contrast with the concrete and socially limited physical risks of flying and traveling by steamship, the new medium of film appeared to lead to less tangible yet much more pervasive cultural dangers as the movies became a popular form of entertainment across the social spectrum. While most Britons and Germans would have agreed that film had the power to pose a cultural danger to contemporary society, they would also have been hard pressed to articulate their unease in unambiguous terms. Film gave rise to an elusive danger – that, at least, is the impression one gains when reading public debates about the new medium. The triumph of film posed puzzling questions. How profound was the social and cultural impact of film, and what was the nature of film's effects on the population? What complicated these questions was the fact that throughout our period no authoritative explanation emerged as to exactly how film captured an audience. Fundamental knowledge problems that touched on the character of film as a technical creation thus underpinned debates about the cultural risks of the movies and maintained a widespread sense that film was a potentially threatening presence throughout our entire period.

CHAPTER 4

# Elusive illusions: the cultural and political properties of film

The rise of film played a central role in the transformation of the British and German media landscapes in the early twentieth century and reshaped both the market for topical news and the world of entertainment. In conjunction with sensationalistic types of journalism, and, since the 1920s, the radio, film contributed to an expansion of the public sphere, and triggered passionate controversies about a new "mass culture" whose appeal to countless, supposedly uncultivated, consumers caused concern. Innovative media, it was claimed, upset existing cultural hierarchies by making "high culture" more widely available, as well as by creating altogether new entertainments that took their place alongside established cultural pursuits. The educated middle class contrasted the commercialism of popular culture with the products of an established "high culture" that, in their view, remained uncontaminated by materialistic concerns. While many historians have stressed the social elitism evident in contemporary debates about "mass culture," few have analyzed in detail how technological developments affected British and German interpretations of popular culture in the late nineteenth and early twentieth centuries.[1] Since technological innovations are usually relegated to the background of current historical writing on popular culture, we know very little about how public evaluations of technology and of popular culture related to one another. This chapter complements existing scholarship on popular culture by uncovering those meanings of film as a *technology* that underpinned contemporary debate about the medium.

Public evaluations of new cultural forms often centered on the veracity of cultural artifacts that owed their existence to mechanical creation processes.

[1] Dan LeMahieu, *A Culture for Democracy: Mass Communication and the Cultivated Mind in Britain Between the Wars* (Oxford, 1988); Corinna Müller, "Der frühe Film, das frühe Kino und seine Gegner und Befürworter," in Kaspar Maase and Wolfgang Kaschuba (eds.), *Schund und Schönheit: Populäre Kultur um 1900* (Cologne, 2001), 62–91; Kaspar Maase, *Grenzenloses Vergnügen: Der Aufstieg der Massenkultur, 1850–1970* (Frankfurt, 1997).

In the case of the gramophone, cultural traditionalists denounced recordings, not so much because they provided imperfect sound documents, but because "canned music" lacked the authenticity that classical works gained in concerts. Others, however, praised the device for its ability to popularize classical music.[2] Evaluations of music records depended on whether it was sound alone (which, in principle, lent itself to mechanical conservation) or the act of performance that critics deemed essential to a musical artwork's aesthetic properties. In short, arguments revolved around the question whether art lent itself to mechanical reproduction, to put it in Walter Benjamin's phrasing. Turning to film, commentators identified different issues relating to this medium, which, rather than popularizing existing works of "high" art, established altogether novel forms of mass entertainment within less than two decades of its invention. First, cinematography's realistic effects provided a recurring source of public uncertainty, since observers never fully understood why the camera and the projector managed to produce visual illusions that possessed the power to transport audiences effortlessly into fictional worlds. A fundamental knowledge problem regarding the functioning of film thus ran through all assessments of the medium. Secondly, many cultural and political commentators feared that irresponsible profit motives drove the film industry to abuse this technological wonder. In this respect, film gave more cause for public concern in the 1920s than the radio, since cultural traditionalists controlled the leading broadcasting stations in Germany and Britain.[3] Finally, uncertainty about how film worked to produce its realistic effects and suspicions about the movie industry combined to generate persistent insecurity about the short- and long-term consequences of frequent film consumption for the numerous audiences attending shows. In brief, many commentators viewed film as a new cultural risk. Of course, everybody would have agreed that the risks associated with film culture differed fundamentally from the physical dangers posed by passenger shipping and, after the Great War, aviation. Nonetheless, contemporaries found it difficult to spell out what they considered to be the exact nature of the perils to which audiences exposed themselves when they watched movies. Uncertainty about the functioning of film, the entrepreneurial energies displayed by film moguls, and its unclear influence on audiences led a host of commentators to view cinematography as a powerful yet potentially untrustworthy tool of deception. Given film's success

---

[2] LeMahieu, *A Culture for Democracy*, 80–99; Dan LeMahieu, "*The Gramophone*: Recorded Music and the Cultivated Mind in Britain Between the Wars," *Technology and Culture* 23 (1982), 372–91.
[3] Karl Christian Führer, "A Medium of Modernity? Broadcasting in Weimar Germany, 1923–1932," *Journal of Modern History* 69 (1997), 722–753; LeMahieu, *A Culture for Democracy*.

in attracting huge audiences from around 1910 on, commentators emphasized that the risks of film – however one wished to define them – were not restricted to specific social groups but affected society at large. In brief, film presented an extensive, rather than a limited, risk imposition of an unspecified nature.

Despite the mechanical nature of film, evaluations of its place in the visual culture of modernity rarely revolved around systematic considerations of the medium's technical properties. Observers usually sought to assess the social, political, and cultural impact of cinematography without reflecting on the medium in explicitly technical terms. Since public exchanges often pushed into the background the simple fact that it was mechanical processes that were responsible for the existence and suggestiveness of the movies, they addressed film as a popular entertainment first and as a technology second, if at all. At the same time, the premise that film possessed extraordinary manipulative powers lay beneath numerous contemporary appraisals. Unclear notions of cinematography as an inexplicably manipulative instrument permeated debates because contemporaries failed to advance authoritative technical explanations detailing how this modern wonder brought its deceptive images into existence in the first place. This fundamental knowledge problem gave rise to a silence about film's technical aspects that had profound consequences, not only for assessments of the supposed risks posed by cinema culture, but also for wider debates about technological innovation. Although it was undeniably the result of technological innovation, film came to be discussed as if it were something other than a technology.

### DANGEROUS ILLUSIONS: FILM AS A DECEPTIVE TOOL

From the moment when, in 1895, the Lumière brothers amazed and alarmed the first film audiences with an image of a train that appeared to hurtle towards the spectators as it filled more and more of the screen, the public credited cinematographic technology with "wondrous" attributes.[4] To begin with, camera and projector miraculously translated a series of still photographic images into lifelike moving pictures. In this context, film's mysterious aura derived from the fact that it proved impossible to *imagine* how the static images that made up a reel gave rise to an impression of motion. In short, even the fundamental technical process that

---

[4] On the effects of early first screenings on audiences, see Stephen Bottomore, "The Panicking Audience? Early Cinema and the Train Effect," *Historical Journal of Film, Radio and Television* 19 (1999), 177–216; Elizabeth Ezra, *Georges Méliès: The Birth of the Auteur* (Manchester, 2000), 1–2.

created moving images resisted straightforward explanation. Moreover, the "intensely exciting, intensely realistic" effects of screen projections stunned observers.[5] It was the inexplicable ability to transport audiences effortlessly into alternative realities, often of an overtly fictional character, that, as we have seen, also prompted contemporaries to pronounce film a modern wonder. For decades, commentators found it impossible to explain convincingly the ease with which movie presentations cast their spell over viewers. As late as 1930, leading British film critic Paul Rotha noted that "there has not as yet been any systematic inquiry" into how films produced their "emotional effects."[6]

For its dramatic visual effects, film relied on a technological paradox: throughout a film's running time, its technical nature became imperceptible, ensuring that audiences concentrated on the content of the illusions they beheld rather than on the tricks that created these illusions. In contrast to aviation and shipping, where celebrations of advances such as increases of speed, size, and power drew attention to engineering feats, film's technological refinement manifested itself in the very ability to conceal the mechanical character of its products. Film possessed the power to render its technological nature invisible – it was a self-effacing medium. Thus film was a deceptive technology in several respects. The "fantasy machine," as one German detractor labeled cinematography in 1931, transformed still frames into moving images, blurred the line separating "truth" and "fiction," and concealed its character as a machine.[7] As a consequence, the British and German public, who attended the movies in their millions, usually overlooked cinematography's technological properties, eventually not thinking of film as a technology at all. The pages that follow trace how film's technological characteristics came to be sidelined in a public debate about this medium, which generated a flood of mechanically produced images in Britain and Germany. The absence of authoritative technical accounts of how the movies achieved their extraordinary make-belief, in turn, fueled profound insecurity about the medium because observers insisted that, as a tool of manipulation, film was a dubious and potentially dangerous element of modernity in Britain and Germany.

Critics began to voice vehement concern about film's deceitful qualities at the beginning of the second decade of the twentieth century. By this time,

---

[5] National Council of Public Morals (hereafter NCPM), *The Cinema: Its Present Position and Future Possibilities* (London, 1917), lxx. The National Council of Public Morals was an ecumenical charitable body including leading Anglican, Catholic, Jewish, and school organizations.
[6] Paul Rotha, *The Film Till Now: A Survey of the Cinema* (London, 1930), 204.
[7] Rene Fülöp-Miller, *Die Phantasiemaschine: Eine Saga der Gewinnsucht* (Berlin, 1931).

film had developed into an industry, with leading companies such as the US-based Motion Picture Patents Company as well as the French Gaumont and Pathé-Frères, all of whom operated on an international scale.[8] While films had initially featured in variety shows and music halls in cities, and in traveling exhibitions at fun-fairs in the countryside, a cinema building boom swept Europe and North America after 1905. According to contemporary estimates, Britain boasted between 5,000 and 7,000 cinemas in 1914, and German figures ranged between 3,000 and 4,500. Because admission prices started as low as 10 to 20 *Pfennige* or 3d, "going to the pictures" quickly developed into a mass phenomenon. Weekly attendance figures allegedly reached 20 million in Britain in the middle of the Great War. During this period of dynamic growth, companies gradually complemented programs consisting of short, factual and fictional "attractions" with hour-long features that became vehicles for the first movie "stars."[9] Many of these were carefully scripted and expensively produced, making a spectacular contribution to the thriving urban commercial entertainment sector, which also included theaters, music halls, variety shows, fairs, sporting events, and a vibrant penny press.

The cultural charges leveled at film after 1910 were launched from predictable social quarters: many of the medium's critics belonged to the small but influential British and German cultural elites who acted as their nations' "public moralists" and commented on current affairs during our period. Their university education, as well as their positions as teachers, state administrators, scholars, scientists, medical doctors, clergymen, lawyers, and judges, lent social authority to their views about film culture, which were often closely linked with professional interests.[10] In the early stages

---

[8] Richard Abel, *The Ciné Goes to Town: French Cinema, 1896–1914* (Berkeley, 1994), 23.

[9] On the characteristics of the "cinema of attraction," see Tom Gunning, "The Cinema of Attraction: Early Film, Its Spectators and the Avant-Garde," *Wide Angle* 8:3 (1986), 63–70. On the number of cinemas, see Rachel Low, *The History of the British Film, 1906–1914* (London, 1973), 22; Corinna Müller, *Frühe deutsche Kinematographie: Formale, wirtschaftliche und kulturelle Entwicklungen* (Stuttgart, 1994), 29; Müller, "Der frühe Film," 64. Attendance figures can be found in Emilie Altenloh, *Zur Soziologie des Kinos* (Jena, 1914), 67; Kaspar Maase, *Grenzenloses Vergnügen*, 108–9; Rachel Low, *The History of the British Film, 1906–1914*, 2nd edition (London, 1973), 17.

[10] On British and German cultural elites, see Stefan Collini, *Public Moralists: Political Thought and Intellectual Life in Britain, 1850–1930* (Oxford, 1993); LeMahieu, *A Culture for Democracy*; Fritz K. Ringer, *The Decline of the German Mandarins: The German Academic Community, 1890–1933* (Cambridge, MA, 1969); Gangolf Hübinger and Wolfgang J. Mommsen (eds.), *Intellektuelle im Deutschen Kaiserreich* (Frankfurt, 1993). Two examples of these critics are psychiatrist Robert Gaupp, who had researched the relationship between "moral insanity" and juvenile delinquency before thundering against the cinema menace, and Karl Brunner, who saw his lobbying activities in favor of censorship as an extension of his work in the Prussian welfare ministry. See Robert Gaupp, "Über moralisches Irresein und jugendliches Verbrechertum," *Juristisch-psychiatrische Grenzfragen: Zwanglose Abhandlungen* 2 (1904), 51–68; Karl Brunner, *Das neue Lichtspielgesetz im Dienst der Volks- und Jugendwohlfahrt* (Berlin, 1920).

of the cinema debate, British and German public moralists exerted considerable influence on political debate because they established the categories that subsequently informed exchanges among politicians that led to censorship in both countries. These middle-class commentators, who had little or no background in engineering, were unlikely to pay attention to the technical properties of film. Instead, they drew on a moralizing vocabulary that had been used to formulate concerns about popular culture since the middle of the nineteenth century. British music halls, for instance, attracted censure not only for their noisy, socially mixed crowds, in which prostitutes were repeatedly found to ply their "trade," but also for the sexual innuendo that ran through many performances and provided what Peter Bailey has termed an "ironic counterpoint to the language of respectability."[11] Similarly, German moral reformers identified cheap serial fiction, known as *Kolportageromane*, as a cultural peril. Peddled on the doorstep and reaching circulations of up to a million, *Kolportageromane* primarily caused alarm because they narrated plots featuring romantic entanglements and criminal conspiracies in melodramatic and sensationalistic styles, thereby "overheating the imagination" of a supposedly unsophisticated readership. Widespread distribution of this cheap fiction among female readers added to the concerns, because women's purportedly precarious rational faculties predisposed them to become particularly uncritical consumers of texts that would unleash unspeakable desires.[12] Moral panic revolving around class and gender anxieties pervaded the expansion of commercial popular culture and provided the framework for discussions in which the movies joined the numerous other manifestations of decadence that allegedly drove "modern" societies to their ruin. By aligning the new medium with other forms of popular culture, public moralists placed the new medium in a familiar light. This worked to push its novel technological character into the background, but left intact the assertion that film posed an unprecedented challenge in breaking visual taboos in front of "mass audiences." Of course, no convincing evidence backed up this assertion, because of the incomplete understanding of how film worked in the first place.

[11] Peter Bailey, "Conspiracies of Meaning: Music-Hall and the Knowingness of Popular Culture," *Past and Present* 144 (1994), 139–70, here 156.
[12] For examples of this debate, see Kaspar Maase, "Einleitung: Schund und Schönheit: Ordnung des Vergnügens um 1900," in Maase and Kaschuba (eds.), *Schund und Schönheit*, 9–28, esp. 12–14. On public fear about pornography, which further radicalized charges against serial literature, see Gary D. Stark, "Pornography, Society, and the Law in Imperial Germany," *Central European History* 14 (1981), 200–29. On concerns about female readers in nineteenth-century France, see Jann Matlock, *Scenes of Seduction: Prostitution, Hysteria, and Reading Difference in Nineteenth-Century France* (New York, 1994), esp. 199–219.

By means of their inexplicable persuasive properties, films, it was held, dissolved the distinction between truth and fiction and induced viewers to imitate the reprehensible kinds of behavior, exhibited on the screen, that threatened public order. Champions of cinema control forecast unknown levels of unlawfulness and violence, quoting criminals who claimed that their misdeeds were inspired by film scenes. Cultural conservatives opposed pictures such as *Motor Mad* (1909) – in which an "insane" man shoots a policeman before being committed to an asylum – not primarily because the plot involved an assault on state authorities,[13] but because, they argued, such films rendered topics *visible* that ought to remain out of sight. Since film supposedly abolished the critical distance between events on the screen and patrons in the auditorium *and* showed events on the screen in "living images," cinematic realism was deemed to have particularly corrosive effects on viewers' morals. One British custodian of morals, therefore, welcomed the decision to restrict cinematic representations of burglary so that audiences could no longer see how locks were picked. After all, he argued, "many of these films prove to be direct incentives to crime. Clever burglaries are exhibited before the eyes of mischievous boys who at once have their attention called to the possibility of the 'expert crackman's life.'"[14] Similar uneasiness informed discussions about "the 'sex' interest," which even a sympathetic investigation "regretted" to find "so dominant" in many programs in 1917.[15] One pamphlet included a letter by a female middle-class patron that bore out how deeply audience members were shocked by a film entitled *Edith, the White Slave*, whose protagonist, a female immigrant to the United States, ends up in a brothel "out of heedlessness."[16] The sight of "matters whose filthy abhorrence had never become so clear to me although I had known of them" made such a strong impression that this writer claimed to have been "haunted and vexed" for two days.[17] Like crime features, this "sex film" broke a visual taboo by showing with unprecedented intensity a topic that, the letter implied, had previously received more measured and less disturbing treatment. The panic about sexual films reached its climax in Germany immediately after the Great War, when the

---

[13] The plot of *Motor Mad* can be found in *1909: General Catalogue of Classified Subjects: "Urban," "Eclipse," "Radios" Film Subjects and "Urbanora" Educational Series* (London, 1909), 17.

[14] H. D. Rawnsley, *The Child and the Cinematograph Show and the Picture Post-Card Evil* (London, 1913), 4–5.

[15] NCPM, *The Cinema*, xxx, xxvii–xxviii.

[16] On white-slave films, which formed a genre in the immediate prewar years, see Hanns Zischler, *Kafka geht ins Kino* (Reinbeck, 1998), 47–60; Robert C. Allen, "*Traffic in Souls*," *Sight and Sound* 44 (1975), 50–2.

[17] Gaupp and Lange, *Der Kinematograph*, 23.

provisional government repealed all censorship laws to institute freedom of expression, thereby creating what some saw as a national danger of the first order.[18] Films quite literally brought into the light topics and actions that many thought should remain altogether hidden from sight because of the persuasive powers of cinematic realism. As the new medium violated *visual* rather than purely narrative taboos, it was charged with exerting direct and lasting effects in the viewer.

The composition of cinema audiences added to the virulence of visual threats associated with the rise of film. As Martin Loiperdinger has pointed out, early "cinemas constituted a gray area of the public sphere which brought together disparate sections of the population in an informal and anonymous setting."[19] Commentators argued that film images developed their full illusionary potential only because many consumers of moving pictures displayed a striking absence of rational faculties that would have modified the emotional impact of events exhibited on the screen, and critics rounded up the usual suspects in this renewed charge against popular culture.[20] Both the social profile as well as the size of "crowds" attracted censure. Alongside the large number of working-class viewers, the presence of women at cinemas caused considerable alarm in public debates and thus continued long-standing suspicions about female participation in popular culture in both countries.[21] The lure of the cinema, commentators suggested, was so strong that women neglected their duties as mothers and wives, thereby threatening the foundations of orderly family life.[22] Nonetheless, critics were convinced that an even bigger danger threatened children and adolescents, since younger audience

---

[18] *Verhandlungen des Deutschen Reichstages* (hereafter *Verhandlungen*), (Berlin, 1919), vol. 328, 1592. In Britain, similar fears survived into the 1930s. See R. G. Burnett and E. D. Martell, *The Devil's Camera: The Menace of a Film-Ridden World* (London, 1932), 9–10.

[19] Martin Loiperdinger, "Das frühe Kino der Kaiserzeit: Wilhelm II. und die 'Flegeljahre' des Films," in Uli Jung (ed.), *Der deutsche Film: Aspekte seiner Geschichte von den Anfängen bis zur Gegenwart* (Trier, 1993), 21–50, 31.

[20] See Gaupp and Lange, *Der Kinematograph*, 3–4; Ernst Schultze, *Der Kinematograph als Bildungsmittel: Eine kulturpolitische Untersuchung* (Halle, 1911), 7; Hermann Lemke, *Die Kinematographie der Gegenwart, Vergangenheit und Zukunft: Eine kulturgeschichtliche und industrielle Studie* (Leipzig, 1911), 18.

[21] On female spectatorship, see Miriam Hansen, "Early Cinema: Whose Public Sphere?" in Richard Abel (ed.), *Silent Film* (London, 1996), 228–44; Gaylyn Studlar, "The Perils of Pleasure: Fan Magazine Discourse as Women's Commodified Culture in the 1920s," in Abel (ed.), *Silent Film*, 263–97; Shelley Stamp, *Movie-Struck Girls: Women and Motion-Picture Culture after the Nickelodeon* (Princeton, 2000); Miriam Hansen, *Babel and Babylon: Spectatorship in American Silent Film* (Cambridge, MA, 1991). On the female as a threatening figure in popular culture, see Andreas Huyssen, "Mass Culture as Woman: Modernism's Other," in *After the Great Divide: Modernism, Mass Culture, Postmodernism* (Houndmills, 1988), 44–62.

[22] Brunner, *Der Kinematograph*, 19. See also *Verhandlungen*, vol. 333, 5173.

members categorically lacked the capacity to distinguish between cinematic "fictions" and lived "reality." One girl allegedly suffered a "heavy nervous shock" (*schwere Nervenerschütterung*) followed by "strong nausea" while watching a religious film that included a dramatization of the massacre of the innocents in Bethlehem.[23] Films purportedly glorifying condemned forms of conduct were said to introduce children and adolescents to the worlds of crime and sexual depravity. Once on the slippery slope of cinematic spectatorship, young people would act out the fantasies they had witnessed on the screen through unlawful behavior ranging from theft to rape and murder. In Britain and Germany, public guardians supported their claims by drawing attention to rising juvenile delinquency, which coincided with the spread of cinemas during the first two decades of the twentieth century.[24]

This steady flood of writings about the dangers of film left the cinema industry relatively unimpressed. Given the sector's economic success, production companies continued to opt for features that promised audiences exactly the scandalous, hair-raising, and farcical adventures to which the cultural Cassandras objected so vociferously. Whether voyeuristically displaying or seriously confronting moral taboos, the film trade invited controversy by choosing topics and settings that were bound to infuriate British and German guardians of public decency.[25] Moreover, the cinema industry wooed its female clientele in ways that invited public argument, since feature films as well as advertising material often represented women in open defiance of prescribed gender roles. Before the Great War, the *Pathé Cine Journal*, for instance, included a weekly column about a self-confident character named "Perfecta Pinch," whose tales actively challenged conventional models of femininity when Perfecta defiantly walked the streets of London in her "new trouser skirt" or when she stood her ground in a male working-class crowd at the Cup Final.[26] The cinematic trend of showing women in self-assertive roles, in sharp contrast to dominant gender hierarchies, continued to fuel unease about the social impact of films in the

---

[23] Gaupp and Lange, *Der Kinematograph*, 11.
[24] Rawnsley, *The Child*, 4–5; *Parliamentary Debates: House of Commons* (hereafter *Parliamentary Debates*), fifth series, vol. 37 (London, 1912), 750–1; vol. 58, 1840, 1846–7; *Verhandlungen*, vol. 330, 3164; vol. 333, 5170, 5173.
[25] Germany reached a high point of acrimony in 1919 when a movie entitled *Anders als die Anderen* (*Different from the Rest*) about same-sex relationships caused violent disturbances during screenings. See Alice A. Kuzniar, *The Queer German Cinema* (Stanford, 2000), 27–30. Advertisements for films with a supposed "sex interest" can be found in Hochschule für Fernsehen und Film München Library, Privatarchiv Eberhard von Berswordt, Verleihprogramme 1921–22, file 5209.
[26] *Pathé Cine Journal*, 11 April 1914, 8; 25 April 1914, 9.

interwar period.²⁷ Finally, the cinema industry caused debate through its aggressive advertising techniques. Since the late nineteenth century the spread of commercial advertising met with aesthetic disapproval as critics complained that the increasing number of billboards and posters defaced the countryside and the urban environment. Furthermore, promotional campaigns, it was argued, enticed consumers to make ill-conceived choices about products because they stoked desire on the basis of partial and misleading information.²⁸ Given the alleged power of advertisements to seduce prospective clients, the public-relations efforts of the film industry came to be regarded with suspicion because they replicated and enhanced the deceptive qualities widely associated with the new medium itself. A caricature from 1912 in *Punch* took up this theme, depicting an elderly lady, surrounded by billboards at a cinema entrance, who declined to see the "moving pictures" themselves because she found the advertisements alone "moving" enough.²⁹ As propaganda campaigns became more elaborate in the mid-1920s, when managements fitted out entire theaters to match film themes and production companies systematically issued press releases, sensationalist effects still counted among the prime aims of cinema promotion in the effort to attract the attention of potential audiences.³⁰

While the film industry and its critics were locked in a vociferous, protracted debate, concern about the new medium culminated in cinema censorship, which strove to establish visual control. Neither public moralists nor parliamentarians pushed for a complete ban on cinema shows, fearing public unrest as a result of such a "radical" measure; but politicians in both countries were convinced that gaining control of this novel element in visual culture was an urgent national question.³¹ Critics of the cinema claimed that they were not fundamentally opposed to this modern form of entertainment. Rather, some had reformist intentions, and pointed towards film's potential as an instructional and educational tool at the same time as they

---

²⁷ See Kuzniar, *The Queer German Cinema*, 34–6.
²⁸ Paul Readman, "Landscape Preservation, 'Advertising Disfigurement' and English National Identity, c. 1890–1914," *Rural History* 12 (2001), 61–83; LeMahieu, *A Culture for Democracy*, 119–20; Gerhard Kratzsch, *Kunstwart und Dürerbund: Ein Beitrag zur Geschichte der Gebildeten im Zeitalter des Imperialismus* (Göttingen, 1969), 259–69.
²⁹ *Punch*, 23 October 1912, 341.
³⁰ See Stiftung Deutsche Kinemathek Berlin, Schriftgutarchiv (hereafter SDKB), file "Metropolis," *Metropolis: Presse- und Propagandaheft*, ed. Presse und Propagandaabteilung (Berlin, no date); SDKB, file SD 1244, *Wings: Ein Paramount Film der Parufamet* (no place, 1929). For examples of temporary cinema redesigns, see articles in *British Gaumont News*, February 1930, 10–11; June 1931, 16–17; August 1931, 16–17; February 1932, 17. Public hostility to advertising stunts is mentioned in Thomas J. Saunders, *Hollywood in Berlin: American Cinema and Weimar Germany* (Berkeley, 1994), 124–5.
³¹ For a judgment that an outright ban of cinemas was unrealistic and "radical," see *Verhandlungen*, vol. 330, 3164.

called for censorship motivated by moral disapproval of the predominant cinematic fare.[32] The institutional arrangements for censorship that both countries arrived at between 1910 and 1920 differed. While the British Board of Film Censors (BBFC) was established as a private review body involving trade representatives, German censorship remained under the control of state organs because neither politicians nor the public credited the cinema industry with the ability to regulate itself.[33] The fact that German Liberals and Socialists, who supported unbridled freedom of expression in the print media as well as in the dramatic and pictorial arts, took the exceptional step of advocating film censorship demonstrates that anxiety about film reached higher levels than that concerning other forms of entertainment.[34] Irrespective of the varying organizational provisions, guidelines for censorship remained vague. The German censorship law from 1920, for instance, called on inspectors to perform a "censorship of effect" (*Wirkungszensur*) by appraising the potential impact of films on audiences' attitudes as well as on their behavior.[35] This procedure was bound to provoke conflict, since, as we saw earlier, it proved impossible fully to account for the effects of films on viewers. German censorship arrangements reproduced the very lack of secure knowledge about the social, political, and moral consequences of

---

[32] See Müller, "Der frühe Film." On German reformers in general, see Kevin Repp, *Reformers, Critics and the Paths of German Modernity: Anti-Politics and Reform Impulses, and the Search for Alternatives, 1890–1914* (Cambridge, MA, 2000).

[33] On the behind-the-scenes negotiations between trade representatives and officials from the Home Office, see Nicholas Hiley, "'No Mixed Bathing': The Creation of the British Board of Film Censors in 1913," *Journal of Popular British Cinema* 3 (2000), 5–19. The foundation of the BBFC is described in detail in Low, *The History of the British Film*, 82–9. See also *The Times*, 23 February 1912, 11; *Parliamentary Debates*, fifth series, vol. 43, 1612. On British film censorship in the interwar period, see James C. Robertson, *The Hidden Cinema: British Film Censorship in Action, 1913–1972* (London, 1989), 6–40; Jeffrey Richards, "The British Board of Film Censors and Content Control in the 1930s: Images of Britain," *Historical Journal of Film, Radio and Television* 1 (1981), 95–116; Gary D. Stark, "Cinema, Society and the State: Policing the Film Industry in Imperial Germany," in Gary D. Stark and Bede Karl Lachner (eds.), *Essays on Culture and Society in Modern Germany* (Arlington, TX, 1982), 122–66; Jan-Pieter Barbian, "Filme mit Lücken: Die Lichtspielzensur in der Weimarer Republik: von der sozialethischen Schutzmaßnahme zum politischen Instrument," in Uli Jung (ed.), *Der deutsche Film: Aspekte seiner Geschichte von den Anfängen bis zur Gegenwart* (Trier, 1993), 51–78; Jan-Pieter Barbian, "Politik und Film in der Weimarer Republik: Ein Beitrag zur Kulturgeschichte der Weimarer Republik," *Archiv für Kulturgeschichte* 80 (1998), 213–45.

[34] See *Verhandlungen*, vol. 328, 1591; vol. 330, 3166; vol. 333, 5173. On the whole, there was widespread agreement across party lines in Germany that the film sector suffered from excessive entrepreneurial greed, which led to the production of objectionable features. See Schultze, *Der Kinematograph*, 135–6; Brunner, *Der Kinematograph*, 3; *Verhandlungen*, vol. 328, 1592. On deeply ingrained anti-capitalism in German intellectual circles, see Rita Aldenhoff, "Kapitalismusanalyse und Kulturkritik: Bürgerliche Nationalökonomen entdecken Karl Marx," in Hübinger and Mommsen (eds.), *Intellektuelle im Deutschen Kaiserreich*, 78–94.

[35] *Das Reichslichtspielgesetz vom 12. Mai 1920: Für die Praxis erläutert von Dr. jur. Ernst Seeger* (Berlin, 1932), 12.

cinematic entertainments that rested on an incomplete understanding of the workings of film in the first place.

Thus the proliferation of film, cultural conservatives argued, exposed broad sections of the population to new cultural and moral perils. In this respect, the medium's social impact differed fundamentally from that of aviation and shipping, whose physical dangers remained restricted to specific groups such as soldiers and travelers. Many contemporaries considered film as a pervasive rather than a limited risk imposition, and this perception triggered successful calls for censorship in the early 1910s. Opponents were convinced that the proliferation of cinematic entertainment increased disreputable and criminal behavior, thereby attributing to film consumption direct, lasting, and detrimental effects. At the same time, it remained unclear how film exerted its influence over audiences, and arguments about the new medium resembled exchanges about earlier popular entertainments that did rely on sophisticated equipment to a much smaller extent. This pattern of debate reinforced the tendency in both countries to view film in cultural rather than technological terms, a trend that was by no means restricted to the medium's detractors.

### FILM AS MODERN ENTERTAINMENT AND MODERN ART

Notwithstanding public moralists' voluble laments that film culture appealed to the base instincts of the lower orders, cinematography exerted a strong appeal across the classes from the outset. In 1896, the audience at the Alhambra variety hall in Leicester Square, London, cheered Edward, Prince of Wales, when he came to watch a brief film that showed the triumph of his horse Persimmon at that year's Derby.[36] William II, who relished newsreels because they widened his public presence, became Germany's most prominent film fan, an enthusiasm the industry acknowledged with a celebratory volume of stills to mark his twenty-fifth anniversary on the throne in 1913.[37] Leading politicians also displayed an appreciation of the new medium. According to a trade journal, Prime Minister Herbert Asquith "laughed heartily" during his first cinema show in July 1914.[38] If the July Crisis did not spoil the fun for Britain's head of government, neither did

---

[36] Roy Armes, *A Critical History of British Cinema* (London, 1978), 21.
[37] Martin Loiperdinger, "The Kaiser's Cinema: An Archaeology of Attitudes and Audiences," in *A Second Life: German Cinema's First Decades* (Amsterdam, 1996), 41–50; The celebratory volume was prepared in advance of the actual anniversary. See Paul Klebinder, *Der deutsche Kaiser im Film* (Berlin, 1912).
[38] *The Bioscope*, 23 July 1914, 321.

political turmoil prevent prominent public figures from attending film premieres during the Weimar Republic.[39] On the local level, lord mayors and councilors played a conspicuous part in opening ceremonies for new cinemas in cities such as Osnabrück, Manchester, and Salford before the Great War.[40]

The acceptance of cinematography among the upper social echelons was not simply a result of censorship regimes that banned films deemed morally and politically objectionable. Rather, the industry itself strove to improve its reputation through a variety of initiatives. Individual proprietors not only claimed to control "misconduct" – that is, amorous activities – among audience members, but also selected films that steered clear of controversy.[41] Furthermore, film's appeal across the social range led to the development of a differentiated infrastructure that took into account the various socio-economic positions of its patrons. Although many badly lit, unventilated, and inadequately heated "fleapits" continued to exist, specially designed "picture palaces," which initially imitated conventional theater architecture, began to proliferate in larger cities in both countries in the 1910s. Charging patrons higher admission prices, up-market movie houses in the West End of London and around the Kurfürstendamm in Berlin cultivated a glamorous atmosphere and attracted a well-heeled clientele. While the audiences that frequented picture palaces were never socially homogeneous, and included patrons of modest means who entered comfortable movie houses for a few hours' escape from mundane domestic environments, the rise of luxurious venues with boxes at steep prices ensured that upper-class film lovers did not jeopardize their social standing by attending a show.[42] Moreover, the styles of new venues self-confidently expressed film's central place in the culture of modernity. In the 1920s, German cinema architects adopted modernist designs, the best-known examples of which were the Titania Palast and the Universum in Berlin. In Britain, meanwhile, the

---

[39] The premiere of *Metropolis* drew a particularly prominent crowd, including Chancellor Wilhelm Marx. See *B. Z. am Mittag*, 11 January 1927, 7. See also Anton Kaes, "Film in der Weimarer Republik: Motor der Moderne," in Wolfgang Jacobsen, Anton Kaes, and Hans Helmut Prinzler (eds.), *Geschichte des deutschen Films* (Stuttgart, 1993), 39–100, here 64.

[40] In Osnabrück, the Lord Mayor opened the Kaiserpalast in 1912. See Anne Paech, *Kino zwischen Stadt und Land. Geschichte des Kinos in der Provinz: Osnabrück* (Marburg, 1985), 39. A year earlier, Manchester's Lord Mayor had presided over the opening of The New Picture House. See Derek J. Southall, *Magic in the Dark: The Cinemas of Manchester and Ardwick Green* (Manchester, 1999), 8. On neighboring Salford, see Tony King, *The History of Salford Cinemas* (Manchester, 1987), 23.

[41] On the control of "misconduct," see NCPM, xxv–xxvi.

[42] On cinema patrons from different classes attending varying venues, see Jeffrey Richards, *The Age of the Dream Palace: Cinema and Society in Britain, 1930–1939* (London, 1984), 16–18; Annette Kuhn, "Cinema going in Britain in the 1930s: Report of a Questionnaire Survey," *Historical Journal of Film, Radio and Television* 19 (1999), 531–41, here 535.

Odeon chain, which opened more than a hundred establishments between 1933 and 1939 throughout the country, embraced Art Deco as its leading style to set itself apart from competitors. Frequently situated in suburban neighborhoods such as Muswell Hill, Surbiton, and Southgate, the new Odeon cinemas attracted a clientele that included the middle class, which had previously steered clear of cheap local cinemas.[43]

The appeal of film to all classes, as well as the stylistic sophistication of its new venues, came to the aid of the medium's defenders, who, especially in the wake of World War I, argued that going to the movies represented a legitimate pastime under the social and psychological conditions of the "modern age." Artist and cultural commentator Curt Moreck, who published a *Moral History of the Cinema* in 1926, went so far as to attribute film's success to technological advances that "stifled [man's] creative potential" in everyday life and made "the invention of cinematography" a "necessity." "A being who overcomes all distances with playful ease with the help of the aeroplane, the car, and the ocean liner . . . who knows no wonders that would be beyond its reach, such a being requires an altogether new means of diversion and entertainment," he argued.[44] Cinematic culture was considered a powerful antidote to the alienating consequences of technological modernity in several respects. Feature presentations, a number of vocal film fans maintained, offered a temporary release from "the increasing strenuousness, complexity, and neurasthenia of modern life."[45] In particular, social and cultural commentators singled out the alienation arising from factory and office work as a factor generating a widespread desire to chase away "exhaustion and boredom" through film shows.[46] The casualness of the movie house also created an enjoyable counterpoint to the discipline and routine that characterized the world of work. "The cinema is primarily a sort of public lounge . . . One can go alone, *à deux, en famille*, or in bands . . . punctuality and decorum are of little importance. One can drop in and out at will," explained a British film fan in 1930.[47] The lack of

---

[43] Allen Eyles, *Odeon Cinemas*, vol. 1: *Oscar Deutsch Entertains Our Nation* (London, 2002). Klaus Kreimeier, *The Ufa Story: A History of Germany's Greatest Film Company, 1918–1945* (New York, 1996), 111–20. On the styles of one of Odeon's competitors, see Allen Eyles, *Gaumont British Cinemas* (London, 1996); P. Morton Shand, *Modern Theatres and Cinemas: The Architecture of Pleasure* (London, 1930).

[44] Curt Moreck, *Sittengeschichte des Kinos* (Dresden, 1926), 63. On Moreck, see Rudy Koshar, *German Travel Cultures* (Oxford, 2000), 84.

[45] Shand, *Modern Theatres and Cinemas*, 1. See also Montagu A. Pyke, *Focussing the Universe* (London, 1910), 7; Iris Barry, *Let's Go to the Pictures* (London, 1926), 7.

[46] Miller, *Phantasiemaschine*, 27–8. A similar point is raised in Paul Rotha, *Documentary Film*, 2nd edition (London, 1939), 31.

[47] Shand, *Modern Theatres and Cinemas*, 9.

established conventions to regulate the theater as a social space, along with continuously running programs and modest prices, allowed pleasure seekers to attend cinemas spontaneously. Moreover, cinema programs offered relaxation and entertainment suited to a range of moods, including a "wish to escape, lassitude, a sense of lack in [one's] nature or [one's] surroundings, loneliness (however passing) and natural frivolity."[48] Film enthusiasts thus commended film for fulfilling a host of desires for diversion.

Commentators who praised the informality and flexibility of cinematic culture as a welcome contrast to the socio-economic constraints imposed on everyday life – in short, as an appropriately "modern" form of entertainment suited to the times – did not always neglect the moral and political questions that had long dominated the public moralists' agenda. Many writers felt obliged to assure readers that a regular intake of films harmed "neither mind nor soul (*Gemüt*)," despite attempts to paint film as a "new monster."[49] Other observers, however, began to disregard contemporary cultural critiques of the movies altogether and self-confidently hailed film entertainment as an end in itself. "We want to be amused and not instructed, intrigued but not edified," declared the author of a book on contemporary cinema architecture.[50] In fact, prominent politicians in the 1930s openly welcomed productions that offered primarily escapism. Addressing the House of Commons in 1937, the President of the Board of Trade, Oliver Stanley, "confessed," as he put it, that Mickey Mouse counted among the characters he admired most while, in a speech to leading members of the German motion picture industry in March 1933, Joseph Goebbels saw the creation of "little amusements" against "daily boredom and worries" as one of film's virtues.[51] The cultural charges that British and German public moralists had leveled against cinematography began to look anachronistic in the interwar period in the light of assessments of movie entertainment as a legitimate and even necessary form of temporary escapism. This new appreciation implied that film represented a praiseworthy and harmless technology that exerted strictly fleeting effects on viewers, a claim that contradicted earlier attacks that attributed to the movies a direct, lasting, and corrupting impact.

---

[48] Elizabeth Bowen, "Why I Go to the Cinema," in Charles Davy (ed.), *Footnotes to the Film* (London, 1937), 205–20, here 205.
[49] Rudolf Arnheim, *Film als Kunst* (Berlin, 1932), 13. See also *Parliamentary Debates*, fifth series, vol. 204, 271; Barry, *Let's Go to the Pictures*, viii.
[50] Shand, *Modern Theatres and Cinemas*, 1.
[51] *Parliamentary Debates*, fifth series, vol. 328, 1171. Joseph Goebbels, speech delivered on 28 March 1933, reprinted in Gerd Albrecht, *Nationalsozialistische Filmpolitik: Eine soziologische Untersuchung über die Spielfilme des Dritten Reiches* (Stuttgart, 1969), 439–42, here 441.

Nonetheless, the cinematic entertainment culture of the interwar years encountered fundamental attacks from public commentators who, in principle, ranked among the medium's most spirited supporters between the wars. Professional film reviewers – whose emergence provides yet another illustration of the growing public significance accorded to the cinema – frequently found fault with the predominant film fare because of its plots as well as its styles.[52] In an overview of the international film scene in 1930, Paul Rotha lamented that most motion pictures provided few "mental stimulants," while Bela Balazs and Rudolf Arnheim castigated the "petit bourgeois" and "narrow-minded bigotry" (*spießbürgerliche Wertmaßstäbe*) they detected in many films at the end of the Weimar Republic. Reviewers identified an abundance of stereotypical, sentimental, and melodramatic stories, which led them to denounce scores of movies as "kitsch."[53] According to these interpretations, most contemporary feature films lacked artistic authenticity, blunted patrons' critical faculties, and pandered to the public's "lazy instincts," as Rudolf Arnheim put it.[54] While some observers, such as Siegfried Kracauer and Bela Balazs, went on to develop sociopsychological readings of film entertainment as "kitsch," other film theoreticians regretfully diagnosed what they considered to be a profound discrepancy between deplorable, conventional content and the medium's innovative aesthetic characteristics.[55] They aimed to promote films that realized cinematography's artistic potential.

The 1920s witnessed the first productions that revealed the movie camera's artistic attributes to contemporaries. The German Liberal and Social Democratic press praised Sergej Eisenstein's controversial *Battleship Potemkin* (1926) as a "new and immortal work of art."[56] This kind of assessment of outstanding productions was by no means unique in the mid-1920s, when reviews of *Metropolis* (1926) singled out Fritz Lang's camerawork for its "photographic beauty," and Walter Ruttmann's *Berlin: Symphony of a City* attracted admiration in 1927 for the way it portrayed a day in a

---

[52] On the rise of film reviewers, see Sabine Hake, *The Cinema's Third Machine: Writing on German Cinema, 1907–1933* (Lincoln, NB, 1993), 13–15; LeMahieu, *A Culture for Democracy*, 115–16.
[53] Rotha, *The Film Till Now*, 9; Arnheim, *Film als Kunst*, 194; Bela Balazs, *Der Geist des Films* (Halle, 1930), 192–3; Fedor Stepun, *Theater und Kino* (Berlin, 1932), 72.
[54] Arnheim, *Films als Kunst*, 194. See also Balazs, *Der Geist des Films*, 212.
[55] For a detailed reading of Balazs's and Kracauer's socio-psychological approaches to film in Weimar Germany, see Hake, *The Cinema's Third Machine*, 212–70.
[56] *Berliner Tageblatt*, 30 April 1926 (evening edition), 4; *Vorwärts*, 1 May 1926 (morning edition), 24. Siegfried Kracauer praised the film in *Frankfurter Zeitung*. A reprint of Kracauer's piece can be found in Siegfried Kracauer, "Die Jupiterlampen brennen weiter," in *Kino: Essays, Studien, Glossen zum Film* (Frankfurt, 1974), 73–6.

world metropolis.[57] While reviewers commended film makers for experimenting with the medium's aesthetic potential, their prescriptions often remained formal and contradictory. Director Andrew Buchanan stressed that "the film must express in its unique way, with moving images, only those things which no other medium can express."[58] Paul Rotha advanced a considerably more nuanced and complex theory of film as "dynamic mental pictorialism" whose "patterns" of "visual images" derived power from careful editing and led viewers to translate events on the screen into "mental images." In keeping with his support for the British documentary film movement, Rotha urged directors to embrace film's "unique faculty for the collective representation of detail" and to promote "realism" in film.[59] While agreeing with Rotha on the importance of editing, Rudolf Arnheim called into question the supposed "realism" of film images and instead stressed the artificiality of good camerawork. Cinematic representations, Arnheim explained, were not only shot in black and white, thereby differing fundamentally from the colorful perception of the human eye; they also manipulated visual perspectives as well as intentionally disrupted conventional temporal and spatial relationships to achieve effects on the screen.[60] Despite diverging definitions of film as art, reviewers tended to agree on which film makers merited consideration as aesthetic pioneers. Works by Pudovkin, Eisenstein, Lang, and Chaplin featured prominently in a canon that betrayed a bias of cultural elitism, since, with the notable exception of Chaplin, these directors only rarely enjoyed success at the German and British box offices.[61] In emphasizing cinematography's capacity to capture the essence of "modern life," reviewers conceived film as an artistic medium on a par with literature, drama, and the so-called "fine arts."

[57] Positive reviews for *Metropolis* include *B. Z. am Mittag*, 11 January 1927, 7; *Berliner Tageblatt*, 11 January 1927 (evening edition), 2; *The Times*, 22 March 1927, 12. It should be noted that these pieces regard the film's sentimental plot as unfortunate. Praise for *Berlin* is found in *B. Z. am Mittag*, 24 September 1927, 5; *Berliner Tageblatt*, 24 September 1927 (evening edition), 3. For further positive reviews, see *Rote Fahne*, 25 September 1927, 14; *Vorwärts*, 25 September 1927, 17; *The Times*, 5 March 1928, 12.
[58] Andrew Buchanan, *Films: The Way of the Cinema* (London, 1932), 88.
[59] Rotha, *The Film Till Now*, 10–11, 246–7. Another theory was advanced by the *Observer*'s film critic in C. A. Lejeune, *Cinema* (London, 1931).
[60] Arnheim, *Film als Kunst*, 27–35. A detailed reading of Arnheim's writing can be found in Hake, *The Cinema's Third Machine*, 271–94. Disagreements between Rotha and Arnheim are briefly spelled out in the English translation of *Film als Kunst*, to which Rotha contributed a preface. Rudolf Arnheim, *Film: translated from the German by L. M. Sieveking and Ian F. D. Morrow, with a preface by Paul Rotha* (London, 1933), ix–xii.
[61] Stepun, *Theater und Kino*, 13; Lejeune, *Cinema*; Arnheim, *Film als Kunst*, 11.

The interwar period, then, witnessed a major transformation of public debate about the movies as a growing number of commentators began to devote serious attention to the place of film in British and German cultural life. While numerous observers emphasized that feature presentations provided an important antidote to the pressures of everyday life in contemporary society, a small group of professional reviewers and theoreticians focused on the medium's artistic possibilities. As film became accepted as an integral part of modern culture with an appeal across social divides in Britain and Germany, two competing interpretations about its social significance emerged: film was judged as a benevolent, largely inconsequential form of escapism on the one hand, or as a developing art form on the other. Despite their fundamental incompatibility, both views downplayed earlier critiques of film as a source of moral danger. As it focused on cinematography's social and cultural significance, praise for the medium continued to eclipse the status of film as a technology and, consequently, did not give rise to systematic considerations about the cultural impact of technology on society. Rather, debate about film developed into a genre of its own and remained largely unconnected to other public assessments of technology. While some now regarded cinematic entertainments as harmless diversions or praiseworthy works of art, in other quarters the suspicion lingered on that frequent exposure to film did leave a lasting and potentially damaging impression on the collective imagination.

## "INDIRECT AND LARGELY UNINTENTIONAL": FILM'S ELUSIVE CULTURAL PRESENCE

Despite a lack of systematic studies about film's social impact as late as the 1930s, a host of commentators insisted that cinematic culture exerted a significant influence on public consciousness during the interwar years.[62] At the same time, many observers modified the claims, so prominent before World War I, that film made a *direct* and *immediate* impact on patrons' beliefs and behavior. Instead, during the 1920s and 1930s public debate espoused the conviction that film exerted a profound yet *elusive* influence on audiences. A lead comment in the *Manchester Guardian* encapsulated this conviction in 1927, maintaining that "the effects of the film upon the minds and imagination . . . are beyond calculation" because "its ways are more insidious and for that reason more penetrating than the formal

---

[62] Rotha, *The Film Till Now*, 204. See also *The Film in National Life* (London, 1932), 58.

methods of the teacher. It moulds the mind, as games do the body, all the more thoroughly for being a pastime and not an exercise."[63] Indeed, commentators across the political spectrum disagreed with the apologists of entertainment films as inconsequential escapism whom we have just encountered, and instead emphasized that movie consumption possessed an intangible influence whose causes and consequences defied analysis.[64] The conviction that the "effect" of cinematic culture "on the mass imagination" was "cumulative" absolved critics from the responsibility of identifying purportedly corrosive individual feature films and opened the door to summary dismissals on vague grounds.[65] These interventions regarded film as a doubly deceptive phenomenon because it transported audiences into fictional worlds and indirectly conveyed subliminal messages of which the recipients remained unaware.

As a consequence, unease about cinematic culture extended beyond opposition to overtly ideological films, such as Soviet productions, which aroused animosity because of their supposedly insurrectional content. The multinational reach of the film industry with its global distribution networks triggered fears for the cultural foundations of Britain and Germany. Some went so far as to proclaim film "the greatest conqueror of all time," whose "internationalism holds the secret of its immense impact."[66] Seeking to uncover the mystery behind the international appeal of motion pictures, a British commentator pointed out in 1928 that "the language of the film is vision, and vision is universal."[67] Film's non-verbal communicative properties on their own, however, did not provide a comprehensive explanation for the medium's allure, since not all motion pictures met with commercial success in more than one country. To secure a worldwide following, a feature film's visual language had to tell a story that was accessible to individuals from varied social and national backgrounds. Hungarian director Alexander Korda, who produced a rare international British hit with *The Private Life of Henry VIII* (1933), explained that "when a film has reached

---

[63] *Manchester Guardian*, 17 March 1927, 6.
[64] See *Report of the Colonial Films Committee*, Cmd. 3630 (London, 1930), 9; *Parliamentary Debates*, fifth series, vol. 328, 1161; Rotha, *The Film Till Now*, 11; Carl Neumann, Curt Belling, and Hans-Walther Betz, *Film-"Kunst", Film-Kohn, Film-Korruption: ein Streifzug durch vier Filmjahrzehnte* (Berlin, 1937), 14. A Communist voice was Willi Münzenberg, *Erobert den Film: Winke aus der Praxis für die Praxis proletarischer Filmpropaganda* (Berlin, 1925), 2.
[65] Alexander Korda, "British Films: To-day and To-morrow," in Charles Davy (ed.), *Footnotes to the Film*, 162–71, here 162.
[66] Moreck, *Sittengeschichte des Kinos*, 53.
[67] Rudolph Messel, *This Film Business* (London, 1928), 292. See also Moreck, *Sittengeschichte des Kinos*, 55.

out to appeal to the Lowest Common Multiple ... in human emotion, then it will succeed all over the world."[68] Where Korda struggled for neutral descriptive language, others couched their statements in overtly pejorative terms. A German critic found that internationally successful movies "addressed the most primitive emotions and desires that are experienced by a Chinese coolie as well as an English merchant and an American farm girl."[69]

Nothing demonstrated the internationalism of film culture more strikingly to contemporary observers than the prominence of the American movie industry in interwar Europe. Evaluations of Hollywood's international success formed part of more general, highly contentious debates in Britain and Germany during the 1920s and 1930s that cast America as a "cipher of unconditional and untrammeled modernity."[70] Whether analysts turned to mass production, innovative advertising practices, large department stores, automobiles, domestic appliances, or radios, American living standards exhibited striking contrasts with those of Britain and Germany. More than anything else, however, the cultural presence of the United States in the "Old World" elicited ambivalent reactions. While jazz and swing musicians as well as dance stars such as Josephine Baker aroused widespread attention, Hollywood movies established the most pervasive contact between European audiences and cultural products from the United States.[71] American films certainly did not meet with universal rejection among reviewers who admired slapstick comedies and commended Hollywood for eschewing the social snobbery that allegedly plagued British feature films.[72] Positive appraisals, however, stood next to a torrent of passionate denunciations that, in keeping with prominent contemporary strands of anti-Americanism, accused Hollywood of

[68] Korda, "British Films," 168.  [69] Miller, *Die Phantasiemaschine*, 54.
[70] Detlev J. K. Peukert, *Die Weimarer Republik: Krisenjahre der Klassischen Moderne* (Frankfurt, 1987), 179.
[71] Victoria de Grazia, "Mass Culture and Sovereignty: The American Challenge to European Cinema, 1920 to 1960," *Journal of Modern History* 61 (1989), 53–87; Mary Nolan, *Visions of Modernity: American Business and the Modernization of Germany* (New York, 1994); Nancy Nenno, "Femininity, the Primitive, and Modern Urban Space: Josephine Baker in Berlin," in Katharina von Ankum (ed.), *Women and the Metropolis: Gender and Modernity in Weimar Culture* (Berkeley, 1997), 145–61; LeMahieu, *A Culture for Democracy*, 91–8, 161–3; Frank Trommler, "The Rise and Fall of Americanism in Germany," in Frank Trommler and Joseph McVeigh (eds.), *America and the Germans: An Assessment of a Three-Hundred-Year History* (Philadelphia, 1985), 333–42; Philipp Gassert, "Amerikanismus, Antiamerikanismus, Amerikanisierung: Neue Literatur zur Sozial-, Wirtschafts- und Kulturgeschichte des amerikanischen Einflusses in Deutschland und Europa," *Archiv für Sozialgeschichte* 39 (1999), 531–561, esp. 532–42.
[72] Saunders, *Hollywood in Berlin*, 171–95; E. W. and M. M. Robson, *The Film Answers Back: An Historical Appreciation of the Cinema* (London, 1939), 57.

spreading primitivism as well as notions of material and sexual excess.[73] Ultimately motivated by fears that American motion pictures would gradually corrode the foundations of British and German national identity, many of these diatribes betrayed a fear of the United States as a new cultural competitor.

In the United Kingdom, where the First World War had seriously disrupted domestic film output, a series of bankruptcies in the production sector spurred a widespread sense of being flooded by American films as the market share of British-made features declined from 25 percent to under 10 percent between 1914 and 1923. An intensive public debate about the necessity to support the domestic industry resulted in the passage of the Films Bill in 1927, which legislated for a quota system under which cinema proprietors were obliged to include in their programs a proportion of British films that was to rise to 20 percent over ten years.[74] By contrast with its attitude to censorship, the state did not shy away from constructing a legal framework to assist the domestic film industry. Given the widespread conviction that film "is to-day the most universal means through which national ideas and national atmosphere can be spread," as Philip Cunliffe Lister, the President of the Board of Trade, argued in his statement introducing the Films Bill, the dominance of the American movie industry threatened to drive a wedge between the British people and their "national culture." One leader article in 1927 urged that it was high time for action, since "the plain truth about the British film situation is that the bulk of our picturegoers are Americanised . . . They talk America, think America, and dream America. We have several million people . . . who are temporary American citizens."[75] As a consequence, cinema shows were hardly ever credited with addressing the themes that were expected to instill national pride and self-confidence into audience members.[76] Parliamentarians passed the Films Bill in the hope that indigenous film production would rectify the lamentable effects on British identity that sprang from America's dominance in the international film business.

Anxieties about Americanization persisted in Britain despite the legislation enacted in 1927. After all, the Cinematograph Films Act primarily

---

[73] Rotha, *The Film Till Now*, 27, 83; Messel, *This Film Business*, 23; Ali Hubert, *Hollywood: Legende und Wirklichkeit* (Leipzig, 1930), 133.

[74] Margaret Dickinson and Sarah Street, *Cinema and the State: The Film Industry and the British Government, 1927–1984* (London, 1985), 10–51.

[75] *Daily Express*, 18 March 1927, 6. Similar complaints could be found in 1930, too. See *The Film in National Life*, 11.

[76] The absence of British films about the First World War rankled tempers in this context; see *Parliamentary Debates*, fifth series, vol. 204, 250, 257–8.

aimed to guarantee British productions a limited domestic market share rather than altogether to eliminate the overall dominance of Hollywood features. Nonetheless, the renewal of the Cinematograph Films Act in 1937 bears out the fact that, despite Hollywood's continuing dominance, wide sections of the British public credited legal initiatives with a salutary effect on the British film sector. In November 1937, Member of Parliament R. C. Morrison was in a position to enumerate British productions such as *The Private Life of Henry VIII* (1933), *Catherine the Great* (1934), and the unabashedly colonialist *Sanders of the River* (1935) in a speech claiming that these features had "put British films on the map of America." Although few went as far as Morrison, who found it increasingly "difficult to define an American or a British film" in terms of quality, movies made in the United Kingdom became a source of national pride during the 1930s.[77] Oliver Stanley, the President of the Board of Trade, had these and other examples in mind when he opened the parliamentary proceedings of the new Films Bill with the words: "I want the world to be able to see British films true to British life, accepting British standards and spreading British ideas."[78] Premised on a modest revival of domestic production, statements along these lines tacitly redefined the national meaning of film in Britain. Rather than viewing the medium as a corrosive means of Americanization, in the 1930s a growing number of Britons appreciated the movies as a tool for the truthful expression and consolidation of national identity.

In Germany, public alarm at America's cinematic presence did not reach the levels of intensity that triggered parliamentary action in Britain, because Hollywood never attained as towering a position in the Weimar Republic as in the United Kingdom. While a ban on film imports had strengthened domestic production during the First World War, the German film industry engaged in a major export drive until the inflationary period came to a close in 1923. It was only in the mid-1920s that American films reached numerical parity with German full-length productions. Hollywood's prominence in German film culture began to wane in the early 1930s with the rise of sound film, which, together with the expense of synchronization, erected a language barrier for German audiences.[79] Nevertheless, individual

---

[77] *Parliamentary Debates*, fifth series, vol. 328, 1193. Other movies that generated national pride were Hitchcock's *The Thirty Nine Steps* (1935) and Herbert Wilcox's *Victoria the Great* (1937). See the reviews collected in the British Film Institute Library from *New Statesman*, 22 July 1935; *Sunday Times*, 9 June 1936; *Monthly Film Bulletin*, June 1935; *The Daily Telegraph*, 10 June 1935, as well as *Daily Herald*, 17 September 1937, 1; *Daily Mail*, 17 September 1937, 6; *Manchester Guardian*, 19 September 1937, 10, 12.
[78] *Parliamentary Debates*, fifth series, vol. 328, 1137.
[79] Saunders, *Hollywood in Berlin*, 51–83, 221–40.

features from the United States did provoke fervent animosity, especially when they devoted themselves to political subjects. The *éclat* over Lewis Milestone's *All Quiet on the Western Front* in December 1930, which led to pitched street battles and systematic disruption by audiences, arose not only from their animosity to the pacifism of Erich Maria Remarque, whose international bestseller provided the movie's plot; the Conservative *Neue Preußische Kreuz-Zeitung* also denounced this "film of wretchedness" (*Film der Erbärmlichkeit*) with particular vehemence because of its "anti-German tendencies," which exceeded those of "all other features based on American scenarios." Goebbels, who personally orchestrated the SA's terror tactics against this film, unsurprisingly cast his anti-Americanism in anti-Semitic terms, labeling it a product of the "Jewish-Bolshevist underworld." Milestone's movie provided the latest episode in a protracted assault on national identity because, in the eyes of the political Right, it insulted the memory of the imperial army.[80] Of course, the German Right was by no means alone in emphasizing the importance of film in advancing political causes, but it was in the Conservative circles of the late empire and early Weimar Republic that a paranoid interpretation of cinematic culture took root that would enjoy prominence beyond 1933.

In the immediate postwar years, the Right came to regard film as a national problem in conjunction with charges that during World War I Allied propaganda had, by secretly securing influence over the German press, gradually eroded popular morale and eventually caused the revolution in 1918. Part of the influential myth that attributed defeat to the "stab in the back" of a victorious German military, this explanation for the outcome of the war ascribed three characteristics to Allied public-relations work between 1914 and 1918. First, the Right erroneously argued that Germany's opponents had employed propaganda with purposeful consistency.[81] Second, figures such as General Erich Ludendorff, Chief of the Supreme Command between 1915 and 1918, emphasized that Allied public-relations measures had achieved their supposed triumph because

---

[80] *Neue Preußische Kreuz-Zeitung*, 6 December 1930 (edition B), 7; *Völkischer Beobachter*, 7/8 December 1930, 1. See also *Völkischer Beobachter*, 9 December 1930, 1.

[81] This point has recently been emphasized in Wolfgang Schivelbusch, *Die Kultur der Niederlage: Der amerikanische Süden 1865, Frankreich 1871, Deutschland 1918* (Berlin, 2001), 256–73. On British film propaganda, which often remained relatively uncoordinated and caused passionate domestic controversy, see Gary S. Messinger, "An Inheritance Worth Remembering: The British Approach to Official Propaganda during the First World War," *Historical Journal of Film, Radio and Television* 13 (1993), 117–27; Nicholas Reeves, "Cinema, Spectatorship and Propaganda: 'Battle of the Somme' (1916) and Its Contemporary Audience," *Historical Journal of Film, Radio and Television* 17 (1997), 5–28.

they worked "silently" (*geräuschlos*) to shape public "opinion in unconscious ways."[82] Ludendorff here expressed conventional wisdom, which held that propaganda proved most compelling when it strove to convey its messages indirectly. Finally, the Right's scenario hinged on the anti-Semitic charge that, as Ludendorff put it, "the German Jews undoubtedly cooperated with our enemies" and thus assisted in the proliferation of corrosive public-relations material in the *Reich*.[83] The absence of any concrete evidence in support of the last accusation only served as proof of the elusive and powerful propagandistic influence of Germany's adversaries during the war.

This paranoid interpretation of the First World War led the German Right to harbor several profound suspicions about popular film during the Weimar years. For one thing, cinematography was widely believed to work in the elusive manner that distinguished effective political propaganda, in general. In fact, it became an internationally acknowledged truism that film was the medium best suited for disseminating political propaganda because, as a British parliamentarian pointed out in 1937, film "is largely indirect and largely unintentional" in its mode of operation.[84] Furthermore, German Conservatives and National Socialists – in strong contrast to British commentators – frequently drew attention to the prominent position Jews occupied in the international film business, and especially singled out American entrepreneurs such as Samuel Goldwyn, Adolph Zukor, and Marcus Loew. While the Jewish presence in the film sector provided a pretext for Conservatives to denounce movie entertainments as corrupted by international capitalism, the Nazis went beyond traditional anti-Semitic tropes.[85]

Hitler and his followers frequently credited the Jews' socio-economic success, as well as their integration into gentile society, to a Protean ability to obscure their "true" identity, disparaging them as "master liars" who lacked an authentic culture of their own because of their supposed compulsion

---

[82] Erich Ludendorff, *Meine Kriegserinnerungen, 1914–1918* (Berlin, 1919), 300.
[83] Erich Ludendorff, *Kriegführung und Politik* (Berlin, 1922), 190–2.
[84] This is the assertion of a British Member of Parliament from 1937. See *Parliamentary Debates*, fifth series, vol. 328, 1161. This belief straddled political divides. See Willi Münzenberg, *Erobert den Film*, 2, 4. For individual articles see the following two left-wing journals: *Film und Volk*, June 1928, 6–7; May 1929, 1–2; November 1929, 5–6; *Arbeiterbühne und Film*, June 1930, 31; July 1930, 23–5; August 1930 22–4; January 1933, 22–4. A detailed recent study is Bruce Murray, *Film and the German Left in the Weimar Republic: From Caligari to Kuhle Wampe* (Austin, 1990). Similar attitudes existed on the British Left, too. John Grierson, "Preface," in Paul Rotha, *Documentary Film*, 2nd edition (London, 1939), 7–10, here 9.
[85] See Miller, *Die Phantasiemaschine*, 17–23; "Der 'Deutsche' Film," *Nationalsozialistische Monatshefte*, December 1931, 550–3; Marian Kolb, "Die Lösung des Filmproblems im nationalsozialistischen Staate," *ibid.*, December 1931, 553–6.

to "imitate."[86] Nazi reasoning, then, posited a threatening affinity between Jewishness and film: the prominence of Jewish businessmen in the film world granted significant influence over a popular, highly effective medium of manipulation to a group that allegedly predicated its existence on strategies of social deception. Although National Socialist tirades against "Jewish-Bolshevist" plots in film circles represented an extreme in Germany, they formed part of a wider opposition to the "'pernicious' influence of Jews on German culture" – a terrain on which "the conservative German bourgeoisie, the traditional academic world, [and] the majority of opinion in the provinces . . . came together with the more radical anti-Semites," as Saul Friedlander has observed.[87] Thus, if German debate about the international reach of film culture displayed less overt anti-Americanism than in Britain, this dissimilarity arose not only from the strength of Germany's own movie industry but also from the fact that the German Right often couched its complaints in anti-Semitic terms that, of course, encompassed anti-American overtones.

Given the National Socialists' virulent anti-Semitism, Hitler's regime initiated the first purge of Jews employed in the German film industry within less than two months after the "seizure" of power, and, by 1937, virtually all "non-Aryans" were excluded from this sector.[88] The expulsion of German Jews from film production represented the first step towards the establishment of a National Socialist film culture. Before 1937, Goebbels's diary entries on this subject often reveal a sense of frustration and indicate that, despite their unchallenged position of power, the Nazis made less progress in implementing their ideas for cinema than they desired. Administrative disorder in the propaganda ministry, Goebbels's "mania to intervene" personally in a wide range of film questions, abrupt adjustments of production guidelines to reflect changing foreign political constellations, and the fact that it was only in the spring of 1937 that the Nazis gained direct control of UFA – the country's largest studio – partly explain the delays in the

---

[86] Adolf Hitler, *Mein Kampf*, 218th edition (Munich, 1936), 331–5, 341. This conviction remained a staple of Nazi thought. See Joseph Goebbels, speech on 18 February 1943, in Helmut Heiber (ed.), *Goebbels Reden*, vol. II: *1939–1945* (Düsseldorf, 1972), 171–208, esp. 181–2. For a statement that Jews had advanced neither the technical nor the artistic side of cinematography, see Neumann *et al.*, Film-"Kunst," 27. Omer Bartov has emphasized that the National Socialists conceived of the Jews as "elusive enemies." See Omer Bartov, *Mirrors of Destruction: War, Genocide, and Modern Identity* (New York, 2000), 104–8.

[87] Saul Friedlander, *Nazi Germany and the Jews*, vol. I: *The Years of Persecution, 1933–1939* (New York, 1997), 107. See also Dietz Bering, "Jews and the German Language: The Concept of *Kulturnation* and Anti-Semitic Propaganda," in Norbert Finzsch and Dietmar Schirmer (eds.), *Identity and Intolerance: Nationalism, Racism, and Xenophobia in Germany and the United States* (Cambridge, 1998), 251–91.

[88] On the effects of this early purge, see Kreimeier, *The Ufa Story*, 210–14.

emergence of feature films that satisfied Goebbels. Last, but not least, the regime's aesthetic directives lacked programmatic clarity. While the propaganda minister acknowledged the cinema as a legitimate source of escapist, apolitical entertainment, he also called upon the film industry to produce "popular (*volkstümlich*) art" with a "National Socialist character" that addressed "National Socialist problems." Although Goebbels demanded films with National Socialist content, he provided little stylistic guidance for directors, since his vision of markedly National Socialist film art regularly included reminders that the ideological content had to remain undetectable, because, in Goebbels's words, "a propaganda that becomes noticeable is ineffective."[89]

Administrative muddles and programmatic vagueness thus help to explain why films of the Third Reich do not present an ideologically homogeneous corpus but show "a division between a self-consciously national cinema with political ambitions and a popular cinema committed to private pleasures and fantasies."[90] Only in the later 1930s did Goebbels bestow unqualified praise on productions that met his unclear requirements for National Socialist film "art," such as Veit Harlan's *Der Herrscher* (*The Ruler*), a 1937 feature about a heavy industrialist who overcomes a mental crisis, family intrigue, and opposition from incompetent managers to lead his enterprise in the service of the *Volksgemeinschaft*.[91] Because it blended family melodrama with central ideological motifs such as anti-bourgeois reflexes, the leadership principle, the notion of struggle, autarky, and industrial strength, this "modern and National Socialist" film, as the

---

[89] Goebbels's "mania" to intervene and the impact of changing foreign political constellations is described in Felix Moeller, *Der Filmminister: Goebbels und der Film im Dritten Reich* (Berlin, 1998), 100–6, 165–71. The programmatic statement is from Joseph Goebbels, speech to the Reichsfilmkammer on 5 March 1937, in Albrecht, *Nationalsozialistische Filmpolitik*, 447–65, here 456.

[90] Sabine Hake, *Popular Cinema of the Third Reich* (Austin, 2001), 21. The field of film studies has recently paid increasing attention to films whose ideological content is not immediately apparent. In addition to Hake's book, see also Eric Rentschler, *The Ministry of Illusion: Nazi Cinema and its Afterlife* (Cambridge, MA, 1996); Karsten Witte, "Film im Nationalsozialismus: Blendung und Überblendung," in Jacobsen, Kaes and Prinzler (eds.), *Geschichte des deutschen Films*, 119–170. As part of the Reich's entertainment policy, American productions were frequently screened during the early years of the regime. Philipp Gassert, *Amerika im Dritten Reich: Ideologie, Propaganda und Volksmeinung 1933–1945* (Stuttgart, 1997), 164–80. On Mickey Mouse as a star in the Third Reich, see Hans Dieter Schäfer, "Das gespaltene Bewußtsein: Über Lebenswirklichkeit in Deutschland 1933–1945," in *Das gespaltene Bewußtsein: Über deutsche Kultur und Lebensbewußtsein, 1933–1945* (Munich, 1982), 114–62, esp. 130. Articles on Joan Crawford and Gary Cooper appeared in *Filmwelt*, 20 September and 11 October 1936.

[91] The mixed reception of early political films such as *Hitlerjunge Quex*, *S. A. Mann Brandt*, and *Hans Westmar* is covered in Moeller, *Der Filmminister*, 158–60. For a reading of *Der Herrscher*, see Linda Schulte-Sasse, *Entertaining the Third Reich: Illusions of Wholeness in Nazi Cinema* (Durham, NC, 1996), 275–81.

propaganda minister termed it, received Germany's annual national film prize in 1937.[92] Other films, as diverse as Leni Riefenstahl's documentary on the Olympic Games and Karl Ritter's militaristic oeuvres, provided further evidence for claims in the party press that "the German film has risen to become a national art."[93]

During the Second World War, when he compared German cinematic propaganda favorably with productions from the United Kingdom in a speech in February 1941, Goebbels went so far as to assert publicly that his policies regarding film reaped success not only with domestic but also with foreign audiences. Unlike British cinematic features, which allegedly tramped noisily along in "heavy boots," the works created under the minister's aegis "assumed a more silent method, seeking to nestle [themselves] into creases in the enemy's skin. [They] disguise [themselves] and thus exert an unimaginable effect on world public opinion that we could hardly anticipate."[94] Given the Nazis' frequent denunciations of Jews as "parasites," Goebbels's self-congratulatory praise for National Socialist propaganda cast the appeal of film in revealingly contradictory terms. Even after, as the Nazis would have put it, Aryans had wrested control of the cinematic medium from the Jews, Goebbels defined the seductive potential of film in terms that, by stressing cinematography's "parasitical" and insidiously deceptive nature, evoked associations of Jewishness. The fact that Goebbels unintentionally broke a taboo by celebrating a triumph of Nazi film policy in terms usually employed to denounce Jewishness illustrates better than anything that film, widely understood to exert a deceitful influence in a subtly elusive manner, retained its reputation as a perplexing and potentially threatening medium as late as the Second World War.

The ascent of film as a popular entertainment gave rise to a wide variety of evaluations between the early years of the century and World War II. In the first two decades of the twentieth century, middle-class critics, with their fears that the new medium exerted a direct and corrupting influence over its numerous patrons, set the tone of public debate in both countries. By contrast, a growing number of defenders, while only rarely

---

[92] The quotation is from Elke Fröhlich (ed.), *Die Tagebücher von Joseph Goebbels: Sämtliche Fragmente*, Teil 1: *Aufzeichnungen 1924–1941*, vol. III (Munich, 1987), entry for 12 March 1937.

[93] This is the headline of a eulogy on Karl Ritter's *Verräter* (*Traitors*), *Völkischer Beobachter*, 17 September 1936, 5. On Karl Ritter see Jay W. Bird, "Karl Ritter and the Heroic Nazi Cinema," in *To Die for Germany: Heroes in the Nazi Pantheon* (Bloomington, 1990), 172–201. On Riefenstahl, see Rainer Rother, *Leni Riefenstahl: Die Verführung des Talents* (Berlin, 2000), 51–115.

[94] Joseph Goebbels, speech on 15 February 1941, in Albrecht, *Nationalsozialistische Filmpolitik*, 465–79, here 468–9.

going so far as to hail the movies as art, considered cinematic entertainments as ephemeral antidotes to the pressures of modern life after World War I. This interpretation conflicted with statements during the interwar years that credited film with the power to shape viewers' beliefs in elusive, indirect, yet lasting ways. In Britain, anti-American rhetoric included anxieties about Hollywood's dominance, while anti-Semitism colored the voluble denunciations of contemporary film culture by the German Right. Although World War I represented a temporal watershed with respect to dominant public interpretations, suspicions about the medium's cultural impact loomed large throughout our entire period and led both countries to establish censorship regimes.

While persistent doubts accompanied the development of film culture into the most successful form of "mass entertainment" at the time, opposition to the cinema did not become strong enough to generate effective summary rejections of film as a medium of modernity. The support of millions of fans who, far from writing lengthy tracts in defense of the cinema, simply attended shows, partly curbed the power of the voices that opposed film culture. Furthermore, while many observers agreed that the movies represented a risk imposition that was much more extensive than the physical dangers engendered by passenger shipping and aviation, the exact nature of the danger of film never came to be defined in clear terms. It was the absence of authoritative accounts of how the movies cast their spell over audiences that was responsible for the uncertain footing of much contemporary criticism of film. As long as the workings of movies remained a mystery, claims about their detrimental effects on audiences had to rest on shaky foundations.

The search for an authoritative explanation detailing the nature and extent of film's influence was hampered not least by the fact that, for a variety of reasons, most observers did not consider the movies from a technical standpoint. First, most of the public moralists who dominated the early debate possessed an educational background that prompted them to consider film in cultural terms and *not* to discuss the movies as a technology. Second, since screen images appeared "lifelike" rather than mechanical – or, to put it differently, since film was a self-effacing medium – contemporary discussion did not lead to a detailed analysis of how the medium's technical properties influenced the content. Instead, while acknowledging that film was in some unspecified way a particularly effective medium for conveying messages, film debate concentrated on the content of individual genres and features. The movies came to be judged in non-technical terms and were included among the subjects of a wider debate about popular culture. The fact that public debate about film devoted hardly any attention to the role technology played in generating new cultural dangers may well have been crucial for the maintenance of a cultural climate conducive to technological innovation in Britain and Germany at the time. After all, had the explicit conviction taken root that the movies as a *technology* generated new, pervasive, and identifiable cultural risks, subsequent fears would easily have amplified the ambivalence that arose from contemporary concerns that the machine was beginning to overwhelm its creators. This theme not only was prominent in films such as *Metropolis* or *Modern Times* but also underpinned the philosophical pessimism about the machine formulated most eloquently by Oswald Spengler. Since the dangers of film were not considered in technological terms, critical concern did not easily connect with existing ambivalent sentiments about the processes of innovation to form what could have presented a potent strand of opposition to technology. As it was, contemporary debate displayed a profound contradiction in public attitudes towards film when on some occasions it celebrated the medium as a technological marvel while constantly disregarding its technological dimensions in the context of the diffuse cultural dangers for which it was held responsible. Depending on the circumstances, public debate managed simultaneously to celebrate and to disavow film as a technology – a stark reminder of how strongly the meanings of technology are related to its contexts.

British and German discussions, then, generated more than one strategy for confronting the risks that accompanied technological innovation. When it came to the physical dangers of passenger boats and aviation, observers emphasized that risks were restricted to small social groups and would diminish over time as designs became safer. While the risk imposition was much more extensive in the case of film, public debate neither pinned down the nature of the dangers of film nor discussed them in explicitly technological terms. If these discussions served to contain technology's dangers by limiting its impact on the public imagination in very different ways, they amounted either to a grudging, tacit acceptance of cultural danger or to an explicit acknowledgment of physical risk as a necessary evil in the process of innovation.

In their overwhelmingly negative assessments of danger, the debates about physical and cultural perils we have covered so far differ dramatically from the spectacular celebrations of risk that form the subject of the next chapter. The British and German public lionized long-distance pilots who gambled with their lives in unreliable, small aeroplanes. In the context of solo aviation, danger created not aversion to, but admiration for, new aeroplanes as it boosted the cult of the solo pilot. Prevalent social fantasies, or widely shared desires, account for this somewhat paradoxical cultural embrace of the aeroplane as a technology that allowed certain individuals to expose themselves to lethal perils. The fantasies that fueled the almost obsessive fascination with solo pilots were powerful indeed: not only did they marshal public support behind life-threatening mechanical contraptions but they also generated applause from political and social quarters that were at loggerheads on most other issues. Solo pilots and their machines became the screens on which were projected compelling dreams that circulated widely in British and German society and bridged deep ideological chasms at the time. While continuing the examination of ideas about risk in public debate, the next chapter will bring into view the way collective desires and dreams stoked the contemporary fascination with technology.

CHAPTER 5

# Pilots as popular heroes: risk, gender, and the aeroplane

Aviators ranked among the most famous and prominent celebrities of the interwar period. While public fascination with people who flew set in before the war in Britain and Germany, it reached unprecedented proportions and peaked between the 1920s and the mid-1930s. Whether it was Hermann Köhl traversing the Atlantic by aeroplane or Alan Cobham returning from one of his flights across the empire, their achievements generated enormous media echoes in both countries. Given the prominence of aviators' achievements, representations of pilots played a crucial role in shaping the public understanding of flight as well as its promotion in the first third of the twentieth century. Beside the colorful depictions of pilots such as Hermann Köhl or Alan Cobham, "genuine" technological experts paled in comparison. Reports about Cobham and Köhl did mention Geoffrey de Havilland and Hugo Junkers, the respective builders of their planes, but these never moved from the margins to the center of public interest. Although press conferences organized by manufacturers informed journalists about the technological virtues of the machines used during daring flights, the media showed little or no enthusiasm for such purely technical "facts."[1] Designers and their staff remained shrouded in mystery as they pursued admirable, but ultimately esoteric, activities in workshops and laboratories, beyond the grasp of most people.

Images of pilots shaped public assumptions about aviation because they were easily accessible to lay audiences. The figure of the aviator ascribed meanings to aviation and established links between specific debates about technological innovation and wider discourses about the individual in contemporary Britain and Germany. As a public icon, the aviator provided a potent symbol that mediated between a complicated machine and

[1] See the detailed manuscript for the press conference held by the Junkers firm after Köhl's flight on 16 April 1928 in Deutsches Museum, Munich, Junkers archive (hereafter Junkers archive), file 0401/16/31. For a newspaper report on this conference see *Vorwärts*, 17 April 1928 (evening edition), 6.

technological laypersons. In these symbolic representations, pilots served as the interface between the developing aeroplane technology and the general public who desired to learn about a specific technological artifact. In a sense, images of airmen and airwomen created public knowledge about the aeroplane without openly talking about the machine itself. The specialist journal *The Aeroplane* praised Alan Cobham for his feats and called him "a great populariser."[2] Cobham himself considered it his mission to convey to "the public . . . the importance of aviation" by arousing "its imagination . . . in support of this good cause."[3] Similar considerations surfaced in connection with Hermann Köhl's Atlantic crossing, which aroused widespread enthusiasm in Germany in 1928.[4]

Although pilots had attracted public attention from the day the Wright brothers took off in their powered flying machine, they became the subject of popular curiosity verging on the obsessive in the aftermath of World War I. Several factors account for the unprecedented elevation of aviators in the public imagination in the interwar years. One was the fact that technological breakthroughs triggered by the demands of flight in war, such as the development of stable airframes as well as more powerful and reliable aircraft engines, put pilots into a position to attempt long-distance feats of a kind that had not been considered realistic in 1914. At the same time, until the mid-1930s, technological advances remained sufficiently limited to lend an aura of unpredictability and drama to long-distance flights. Put differently, from the 1920s to the mid-1930s aviation was at a stage where technological developments allowed pilots to push beyond existing frontiers but simultaneously endowed these endeavors with significant physical risks, as aeroplane engines, though more sophisticated than before the Great War, still had a tendency to stall, navigation aids remained primitive, and radio equipment was too heavy to carry on long flights.[5] Furthermore, during the immediate postwar years, numerous former war pilots could not find regular employment, as airline jobs remained rare and air forces

---

[2] *The Aeroplane*, 24 February 1926, 271.
[3] Alan Cobham, *Australia and Back* (London, 1926), 124.
[4] Günther von Hünefeld to Hugo Junkers, 11 February 1928, in Junkers archive, file 0401/16/7.
[5] Peter Fearon, "The Growth of Aviation in Britain," *Journal of Contemporary History* 20 (1985), 21–40; idem, "Aircraft Manufacturing," in Neil Buxton and Derek Aldcroft (eds.), *British Industry Between the Wars: Instability and Development* (London, 1979), 216–240; David Edgerton, *England and the Aeroplane: An Essay on a Militant and Technological Nation* (Basingstoke, 1991), 19–37; Walter Rathjen, "Historische Entwicklung des Flugzeuges im Überblick," in Ludwig Bölkow (ed.), *Ein Jahrhundert Flugzeuge: Geschichte und Technik des Fliegens* (Düsseldorf, 1990), 8–51, esp. 26–35; Horst Culmann, "Verkehrsluftfahrt," in Bölkow (ed.), *Ein Jahrhundert Flugzeuge*, 334–59; Harold James, "Die Frühgeschichte der Lufthansa: Ein Unternehmen zwischen Banken und Staat," *Zeitschrift für Unternehmensgeschichte* 42 (1997), 4–13.

either went through a phase of retrenchment (as in Britain), or were completely dissolved and banned (as in Germany). Overcrowded job markets induced courageous, underemployed airmen to engage in dangerous flights that generated publicity and promised ample financial rewards. After his release from the armed services, Cobham initially worked alongside other demobilized war pilots for a flying circus, which teetered on the verge of insolvency, before he embarked on his empire flights in 1925 and 1926. Hermann Köhl, a former fighter pilot, found himself reduced to taking a string of temporary jobs; his luck changed only in 1928, when he managed to talk Hugo Junkers into selling him an aeroplane for his transatlantic attempt. Both pilots stood to benefit substantially from risky flights, as did many others. Finally, the war itself had lent aviators unprecedented visibility as the German and British public began to venerate the aces of the air. Unlike in France, where Louis Blériot had galvanized national attention with his flight across the Channel as early as 1909, in Germany and Great Britain fighter pilots such as Oswald Bölcke, Max Immelmann, Manfred von Richthofen, and Albert Ball (all of whom will receive close attention in a later chapter) were the first undisputed national heroes of heavier-than-air aviation. From the 1920s to the mid-1930s, the popular fascination with dangerous long-distance flights perpetuated the cult of the individual airman, though now he wore a civilian guise.

A combination of technical, economic, and cultural factors, then, created the conditions under which aviators became screens for the projection of social fantasies of personal stardom, thus shaping popular technological knowledge that rendered the aeroplane intelligible to wide audiences. Public enthusiasm for pilots transcended even the political divisions created by contemporary nationalism. While propagandists of technological modernity pointed out how nations might benefit from the military, political, colonial, and economic potential of innovative technologies such as flight, fascination with aviators extended even to those sections of the British and German public that distanced themselves from national or imperial clamor. The German Communist daily *Rote Fahne* may have condemned the "militaristic" overtones of Köhl's transatlantic attempt, but the paper respected "highly the aviators' personal courage." Similar British voices, albeit from a less radical political background, made themselves heard following Cobham's return from Australia.[6] In addition, some pilots, such as

---

[6] *Rote Fahne*, 14 April 1928, 1. For a less radical critique of aerial militarism in the Social Democratic press see *Vorwärts*, 19 June 1928, 5. For a pacifistic appreciation of Cobham's flight, see *Daily Herald*, 2 October 1926, 4.

Charles Lindbergh, were elevated not just to national but to international stardom throughout Europe.[7]

This chapter explores how Britons and Germans imagined aviators, and how social fantasies, or widely shared social dreams, surrounding aviators contributed to a cultural climate conducive to technological innovation between the early 1900s and the middle of the 1930s. Paying particular attention to notions of masculinity and femininity that informed contemporary celebrations of aerial heroism, we shall learn how the early aeroplane was embraced not despite, but *because of*, the dangers contemporaries associated with flight. Pilots came to be seen as an elite consisting of individuals endowed with the outstanding characteristics deemed necessary to meet the challenges of flying aeroplanes successfully. Public writings by and about individual aviators treated the issue of physical risk in a fundamentally different manner from debates about accidents and the rise of film that we have encountered earlier. Rather than searching for ways that would have limited technology's perils, numerous articles about solo pilots trumpeted aviation's dangers. Depictions of aviators thus reveled in the insecurity of flight and, at the same time, promoted the aeroplane as a high-risk technology, until the late 1930s. Only in the years immediately preceding World War II did this specific mode of celebrating the aeroplane lose cultural currency.

### PILOTS AS A MODERN ELITE

Highlighting the social exclusiveness of pilots was one of the most basic conventions that invested them with exceptional status, as autobiographies and press reports stressed the aristocratic and upper-middle-class family backgrounds of some, while remaining silent about the less prestigious pedigrees of others. For instance, the British press informed the public that aviation pioneer Charles Rolls was the third son of Lord Llangattock, with a distinguished educational record at Eton and Trinity College, Cambridge.[8] At the same time, silence reigned about the parents of pilots whose unexceptional pedigree would have marred the elitist image of the

---

[7] On Lindbergh see Terry Gwynn-Jones, *Farther and Faster: Aviation's Adventuring Years, 1909–1939* (Washington, 1991), 214–18; Charles A. Lindbergh, *We: The Famous Flier's Own Story of His Life and His Transatlantic Flight* (New York, 1927).

[8] *The Times*, 13 October 1910, 12. German journalists played up the fact that Hellmuth Hirth came from a family of wealthy entrepreneurs. See *Frankfurter Zeitung*, 30 June 1911 (evening edition), 2. Hirth himself did not remain silent about his pedigree in his autobiography; Hellmuth Hirth, *20,000 Kilometer im Luftmeer* (Berlin, 1913), 11–12, 25.

aviator.⁹ Elaborate descriptions of contacts with the social and political elite also endowed aviators with elevated status. Both before and after the war, successful pilots were showered with social favors, including formal visits to heads of state, luxurious dinners in august places, and elevation to peerages and knighthoods. When Köhl returned from his transatlantic flight, it was a matter of course that he was given a glamorous reception in Berlin, during which he met President Hindenburg.¹⁰ Harry Hawker, who had to abandon a transatlantic attempt in 1919, was not the only pilot to be formally received by George V. His more fortunate colleague John Alcock, the first pilot successfully to cross the Atlantic later that year, was both presented to the monarch and knighted – as was Alan Cobham.¹¹

Stressing pilots' proximity to the "best circles" helped women aviators, in particular, to step into the limelight because elite women could take up a range of public roles without endangering their respectability.¹² Female pilots' self-characterizations, therefore, underlined the social exclusiveness of aviation. Marga von Etzdorf, daughter of a Prussian general staff officer with a landed estate, who flew solo to Japan in 1931, was not the only woman aviatrix to play up her blue-blooded relatives.¹³ Accounts of the aerial exploits of the Duchess of Bedford and Lady Heath also emphasized these women's prominent social status.¹⁴ Finally, encounters with royalty bestowed the ultimate stamp of social distinction on Britain's most famous aviatrix Amy Johnson, in the 1930s.¹⁵ Thus associations with the political and upper-class elite invested aviators with an aura of remoteness and created a firm distinction between pilots on the one side and the majority of the population on the other.

The physical and mental qualities of those who flew added to this imagery of exceptionality. Time and again, press reports and memoirs stressed that aviators had to be well-trained and fit to meet the potential challenges

---

⁹ See the report about William Leefe Robinson, who received the Victoria Cross for shooting down the first Zeppelin airship over Britain. It failed to mention his father's occupation, while detailing his grandfather's employment as "Chief Naval Constructor in Portsmouth Dockyard." See *The Times*, 6 September 1916, 9.
¹⁰ *Neue Preußische Zeitung*, 21 June 1928 (morning edition), 1; *Berliner Tageblatt*, 20 June 1928, 2.
¹¹ *The Times*, 23 June 1919, 13. Harry Hawker, who had failed in a similar attempt to cross the Atlantic, was presented to the monarch nevertheless. See *The Times*, 29 May 1919, 13. Cobham met the King and was knighted in October 1926. *The Times*, 6 October 1926, 9; *The Times*, 12 October 1926, 17.
¹² Leonore Davidoff, *The Best Circles: Society, Etiquette and the Season* (London: Cresset Library, 1986); Dolores L. Augustine, *Patricians and Parvenus: Wealth and High Society in Wilhelmine Germany* (Oxford, 1994), 106–18.
¹³ Marga v. Etzdorf, *Kiek in die Welt: Als deutsche Fliegerin über drei Erdteilen* (Berlin, 1931), 11–12.
¹⁴ See *The Times*, 1 May 1930, 17; 18 May 1928, 16; *Daily Mail*, 18 May 1928, 11. See also Lady Heath and Stella Wolfe Murray, *Woman and Flying* (London, 1929), 125.
¹⁵ *The Times*, 20 October 1934, 12; *Daily Mail*, 20 October 1934, 11.

of an aerial journey successfully. While many pilots agreed that steering an aeroplane in fine weather required relatively little physical strength or skill, bad weather and long hours of flying presented serious tests of toughness.[16] Spectacular coverage of attempts at records established the topos of "physical endurance" in aerial iconography, as pilots rode out the storm in open cockpits despite sleep deprivation.[17] Female pilots also emphasized the physical aspects of their achievements. Marga von Etzdorf proudly remembered that her record journey from Berlin to Tokyo required a vigor that exceeded the usual limits of the human frame. After days and nights of flying without sleep, she began to suffer from severe sunburn, excruciating back pains and double vision. As her "fatigue gradually reached a stage close to complete apathy," she held out and arrived in Tokyo.[18]

Contemporaries expressed surprise at the fitness that pilots displayed in the immediate aftermath of exploits. Meeting Alcock and Brown, who had just achieved the first crossing of the Atlantic in a converted bomber in 1919, a reporter from *The Times* was amazed by their good physical condition. "Captain Alcock said with a laugh 'I am not at all tired,' but Lieutenant Brown confessed, 'I am a bit fagged out.' His eyes were a wee bit bloodshot, but otherwise the men looked as if they had not traveled 100 miles."[19] A similar sense of astonishment welcomed Amy Johnson over a decade later when she returned from her expedition to Australia in August 1930. The "'poor girl' whom everybody had been pitying for her unpleasant and exhaustive journey" emerged from the Imperial Airways passenger aeroplane, which had borne her to Britain in stormy conditions, and, to "their astonishment . . . looked as fresh and trim as if she were ready for a party."[20] Pilots of both sexes were referred to as sportspersons, to underscore their physical prowess.[21]

Mental robustness also featured prominently among the prerequisites for a successful flying career, and many accounts emphasized the extraordinary

---

[16] Hirth, *20,000 Kilometer*, 33–4; Claude Grahame-White, *Aviation* (London, 1912), 119; Etzdorf, *Kiek in die Welt*, 20–2.
[17] A particularly sensationalistic example can be read in James A. Mollison, *Death Cometh Soon or Late* (London, 1932), 163.
[18] Etzdorf, *Kiek in die Welt*, 154. For additional accounts of the physical challenge of aviation, see André Beaumont, *My Three Big Flights* (London, 1912), 92; *The Times*, 2 October 1926, 10; Elly Beinhorn, *Ein Mädchen fliegt um die Welt* (Berlin, 1932), 26, 118; Charles Dixon, *Amy Johnson – Lone Girl Flyer* (London, 1930), 7, 87.
[19] *The Times*, 16 June 1919, 13. The identical report can be found in *Daily Mail*, 16 June 1919, 5.
[20] *Manchester Guardian*, 5 August 1930, 9.
[21] *The Times*, 5 August 1930, 10; *Manchester Guardian*, 2 October 1926, 14; *Frankfurter Zeitung*, 30 August 1931 (morning edition), 1; Hirth, *20,000 Kilometer*, 16.

personal "courage" of airmen and airwomen.[22] Yet pluck alone did not suffice to make a successful pilot, since unbridled courage might turn into recklessness and so unnecessarily endanger both aviator and machine. Self-restraint had to temper boldness, or, put in contemporary terminology, a pilot needed good "nerves" to remain calm in difficult situations.[23] It was no accident that the psychological aspects of pilots' personalities attracted attention during the interwar period.[24] In cultural environments that had turned the "preoccupation with nervous ailments . . . into an obsession," pilots appeared at the pinnacle of mental health.[25] Germans and Britons publicly and privately worried about the psychological effects of "modern" life on men, women, and their offspring, frequently fearing that "nervous degeneration" would affect society on a broad scale and result in the proliferation of mental illness.[26] The horrors of twentieth-century warfare added further disconcerting facets to this debate as large numbers of soldiers returned from the battlefields in states of "shell shock."[27] In this context, male pilots, in particular, came to be seen as individuals who confronted the psychological challenges of the "modern" world in exemplary ways. While it was often assumed that women had little voluntary control over their nerves, men were held to command stronger willpower. Uniform in their praise of the adventurers of "modern existence," German and British

---

[22] For some statements stressing the importance of courage, see Beaumont, *My Three Big Flights*, 135; B. Wherry Anderson, *The Romance of Air-Fighting* (London, 1917), 24; *The Times*, 3 May 1915, 17; Hirth, *20,000 Kilometer*, 44; Manfred Freiherr von Richthofen, *Der rote Kampfflieger* (Berlin, 1917), 21; *Frankfurter Zeitung*, 14 April 1928 (evening edition), 1.

[23] For descriptions of pilots' good "nerves" before 1918, see *The Times*, 27 December 1913, 6; Beaumont, *My Three Big Flights*, 92; *Daily Mail*, 4 June 1910, 4. Richthofen referred to flying as a *Nervenkitzel* (thrill). Richthofen, *Der rote Kampfflieger*, 46, 67. For further German praise of nervous strength during World War I, see Prof. Dr. Johannes Werner, *Briefe eines deutschen Kampfpiloten an ein junges Mädchen* (Leipzig, 1930), 18; Günther Plüschow, *Der Flieger von Tsingtau* (Berlin, 1927), 48. The British example is from Anderson, *Romance*, 23–4.

[24] *The Times*, 20 May 1919, 13; Dixon, *Amy Johnson*, 95, 119; *Daily Mail*, 19 December 1932, 11; *Frankfurter Zeitung*, 14 April 1928 (evening edition), 1.

[25] Peter Gay, *The Cultivation of Hatred: The Bourgeois Experience, Victoria to Freud* (London, 1995), 506.

[26] Janet Oppenheim, *Shattered Nerves: Doctors, Patients, and Depression in Victorian England* (New York, 1991); Elaine Showalter, *The Female Malady: Women, Madness, and English Culture, 1830–1980* (London, 1987); Wolfgang Radkau, "Die Wilhelminische Ära als nervöses Zeitalter, oder: Die Nerven als Netz zwischen Tempo- und Körpergeschichte," *Geschichte und Gesellschaft* 20 (1994), 211–41; Wolfgang Radkau, *Das Zeitalter der Nervosität: Deutschland zwischen Bismarck und Hitler* (Berlin, 2000).

[27] Eric J. Leed, *No Man's Land: Combat and Identity in World War I* (Cambridge, 1981), 163–92; Joanna Bourke, *Dismembering the Male: Men's Bodies, Britain, and the Great War* (London, 1996), 107–23; Robert Wheldon Whalen, *Bitter Wounds: German Victims of the Great War, 1914–1939* (Ithaca, 1984), 39–67; Paul F. Lerner, *Hysterical Men: War, Psychiatry, and the Politics of Trauma in Germany, 1890–1930* (Ithaca, 2003).

accounts nonetheless tended to differ in their assessments of the ways in which male pilots tackled the nervous crises that aviation brought on.[28]

British tales made explicit the strains that flying exerted on aviators' mental frames after 1918. Jim Mollison, famed for his transatlantic and imperial flights, admitted to being temporarily "scared stiff" during his adventures, and made no secret of the fact that he was "frightened to give free rein to meditations on the future" because he regularly risked his life.[29] Even Britain's most famous aviator, Alan Cobham, described his struggles with psychological problems in detailed accounts in the *Daily Mail* and an autobiographical monograph. In July 1926, Cobham embarked on his exploratory journey to Australia "suffering from both mental and physical exhaustion," which left him "miserably depressed" and "not enthusiastic" at all. "Perhaps we had overworked ourselves in the preparation of the flight," he reflected. On the second day of the flight, Cobham was ordered to rest in bed by a doctor in Athens. A month into the flight, Cobham still suffered from violent nightmares that required "a lot of pacifying" by his mechanic. When his engineer Arthur Elliott was killed a few days later, Cobham became "so depressed that [he] felt it was hardly worth while carrying on with the flight." Public encouragement, however, convinced him to persist in his efforts, but he did not regain a balanced mind for weeks.[30] Cobham and other British aviators, then, did not assert that pilots' psyches remained unaffected by the demands of aviation. Rather, they maintained that mental robustness helped them to overcome the severe psychological challenges brought on by flying. In keeping with a long-standing model of exemplary British masculinity that cast (especially) colonial heroes in the mold of the male who bore the "White Man's burden," pilots were by no means the only icons to narrate heroism in a language of suffering. *The Seven Pillars of Wisdom* by T. E. Lawrence, a book Jim Mollison admired profoundly, frequently portrayed its protagonist on the brink of nervous collapse and full of remorse about his conduct during battle.[31] Thus, like other British popular adventurers, British aviators displayed examples of masculinity in crisis while fashioning an iconography of the heroic male.[32]

[28] For a description of Cobham as an adventurer of modern existence, see *The Times*, 1 July 1926, 17.
[29] Mollison, *Death Cometh*, 218–19. An admission of being scared stiff can be found in *Daily Express*, 31 October 1936, 1.
[30] Cobham, *Australia and Back*, 6, 7, 27, 29, 72; *Daily Mail*, 1 October 1926, 9–10.
[31] James Mollison, *Playboy of the Air* (London, 1937), 9.
[32] The "White Man's burden," of course, refers to Kipling's eulogy of imperial sacrifice. See Rudyard Kipling, *Gunga Din and Other Favorite Poems* (New York, 1990), 52–3. On the fashioning of India as "a land of regrets" by Kipling and other writers, see B. J. Moore-Gilbert, *Kipling and "Orientalism"* (London, 1986), 42–7. For a wealth of narratives of imperial suffering by colonizers, see Elizabeth

German accounts evaluated the impact of flight upon the individual's mental frame in different ways. Unswerving toughness and strength provided important motifs in depictions of German aviators in the interwar period as the Right dominated German aviation discourse. Although Hermann Köhl, a member of the right-wing *Stahlhelm* veterans' association, experienced anxiety before setting out across the Atlantic, he considered this state of mind on that occasion as unrepresentative. At dawn, he recalled, "my heart became quiet again. The fresh morning air and the first glimmer of the . . . morning made me feel freer and happier than ever before. The gloomy hints . . . were all swept away."[33] A volume published under the auspices of Conservative ministerial supporters of the German aerial cause in 1925 called on pilots to set national examples as "hard, firm men, free from personal fear," who kept cool at "critical moments."[34] Ernst Jünger's writings, which intellectualized technology, were in step with such reasoning, and celebrated pilots' "wide awake and icy clarity of intellect, which is required to master complicated engines of many hundred horse powers."[35] Unlike in Britain, the German aviator appeared as a man of unmitigated strength and toughness who marginalized or entirely suppressed mental crises. Given the dominance of the political Right in German aviation circles, the image of the pilot featured a form of virility that disavowed associations between masculinity and weakness or suffering, because of their overtones of femininity, failure, and defeat.[36] Accordingly, German eulogies conveyed the conviction that flying toughened the nerves, thus creating individuals who were immune even to the most extreme pressures of "modern" life.

While British and German assessments of aviators' mental strengths differed, it was agreed that pilots embodied a new type of person, who was particularly well suited to withstand the demands of "modern times." In both countries, observers stressed that youth was a prerequisite for a successful flying career, although some pilots were actually in their thirties

---

Buettner, *Empire Families: Britons in Late Imperial India* (Oxford, 2004). On T. E. Lawrence, see Graham Dawson, *Soldier Heroes: British Adventure, Empire and the Imagining of Masculinities* (London, 1994), 195–201.

[33] Hermann Köhl, James C. Fitzmaurice, and Günther v. Hünefeld, *The Three Musketeers of the Air: Their Conquest of the Atlantic from East to West* (New York, 1928), 53–5.

[34] Johannes Poeschel, "Die Stellung des Staates zur Luftfahrt: Luftpolitik," in Johannes Poeschel (ed.), *Einführung in die Luftfahrt* (Leipzig, 1925), 160.

[35] Ernst Jünger, "Preface," Ernst Jünger (ed.), *Luftfahrt ist not!* (Leipzig, 1929), 12.

[36] George L. Mosse, *The Image of Man: The Creation of Modern Masculinity* (New York, 1996), 154–80. For a controversial psychoanalytic interpretation, see Klaus Theweleit, *Male Fantasies*, 2 vols. (Cambridge, 1987–1989). See also Barbara Spackman, *Fascist Virilities: Rhetoric, Ideology, and Social Fantasy in Italy* (Minneapolis, 1996).

*Pilots as popular heroes* 125

and forties. Well beyond his twenties in 1926, Alan Cobham was hailed as personifying the "eternal spirit of youth."[37] Aged twenty-seven when she completed the flight to Australia that launched her career in 1930, Amy Johnson appeared as an example of "youthful innocence" and, consequently, received a tribute from the youth of Britain on her return from Australia.[38] To make a similar point, German publications stressed that candidates for a commercial flying license were not to exceed the age limit of twenty-two.[39] Stella Wolfe Murray, one of the female aerial pioneers, summarized what many considered a truism in 1929 when she wrote "that you need youth with its energy, enthusiasm, and endurance for non-stop flying."[40]

In sum, popular images of aviators rested on the assumption that youth, physical fitness, and mental strength were essential for a successful flying career. The characteristics ascribed to airwomen and airmen not only shaped the social understanding of them, but also created public knowledge about contemporary aerial technology. If steering an aeroplane required vigor, self-control, strong nerves, and youthful vitality, flying had to be a demanding and, therefore, unsafe enterprise. Thus the public image of aviators constructed the aeroplane as a technology beset by high risks and dangers because only exceptional characters could dare to fly these menacing contraptions. In comparison with British representations, German accounts tended to place a particularly strong emphasis on the high-risk nature of contemporary aviation, arguing that flying was only for those with unshakeable nerves. Of course, many of these attributes were not the exclusive preserve of aviators. In terms of their abilities, pilots resembled sports icons such as cricketers, footballers, and athletes, who also prospered in competitive and challenging situations through a combination of youthful energy, skill, fitness, and the capacity to overcome exceptional physical strain. Pilots, however, represented the most extreme version of the sports hero because, in contrast to ordinary sports stars, who restricted themselves to playing games, they engaged in "daring gamble[s]" with their lives.[41] In this respect, aviators ranked among the older socio-cultural species

---

[37] This statement is from the preface by Australian Prime Minister Bruce in Cobham, *Australia and Back*, iv.
[38] Dixon, *Amy Johnson*, 101; *The Times*, 7 August 1930, 8; *Manchester Guardian*, 5 August 1930, 8.
[39] J. B. Malina (ed.), *Luftfahrt Voran! Das deutsche Fliegerbuch* (Berlin, 1932), 121; *Aviaticus: Jahrbuch der deutschen Luftfahrt 1931* (Berlin, 1931), 2.
[40] Heath and Wolfe Murray, *Woman and Flying*, 10. Similar judgments can be found in very different contexts. Wherry Anderson, *The Romance of Air-Fighting*, 5; Plüschow, *Der Flieger von Tsingtau*, 5; Jünger, "Preface," 12; *The Times*, 20 December 1919, 13.
[41] This was how a transatlantic flight was characterized in a British tabloid. See *Daily Mail*, 8 June 1933, 11.

that was composed of death-defying adventurers such as the mountaineer and the explorer, who commanded admiration for risking their lives in spectacular quests.[42] Unlike climbers and explorers, however, pilots used innovative machinery in their exploits, a fact that contributed to aviators' unique aura.

### THE AEROPLANE AS AN INSTRUMENT OF TRANSCENDENCE

Pilots celebrated their first solo flight as a rite of initiation that established their own distinctiveness from the rest of mankind.[43] British flying ace James McCudden may have written ironically about the emphasis that pilots placed on their first flight, but he did not exclude himself: "Oh! That feeling when one has done one's first solo. One imagines oneself so frightfully important."[44] He probably wrote with an eye on colleagues such as Canadian war pilot William Bishop, who proclaimed patronizingly that the "first solo . . . is . . . the greatest day in a flying man's life. I felt a great and tender pity for all the millions of people who never have a chance to solo."[45] Although candidates completed several flights under the guidance of an instructor, the first solo flight forced them to assume sole responsibility for themselves and their machines, a burden that weighed heavily upon them. Seeking to keep stage fright at bay, novices had to remain calm and master their fear – anything but a simple task, as the vivid account by an early devotee reveals:

Learning to fly is a perplexing business. The beginner finds it very difficult indeed to keep his head. He first sits in his machine, with a great engine roaring . . . To make things worse . . . his breath is almost taken away by the rush of wind from the propeller . . . Then, when the men who are holding back the machine let go, it darts away across the ground, providing the now worried novice with a new sensation. He should remember, of course, all that has been dinned into him before starting; but the chances are he does not.[46]

---

[42] On mountaineering, see Peter Hansen, "Albert Smith, the Alpine Club, and the Invention of Mountaineering in Mid-Victorian Britain," *Journal of British Studies* 34 (1995), 300–24; Rainer Amstädter, *Der Alpinismus* (Vienna, 1996); George L. Mosse, *Fallen Soldiers: Reshaping the Memory of the Wars* (New York, 1991), 107–25. On adventurers, see Dawson, *Soldier Heroes*; Joseph Bristow, *Empire Boys: Adventures in a Man's World* (London, 1991); Richard Phillips, *Mapping Men and Empire: A Geography of Adventure* (London, 1996).
[43] I am using the term "initiation" in a wider sense than van Gennep. See Arnold van Gennep, *The Rites of Passage* (Chicago, 1960), 65–115.
[44] James T. B. McCudden, *Five Years in the Royal Flying Corps* (London, 1919), 116.
[45] W. A. Bishop, *Winged Warfare: Hunting the Huns in the Sky* (London, 1918), 18. For similar depictions, see S. H. Long, *In the Blue* (London, 1920), 14; Mollison, *Death Cometh*, 43.
[46] Cecil S. Grace, "The Human Factor in Flying," in Claude Grahame-White and Harry Harper (eds.), *The Aeroplane: Past, Present and Future* (London, 1911), 227.

In retrospect, German fighter ace Manfred Richthofen also found it problematic to keep his nerves under control and admitted that, upon learning of his imminent first solo, "I had to swallow my cowardice[,] take a seat in the machine," and adopt a "daredevillish" attitude of "death defiance."[47] As pilots willingly faced the dangers of aviation and struggled to control its risks, they fashioned a language of personal challenge to describe the first solo as a transformative stage of identity formation. Because it disturbed pilots' mental stability, the "first time" tested and threatened to shatter a candidate's sense of self. After all, Richthofen referred to himself as a coward, while other prospective pilots became anxious that they might completely forget how to act.[48]

The unbalancing effect the aerial rite of initiation produced in novices provided opportunities for aviatrixes to enter the overwhelmingly male world of flying. Women took advantage of the re-casting of identity generated by first flights and managed to establish themselves as female pilots in an environment often suspicious of their aspirations. Marga von Etzdorf, the Prussian general's daughter whom we have already encountered, embarked on "the greatest and most fascinating moment in the life of a pilot," hoping that she would escape the testing eyes of the all-male and misogynous pupils from the nearby German Commercial Aviation School who had repeatedly teased her. Battling the usual stage fright before take-off, her

> biggest fear was that [the male trainees] would notice that I was about to solo for the first time. I could do without an audience on this occasion, especially if it was critical. But I had seen it coming – just when I wanted to speed away, the entire class turned around and watched me with amused smiles [*mit vergnügtem Grinsen*].

As soon as she had left the ground, she found that it was "wonderful to be in the air on your own." While landing turned out to be slightly problematic, Etzdorf met with unexpected positive acknowledgment after her flight. Having missed the runway on her first attempt to bring the aeroplane down safely, she had to approach the airfield again. She remembered sitting in her cockpit

---

[47] Richthofen, *Der rote Kampfflieger*, 67.
[48] Following Victor Turner, we can thus see the first flight as a "liminoid" activity whose "fragmentary and experimental" character indicates the recasting of individual identities in industrial societies. See Victor Turner, "Liminal to Liminoid in Play, Flow, and Ritual: An Essay in Comparative Symbology," in *From Ritual to Theatre: The Human Seriousness of Play* (New York, 1982), 20–60. For a brief critical appreciation of Turner's work on the liminoid, with further literature, see William G. Doty, *Mythography: The Study of Myths and Rituals* (Tuscaloosa, 1986), 91–8.

preaching to myself: be calm, be calm. I know that I shall never again be as proud as in the moment when I had brought down the machine safely, albeit with a few wrong bumps. And I do not think that the most elaborate floral bouquet will ever give me more pleasure than the hastily gathered bunch of wild flowers with which I was welcomed by a delegation from the German Commercial Aviation School.[49]

During her aerial initiation Etzdorf met a twofold challenge. She mastered the vicissitudes of flight and, simultaneously, disproved the skepticism concerning her suitability as an aviator that rested on notions of gender. Successfully passing through the pilots' rite of initiation dispelled such doubts and established Etzdorf as a pilot. The somewhat rustic gift indicated not only that she was recognized and respected by the group of pilots, but also that she was included among them. To be sure, the floral gift she received demonstrated that it was as a woman that she was welcomed into the flying community, since male aerial candidates were, predictably, never greeted with flowers. Yet her sex no longer barred her from membership in the community of aviators once she had completed her first solo flight. Thus a subtext about the reworking of notions of gender that occurred during the first solo informs Etzdorf's narrative. With even "critical" men acknowledging Etzdorf's status as a pilot, she asserted that she had successfully crossed a gender barrier by passing through the aerial rite of initiation. In Etzdorf's view, pilots consequently eluded straightforward classification along gender binaries. Thus narrating her aerial initiation provided Etzdorf with an opportunity to write herself into the community of aviators. Of course, this does not mean that public images of aviators relinquished gender bias, as we shall see below. Still, self-representations of pilots' uniqueness fed on notions that pilots of both sexes had moved beyond established gender roles, albeit, of course, from vastly different points of departure. Pilots were extraordinary men *and* women.

Upon entering the community of fliers, aviators strove to consolidate images of their uniqueness in accounts of their aerial experiences. From the outset of aviation, pilots invested themselves with a mystique of freedom and claimed to be revitalized by the liberty they found in the air. In 1911, a British airman celebrated the "sense of absolute freedom" he found among the clouds, while Hellmuth Hirth wrote in 1913 that the "particularly fascinating effect of flying" left him "totally exalted over earthly existence."[50]

[49] Etzdorf, *Kiek in die Welt*, 24–25.
[50] Claude Grahame-White, "The Fascination of Flying," in Grahame-White and Harper (eds.), *The Aeroplane*, 224; Hirth, *20,000 Kilometer*, 173. A similar point is conveyed in Richthofen, *Der rote Kampfflieger*, 50.

Furthermore, before and after the Great War, aviators asserted that aerial liberty had an energizing effect. Pioneers Claude Grahame-White and Charles Rolls referred to the "sense of power" and the "exhilarating experience" that flight gave them, while Alan Cobham relied on flying to "renew his vitality" in states of exhaustion.[51] Amy Johnson went one step further in 1930 and described the euphoria induced by successful aerial journeys in terms verging on the ecstatic: "Hurrah! I shouted. I caught the sides of [my aeroplane,] shook them, slapped them, and rocked in my seat cheering as though the world could hear ... I could not contain myself."[52] Some pilots drew on the language of medical pathology to address the uncanny side of aviation's liberating and invigorating effects, since flying could easily have the effects of a drug. Alan Cobham explicitly referred to flying as an addiction: "Once a man is really bitten with the fascination of flying he is never able to shake off that irresistible lure."[53] Richthofen claimed to have caught the flying bug right after his first ascent, which left him so "fascinated" that he "could have sat in the aeroplane all day," and he immediately "counted the hours to the next start."[54] Marga von Etzdorf also fell prey to flight on her first solo because "it was on that day that it really grabbed me, never to release me again, the feeling of unlimited, three-dimensional freedom, which only flying can give you."[55]

Narrating aerial liberation in the pathological language of addiction cast pilots in ambivalent terms. The suggestion that aviators lost command of themselves potentially conflicted with images that emphasized their outstanding discipline and "nerves." The firm association between flight and freedom, however, ensured that this tension was not further explored publicly. Instead, pilots claimed to have realized a collective dream of millions of Britons and Germans. Taking to the air, they purportedly managed to divest themselves of the "reins" of civilization, which so many contemporaries volubly denounced at that time.[56] Moreover, airmen and airwomen claimed that flying not only liberated but also invigorated and energized

---

[51] Grahame-White, "The Fascination of Flying," 224; *Daily Mail*, 3 June 1910, 7. For similar accounts of the energizing effects of flight, see Briscoe and Stannard, *Captain Ball*, 47; McCudden, *Five Years*, 328–9; Cobham, *Skyways*, v; *idem, Australia and Back*, 6.
[52] *Daily Mail*, 26 May 1930, 13.     [53] Cobham, *Skyways*, 75.
[54] Richthofen, *Der rote Kampfflieger*, 47.     [55] Etzdorf, *Kiek in die Welt*, 26.
[56] For an excellent foray into the British world of the critics of contemporary social life, see Frank Trentmann, "Civilization and its Discontents: English Neo-Romanticism and the Transformation of Anti-Modernism in Twentieth Century Western Culture," *Journal of Contemporary History* 29 (1994), 583–625. The best interpretation of contemporary German cultural debates remains Detlev J. K. Peukert, *The Weimar Republic: The Crisis of Classical Modernity* (New York, 1991), 164–90.

them. While contemporary debate frequently pointed to disconcerting depletions of social and individual energy levels due to the demands of "modern" life, aviators prided themselves in having found a spring of rejuvenation.[57] On the basis of their newly gained strength, pilots considered themselves prepared to meet the ultimate challenge of death by risking their lives.

Alongside revitalization, death featured as a prominent motif in British and German aviation literature. When pilots perished, the language of mourning and sacrifice invested their deaths with meaning. An obituary of John Alcock, who died in a crash a few months after his spectacular transatlantic flight in 1919, stated that "flight . . . demands its sacrifices from youth."[58] This conviction rested on the dialectical understanding of technological progress, which argued that improvement was predicated on costly errors which, in turn, would spur better approaches to engineering problems. In Germany, aviation journals of the interwar period featured notices of the deaths of aviators that invoked sacrifice with overtones of military combat when civilian pilots died their untranslatable "*Fliegertod.*"[59] The term "*Fliegertod*" carried associations of "heroic death" in war, "*Heldentod.*" Even in peacetime, pilots' willingness to give up their lives resembled that of soldiers, as the term implied.

Those who lived to tell of their close shaves with death were even more interesting than the deceased, however, because they had tempted fate successfully. Survivors of risky flights made it clear that they had entered into and emerged from a death zone. Such tales became particularly prominent after 1918 as improvements in aeroplane design provided individual pilots with the means to create and tackle novel challenges in remote corners of the globe. Elly Beinhorn recalled her first flight over the Sahara in the following manner:

It is strange what it means: For the first time you have to fly completely on your own over the desert for 450 miles. There is nothing, no house, no rivulet, not the tiniest path, absolutely nothing. Only sand . . . After some time you get sound hallucinations [from the engine] . . . I frequently had to think of the Atlantic flyers, who flew for one whole day and one night over water, nothing but water beneath

---

[57] See Anson Rabinbach, *The Human Motor: Energy, Fatigue, and the Origins of Modernity* (Berkeley, 1992); Maria Osietzki, "Körpermaschinen und Dampfmaschinen: Vom Wandel der Physiologie und des Körpers unter dem Einfluß von Industrialisierung und Thermodynamik," in Philipp Sarasin and Jakob Tanner (eds.), *Physiologie und industrielle Gesellschaft: Studien zur Verwissenschaftlichung des Körpers im 19. und 20. Jahrhundert* (Frankfurt, 1998), 313–46.
[58] *Daily Mail*, 20 December 1919, 4; *The Times*, 20 December 1919, 13.
[59] *Der Luftweg*, 10 March 1928, 45.

themselves. I could understand them a little. Every now and then you saw a "flying Dutchman" on the coast – a big deserted ship which had become completely rotten over time.[60]

Beinhorn expressed a feeling of existential loneliness in a spatial void. Flying alone over monotonous, uninhabited, and barren scenery turned the environment into a landscape rife with threatening symbols of disorientation, peril, and disaster. She left her readers in no doubt as to the reality of the dangers she was courting. The "flying Dutchman" aside, "two aeroplanes whose crews were never heard of again" were the only landmarks by which to orientate oneself in the desert.[61] Jim Mollison's autobiographical narrative, aptly entitled *Death Cometh Soon or Late*, also reveals that he frequently felt close to death. "I have ridden more than once with death, and I have been stricken with fear for my life, for I have, unfortunately, too much imagination to be iron-nerved in the face of peril."[62]

During the interwar years, transatlantic flights proved particularly fascinating. In addition to their difficulty and audacity, such feats commanded an aura of mystery because pilots were cut off from communication for over twenty-four hours while airborne. Restricted by weight limitations, pilots preferred to carry additional fuel supplies rather than heavy radio equipment when they saw a chance to extend the range of their craft. Transatlantic flights thus launched pilots into silent spaces from which they – and only they – could report back authoritatively. Aviators effectively exploited this opportunity for self-promotion by elaborately describing the dangers of their feats. The autobiographical account of Hermann Köhl's successful attempt in April 1928 revolved around the omnipresence of perdition and cast the "battle" with the elements in military language. He emphasized that it required only "a small mishap, some trouble in the water, oil, or gasoline system, and death would be unavoidable."[63] Some hours into the flight, he and his companions set out to cross a storm front in a mood of "utter seriousness. As we shook hands, our eyes met, and without words we promised each other faithful comradeship and to fight bravely, come what may, to the very end."[64] Battered by torrential rain and high winds, the

poor machine literally trembled in all her parts. The wings vibrated and bent, the rudder received shocks of incredible violence. Deep troughs of waves alternated with rolling mountains of water, the caps of which collapsed in their white abyss. I must confess this sight did not cheer me very much. It seemed to be too audacious

---

[60] Beinhorn, *Ein Mädchen fliegt um die Welt*, 18.   [61] *Ibid.*, 17.
[62] Mollison, *Death Cometh*, 25.   [63] Köhl, *The Three Musketeers of the Air*, 53.   [64] *Ibid.*, 79.

for small frail man to fight against these powers of nature. How would it all end? It was a fight for life and death. But so long as there was life in us, we intended to struggle.[65]

With the sea beckoning to receive the pilots to their graves and the machine under enormous strain, Köhl's description portrayed his enterprise as taking the highest risks. After arriving in Newfoundland, he remembered those who had been less fortunate: "I thought of the fliers who had [previously] attempted to reach the goal. Now I know where death lies in wait; where they disappeared."[66]

While Köhl was in transit, confusion reigned in Germany. Once the aeroplane had "disappeared," nobody knew what was happening, and all the public could do was wait.[67] When no news of Köhl's whereabouts became available within twenty-four hours, concern about the aviators' fate led to rumors as first reports that the machine had been sighted turned out ill-founded. Was the aeroplane "missing in action (*verschollen*)?" asked the Social Democratic daily *Vorwärts*.[68] When Köhl eventually "reappeared," German newspapers were at loggerheads about the national significance of his achievement. Yet, from the Left to the Right of the political spectrum, they agreed that the pilots had challenged and overcome death. One popular magazine reminded readers that "an adventure which consciously faces a ninety per cent risk of death . . . is . . . a great feat, if only for its incredible command of nerve."[69] The progressively liberal *Berliner Tageblatt* published the most direct comment about the pilots' defiance of death: "[They] have gone through death. There is no doubt about this. Many went missing this way, swallowed up by nothingness."[70]

Britain too did not lack aerial heroes who were celebrated for achieving an unlikely escape from perdition. As Robert Graves remembered in the late 1920s, the story of Harry Hawker's transatlantic flight and rescue was a focus of public interest in 1919.[71] Having set out on a fine day in May from Newfoundland, Hawker and his co-pilot Kenneth Mackenzie-Grieve disappeared over the Atlantic without a trace. Unconfirmed reports assumed that they had crashed into the sea 40 miles off the Irish coast.[72] When no

---

[65] Ibid., 82.   [66] Ibid., 100.
[67] On the "disappearance" of the machine, see *Rote Fahne*, 13 April 1928, 1; *Vorwärts*, 14 April 1928 (morning edition), 1; *B. Z. am Mittag*, 12 April 1928, 1.
[68] *Vorwärts*, 13 April 1928 (evening edition), 1.
[69] *Berliner Illustrierte Zeitung*, 29 April 1928, 771.
[70] *Berliner Tageblatt*, 20 June, 1928, 1. For further invocations of death in this context, see *Neue Preußische Zeitung*, 14 April 1928 (evening edition), 1; *Rote Fahne*, 14 April 1928, 1.
[71] Robert Graves, *Goodbye to All That* (Harmondsworth, 1960 [1926]), 236.
[72] *The Times*, 20 May 1919, 13.

news of the pilots had been received forty-eight hours after the start, "fear for Hawker and his navigator passed . . . so nearly into certainty that only a shred of hope remain[ed]."⁷³ While Mrs Hawker was publicly praised for her unstinting belief in her husband's ultimate rescue, as well as for her refusal to accept financial help, the press began to publish obituaries commending the "sacrifice" of this "born hero of romance."⁷⁴ For the British public, Hawker and his companion had died.

No one could know that a trawler that lacked radio equipment had fished the airmen out of the sea after a forced landing caused by engine trouble. When word of the unexpected rescue reached London a week after the aviators' start, the press found it difficult to keep up with ensuing events, let alone interpret them. *The Times* reported that initial "incredulity changed into delight . . . and in the *cafes* there were cheers and joyous demonstrations, and the health of the airmen was drunk with great enthusiasm. The good news was shouted from the tops of motor-omnibuses, discussed in the tubes, and passed on from stranger to stranger in the streets."⁷⁵ The eventual return of Hawker and Mackenzie-Grieve to central London – which was packed with spectators – generated "scenes of extraordinary enthusiasm" crowned by "an unprecedented exhibition of mingled joy and pride."⁷⁶ At King's Cross station, the heroes "were not so much received as absorbed . . . into a human cauldron, boiling with enthusiasm, seething with demonstrable energy." One observer gasped that he had not seen anything similar since Queen Victoria's funeral.⁷⁷ Faced with this "spontaneous and unrehearsed" public outburst, *The Times* was struck by the "almost selfish delight in [the] intense relief that comes from the cessation of pain. There is no revulsion as that which comes from a hope gradually and miserably given up and then on a sudden triumphantly realized."⁷⁸

In May 1919 such language was typical of stories about the many war victims the circumstances of whose death had never been ascertained. The survival of Hawker and Mackenzie-Grieve was like a return from the dead. Aviators appeared as individuals who miraculously reappeared after encounters with death at a time when British and German society sought to come to terms with the bereavements of the First World War in a variety of ways. Rituals of mourning and the creation of personal as well as public memory lent expression to the conviction that the victims of war had not died in vain

---

⁷³ *Ibid.*, 21 May 1919, 14.
⁷⁴ *Ibid.* For reports on the concomitant female heroism of Hawker's wife see *Daily Mail*, 20 May 1919, 7; 23 May 1919, 12; *The Times*, 27 May 1919, 13.
⁷⁵ *The Times*, 26 May 1919, 14.  ⁷⁶ *Ibid.*, 28 May 1919, 13.
⁷⁷ *Daily Mail*, 28 May 1919, 5.  ⁷⁸ *The Times*, 28 May 1919, 3.

and that their sacrifice remained meaningful in the historical present.[79] At the same time, contemporaries endeavored to learn about the "other world" to which they believed the dead had passed on. While spiritualism provided one such source of knowledge, pilots added to this knowledge as improving aeroplane technology put them in a position to undertake life-threatening feats during which they appeared to penetrate death zones.[80] When pilots re-emerged, they provided listeners and readers with information about a sphere that millions had entered during the Great War without returning. In a sense, pilots became intermediaries between this world and the realm of death.

The prominence of tales in which aviators tempted death had far-reaching implications for contemporaries' evaluation of aerial technology. Through stories highlighting the risks pilots faced in flight, aeroplanes came to be widely seen as a highly unreliable technology. After all, pilots were in a position to lay claim to close shaves with death only when they could convincingly argue that they stood a decent chance of losing their lives while airborne. The image of the death-defying aviator crucially shaped contemporary perceptions of the aeroplane as an instrument that courted, challenged, brought on, and defeated death. The occasional criticism of spectacular flights, such as Köhl's, bears this out. While the Social Democratic *Vorwärts* condemned his transatlantic enterprise as a mindless "Russian roulette," the paper never doubted the pilot's ability. Instead, it was convinced that Köhl's technological equipment was not yet sufficiently advanced or safe to warrant the enterprise.[81] Iconic pilots thus promoted aviation as a high-risk technology because it provided a means both to tempt *and* to conquer death.

The potent public associations between flight and death evoked an ambivalent reaction on the part of the British and German aviation industries. While pilots gained popularity from stunts deemed uncertain and daring, airlines and aircraft manufacturers were aware that fatal outcomes of flights could result in damaging press coverage about the technological imperfections of individual designs. Throughout the period, the

---

[79] On the impact of the Great War on British and German attitudes to death, see Jay Winter, *Sites of Memory, Sites of Mourning: The Great War in European Cultural History* (Cambridge, 1996), 13–116; Mosse, *Fallen Soldiers*, 70–106; David Cannadine, "War and Death, Grief and Mourning in Modern Britain," in Joachim Whaley (ed.), *Mirrors of Mortality: Studies in the Social History of Death* (London, 1981), 187–242, esp. 202–12, 227–31.

[80] On spiritualism, see Janet Oppenheim, *The Other World: Spiritualism and Psychological Research in England, 1850–1914* (Cambridge, 1988); Jenny Hazelgrove, *Spiritualism and British Society Between the Wars* (Manchester, 2000).

[81] *Vorwärts*, 15 April 1928, 2.

commercial aviation sector attempted to project images that stressed the safety and reliability of aerial technologies to counter narratives focusing on aviation's dangers. For several reasons, however, these initiatives achieved limited success before the late 1930s. First, financial shortages hampered the public-relations efforts through which individual firms tried to achieve control over the popular image of aviation. Until the second half of the 1930s, the British and German aviation sectors mainly consisted of relatively small companies that could not afford large and costly advertising departments. Just when the technological potential of flight expanded, German and British companies faced an especially precarious economic environment. In both countries, policies of fiscal retrenchment as well as the slump presented obstacles to a consolidation of the sector and left many aviation firms fighting for survival. In the case of Germany, regulations imposed by the peace settlement added to the economic pressures before 1925. Unlike the highly capitalized shipping industry, which created and controlled the public image of passenger vessels as luxurious "floating palaces" (as we shall see in the next chapter), aviation companies were in no position to marginalize or silence the public interventions of pilots, which enhanced representations of flight as dangerous. The economic fragility of the German and British aviation sectors thus indirectly supported the prominence of a public imagery that cast pilots as aerial heroes who exposed themselves to high risks.

Second, the German and British industries themselves could not resist the lure of promoting their products by invoking the dangers of flying. The maneuvers of the Junkers works, from which Köhl purchased the plane for his transatlantic adventure, illustrate how the industry indirectly supported media campaigns that portrayed flight as a high-risk technology. On the one hand, the managers at Junkers insisted on publicly dissociating themselves from the planning stages of the enterprise out of fear of a public-relations disaster.[82] The management anticipated that "the dear yellow press as well as wide sections of the public will hound us" if the aeroplane were to come down.[83] Köhl's financial supporter, Baron von Hünefeld, acknowledged the difficult position the Junkers works were in, and praised Hugo Junkers's generosity in handing over "a machine that is most intimately connected with your universally appreciated name."[84] A month before take-off, Junkers personally implored Köhl to postpone the flight from the spring

---

[82] Agreement between Hugo Junkers and von Hünefeld, 19 January 1928, Junkers archive, file 0401/16/2.
[83] Dr. Bruhn to Herta Junkers, 11 February 1928, Junkers archive, file 0401/16/6.
[84] Von Hünefeld to Hugo Junkers, 11 February 1928, Junkers archive, file 0401/16/7.

to the summer, when weather conditions would be more favorable.⁸⁵ On the other hand, the company secured the right to exclusive "propagandistic exploitation" (*propagandistische Auswertung*) if Köhl's hazardry came to a successful completion.⁸⁶ Junkers thus pursued a self-contradictory public-relations strategy in that it strove to detach itself from a perilous enterprise while it seemed set to affirm the dangers of flight, but was simultaneously anxious to profit from the enthusiasm that followed Köhl's triumph. As a result, Junkers did little to reverse the well-known tales about the high-risk nature of flying.

Since the British and German aviation industries did not mount effective independent public-relations efforts, pilots gained the opportunity to establish themselves as celebrities with the help of carefully crafted media campaigns that informed the public about their dangerous adventures. An alliance between the press and the stars worked to mutual advantage, since sensationalist front-page coverage of daring exploits increased circulation while press exposure enhanced a star's marketability. Given these economic stakes, it is hardly surprising that the public image that the stars of aviation presented to their fans was anything but "authentic," as they went to great lengths to stage good "shows" for reporters. To provide attractive promotional material, stars agreed to participate in film and photo shoots several days before their exceptional feats, and these pictures subsequently circulated as faithful records. The scenes in a German newsreel about Elly Beinhorn's start to a flight across Africa in 1933, for instance, had been filmed a day before the beginning of the journey, because at the actual take-off, set for an early hour, there would be insufficient light.⁸⁷ In Britain, where a fierce circulation war broke out among the popular newspapers at the beginning of the 1930s, the press landscape offered ample opportunity to aerial heroes and heroines keen on self-promotion as individual newspapers competed for scoops.⁸⁸ Amy Johnson – the most famous aviatrix in Europe by a wide margin – owed her stardom to a protracted campaign in the *Daily Mail*, which gained exclusive coverage of her journey upon payment of £10,000 as the previously unknown pilot was making her way

---

⁸⁵ Express letter, Hugo Junkers to von Hünefeld, 15 March 1928, Junkers archive, file, 0401/16/12.
⁸⁶ Agreement between Hugo Junkers and von Hünefeld, 19 January 1928, Junkers archive, file 0401/16/2.
⁸⁷ Elly Beinhorn, *180 Stunden über Afrika* (Berlin: Scherl, 1933), 7. For a British example, see *Daily Express*, 9 June 1933, 11.
⁸⁸ James Curran and Jean Seaton, *Power Without Responsibility: The Press and Broadcasting in Britain* (London, 1981), 44–85; Robert Graves and Alan Hodge, *The Long Weekend: A Social History of Great Britain, 1919–1939* (London, 1991 [1940]), 290–3.

to Australia in 1930.⁸⁹ In subsequent years, Johnson consistently sold the press rights for her long-distance flights to the highest bidder, and hired William Courtenay, an "aviation correspondent" for several London dailies, as her press secretary. He was to negotiate with editors, arrange industrial sponsorship deals, handle fan mail, edit her autobiographical writing, and design the storylines to be fed to the press.⁹⁰

Like sports heroes and film stars of their day, famous pilots belonged to a relatively new socio-cultural group that struck a host of observers as quintessentially "modern." They were all media celebrities who owed their prominence to exposure in the press, movies, and radio shows designed for an urban "mass" clientele.⁹¹ Aviators' image of stardom crystallized around several components that underlined their uniqueness in contemporary celebrity culture. In addition to mastering the first flight as a demanding rite of initiation, pilots claimed that, once in the air, they found freedom and revitalizing energy as they risked their lives in potentially lethal flights. By casting the aeroplane as a tool to tempt and defy death, contemporary debate endowed the aerial technology with an aura of transcendence. While Antoine de Saint-Exupéry provided the most eloquent and philosophically refined interpretation of the airborne pilot as a transcendent figure suspended between life and death in interwar Europe, less sophisticated, and probably more accessible, variations on this theme propelled successful aviators to exceptional stardom in Britain and Germany.⁹² Given the intimate cultural links between danger and flight, pilots were widely credited with attributes that most contemporaries considered to be typically male. After all, they had to be fit, self-disciplined, courageous, and, last but not least, willing to expose themselves to the danger of death. Nevertheless, women as well as men gained a high public profile in the world of aviation after the First World War, and consequently we need to examine the relationship between notions of femininity and the public image of flight as a high-risk technology. The United States aircraft industry employed pilots of the "weaker sex" to demonstrate the reliability and safety of machines and thus to boost the confidence of potential customers of air travel, implying that aeroplanes were becoming so safe that "even"

---

⁸⁹ On the negotiations between Johnson's father, who acted as her representative, and the *Daily Mail*, see Constance Babington Smith, *Amy Johnson* (London, 1967), 214–17.
⁹⁰ See William Courtenay, *Airman Friday* (London, 1937), 98–100, 118–23, 251, 259–69.
⁹¹ On the phenomenon of the star, see John Ellis, "Stars as a Cinematic Phenomenon," in Jeremy G. Butler (ed.), *Star Texts: Image and Performance in Film and Television* (Detroit, 1991), 300–15; Richard Dyer, *Heavenly Bodies: Film Stars and Society* (London, 1985), 10–15.
⁹² Antoine de Saint-Exupéry, *Night Flight* (New York, 1938).

women could be entrusted with their operation.[93] In the United States, a female presence in the world of flight served to discount the danger of aviation. Although, as we have seen, this was not the case in either Britain or Germany, aviatrixes rose to prominence in both countries. After all, Amy Johnson owed her fame to a perilous attempt to reach Australia, and Elly Beinhorn crossed the Sahara desert on her own. If British and German celebrations of female pilots affirmed the dangers of flight, they also needed to credit them with qualities such as death-defying courage and self-discipline, attributes that were conventionally regarded as inherently male. How, then, are we to locate prominent, dare-devillish aviatrixes within contemporary gender notions, whose conventions ascribed to women less assertive roles?

### FEMALE STARS OF THE AIR

The stardom of female pilots often remained fraught with problems during the interwar years because their arrival on the public stage highlighted the widespread confusion about masculinity and femininity as well as their hierarchical relations in Britain and Germany.[94] As in France, where, in Mary Louise Roberts's words, contemporary "debate about gender identity became a primary way to embrace, resist, or reconcile oneself" to perceived changes, the British and German public spheres abounded with discussions that attempted to consolidate, dissolve, and recast conventional gender roles between the wars.[95] Public evaluations of aviatrixes produced a fractured discourse that testifies to the culturally disorienting effects of these prominent, active females. To some degree, the difficulties of locating these women discretely within contemporary social and cultural frameworks arose from differences between their own autobiographical accounts and texts penned by male journalists. Furthermore, the distinct political climate in Britain and in Germany contributed to the ambiguities surrounding aviatrixes in

---

[93] Joseph C. Corn, *The Winged Gospel: America's Romance with Aviation, 1900–1950* (New York, 1983), 71–90. Recent work, however, has examined narratives by American aviatrixes as tales of adventure. See Sidonie Smith, *Moving Lives: 20th-Century Women's Travel Writings* (Minneapolis, 2001), 87–106.

[94] For popular overviews of the histories of female pilots see Mary Cadogan, *Women with Wings: Female Flyers in Fact and Fiction* (London, 1992); Judy Lomax, *Women of the Air* (London, 1986).

[95] Mary Louise Roberts, *Civilization Without Sexes: Reconstructing Gender in Postwar France, 1917–1927* (Chicago, 1994), 5. On British negotiations of gender, see Susan Kingsley Kent, *Making Peace: The Reconstruction of Gender in Interwar Britain* (Princeton, 1993); Barbara Caine, *English Feminism, 1780–1980* (Oxford, 1997), 173–217. On Germany, see Ute Frevert, *Women in German History: From Bourgeois Emancipation to Sexual Liberation* (New York, 1989), 168–204; Atina Grossmann, "*Girlkultur* or Thoroughly Rationalized Female: A New Woman in Weimar Germany?" in Judith Friedlander et al. (eds.), *Women in Culture and Politics: A Century of Change* (Bloomington, 1986), 62–80.

nationally specific ways that highlight the different gender orders in these countries. Because their quests for fame easily embroiled women pilots in the culture wars conducted between advocates and opponents of an expanding female presence in the public sphere, aviatrixes met with significantly differing receptions in Britain and in Germany.

The heterogeneous imagery of aviatrixes sprang partly from overtly misogynous tendencies within a flying community that made no secret of its overwhelming hostility to women's aspirations. Before World War I, Hellmuth Hirth was convinced that women were incapable of outstanding aerial achievements for physical reasons; flying, he claimed, required the "strength of a man."[96] Distrust of female pilots went so far as to lead to an international ban on them in commercial aviation between 1919 and 1926.[97] Even after this regulation had come to an end, aviatrixes stressed how difficult it was to make a living from flying. Marga von Etzdorf was convinced that her application to join the German Commercial Aviation School to train for a commercial flying certificate was rejected on grounds of her sex. She prepared for the necessary examination independently, only to find that she had to take the final test under conditions that put her at a disadvantage. While men were never examined individually, she had to face the "five officers from the aerial police entirely on my own (*mutterseelenallein*) for three hours. It is much easier to be examined in a group, but I was unfortunately a woman, and since the possibility of a female candidate had been anticipated neither by the commercial flying school nor by its regulations, I was examined on my own."[98] Etzdorf's account emphasized that gender bias seriously hampered her ambitions. While Etzdorf was passing her test, Lady Heath complained that "it was the fashion . . . for a section of the aeronautical community to decry as 'publicity seekers' all those women who interested themselves in aviation."[99]

In response to this misogyny, women pilots adopted several strategies to assert the legitimacy of their endeavors against "the smiling skeptics who could not visualize a woman's activities outside the four walls of her house."[100] Aristocratic pilots such as Lady Heath and the Duchess of Bedford emphasized their social rank and contributed to the creation of a European version of what Joseph Corn called the "lady-flier stereotype" in an American context.[101] The Duchess of Bedford was accompanied on her flights by a pilot as well as a mechanic. Taking only "several short spells

---

[96] Hirth, *20,000 Kilometer*, 57. See also his disparaging remarks about female passengers, *ibid.*, 222–3.
[97] Lomax, *Women of the Air*, 38–9.   [98] Etzdorf, *Kiek in die Welt*, 56.
[99] Heath and Wolfe Murray, *Woman and Flying*, 22.   [100] *Ibid.*, 23.
[101] Corn, *The Winged Gospel*, 76.

at the controls" and leaving most of the piloting to her male employees, she conformed to the conventional model of an aristocratic lady who was driven by her chauffeur on most of her trips.[102] Although Lady Heath flew solo and would suffer no male aviator in her cockpit, she labored to demonstrate that piloting did not compromise her standing as an elegant, upper-class woman. She pointed out that it was "really absurdly easy" to steer an aeroplane, and ostentatiously illustrated her assertion in the way she dressed.[103] Returning from South Africa, she "was not wearing the usual flying kit, but stepped from her machine attired in an ordinary dress, with fur coat, silk stockings and a straw hat. 'I find my Avian is as comfortable as a motorcar,' she explained; 'any special flying clothes are not necessary.'" She even claimed to have changed her stockings in mid-air.[104] If clothes made the lady, she further underlined her social status by downplaying her technical abilities, which carried connotations of filth and physical labor widely associated with working-class masculinity. According to a popular book written by a male journalist, Lady Heath spoke "candidly of her lack of knowledge of the little Cirrus engine at a Savoy banquet given in London in her honor [and] said, 'I never troubled myself much about my engine as regards overhauling. Each morning before starting the next stage I simply stood on a box and did the usual mysterious things inside my engine.'"[105] By emphasizing the simplicity of steering and maintaining an aeroplane, the "lady fliers" remained silent about the dangers widely associated with flight. Other aviatrixes, however, were unwilling or unable to claim the status of a "lady pilot." In fact, they chose to portray aviation as a dangerous occupation that required considerable training, not least of a technical nature.

Detailed accounts of engine services expertly performed highlighted how women proved themselves in the predominantly male world of aviation. Reports by male pilots hardly ever mentioned maintenance duties such as the overhaul of motors, because their technological expertise was tacitly acknowledged. In contrast, aviatrixes explicitly described the way they serviced their engines in order to write themselves into a male environment on

---

[102] *The Times*, 1 May 1930, 17.    [103] Heath and Wolfe Murray, *Woman and Flying*, 25.
[104] *The Times*, 18 May 1928, 16. See also *Daily Mail*, 18 May 1928, 11. The photographic portrait of Lady Heath that featured in the book on her return flight to South Africa repeated the "lady flier" stereotype. It showed a smiling woman in a fur coat that was casually draped around her shoulders, wearing a white blouse and a fashionable dress. Lady Heath's tale as well as her camera portrait implied that piloting agreed with patterns of dress that were in accordance with uncontroversial notions of upper-class femininity. For the photograph, see Heath and Wolfe Murray, *Woman and Flying*, plate facing 122.
[105] Lady Heath, quoted in Dixon, *Amy Johnson*, 2–3.

egalitarian terms. Marga von Etzdorf claimed that she derived "the greatest pleasure from working on the engines":

> I had to learn all the things from scratch which every ten-year-old boy is familiar with these days anyway. Cylinder and crankshaft, vents and spark plugs were entirely strange words to me. But I quickly found my way around, and it wasn't long before I laughed with my pals at our most recent pupils after we had played an old and popular joke on them ... Under the master's strict eyes, I had to do everything on my own, from vent cutting to taking apart magnetos.[106]

Etzdorf here tells how she was gradually included in an all-male community as she was acquiring the technological competency required to qualify as a pilot. Other women employed similar stories of female technological expertise to demonstrate that they could develop the same engineering skills as men. Elly Beinhorn's account of an emergency landing in Persia due to engine trouble may serve as a case in point. She set herself to work on the engine and found it easy to ascertain the cause of the problem: "I had been fortunate. The whole mess looked pretty promising. On opening the fuel filter and the carburetor such amounts of filth came out that it was no surprise that the engine had gone on strike."[107]

In addition to playing up their ability to acquire and apply technological expertise, female pilots stressed that women could tolerate, and even thrive in, the rough everyday culture of the workshop. Amy Johnson did not complain about the inauspicious welcome she received when arriving late one morning:

> I slip round the hangar door. Up comes a huge gum-boot and my leather coat behind takes on a slightly muddier look than usual. But – I'm inside. On with the overalls – slightly soiled, I'm afraid – and up the steps I go, armed with grease gun, oil squirt, spanner, and the like, for an engine waits my personal attention. First I wash it down – with every minute the engine grows cleaner and I dirtier.[108]

Johnson depicts herself as a female stoic who remains indifferent to physical abuse by a man and gets on with her job no matter how filthy it may be. Her next assignment, "to scrape carbon off a cylinder head," verges on the frustrating: "It is dinner time before I have made any noticeable impression on its blackness. (The blackness, incidentally, has made a most notable impression on me!)"[109] Thus Johnson portrays her work environment as

---

[106] Etzdorf, *Kiek in die Welt*, 20–1.
[107] Beinhorn, *Ein Mädchen fliegt um die Welt*, 72. For a narrative of how Beinhorn took care of her engine, see *ibid.*, 70.
[108] Dixon, *Amy Johnson*, 25–6. (This passage in Dixon's biography is a reprint of an article Johnson wrote for the journal *Air*.)
[109] *Ibid.*, 28.

dominated by demands and behavioral patterns widely associated with working-class masculinity. She depicts the workshop as a place of dirt, where toughness and devotion to physical labor take precedence over manners and cleanliness. Johnson is at pains to convey the pleasure as well as the sense of belonging that she derives from working in this environment. She enjoys "joking and talking shop" with her colleagues over lunch, and emphatically concludes her account with the words, "A hard life, but by Jove – it's a good one!"[110]

While aviatrixes claimed inclusion on terms of equality in the unwelcoming and practically all-male surroundings of the workshop as well as in the wider aviation community, the claim made it difficult for them to formulate gendered identities in unambiguous terms. Writings by women pilots remained riddled with tensions between notions of femininity on the one side and the perceived requirements of flying on the other. Elly Beinhorn's preparations for her reception at the Siamese Royal Court reveal her apprehensions about the effects of piloting on female identity. Before she met the monarch, "it had been drummed into my head that it was up to me to show through a decent curtsey that flying does not ruin the good manners of the female sex."[111] Beinhorn stresses that she sailed through the ritual and thereby disproved contemporary fears that aviatrixes necessarily compromised conventional notions of feminine respectability. Nonetheless, her tale subverts her own endeavors when she proudly refers to herself as a cigarette-smoking, whiskey-drinking, short-haired deer hunter.[112] In fact, a photograph of Beinhorn at the helm of a transatlantic liner, clad in white shirt, white trousers and a captain's hat, suggests that she consciously set out to cross conventional gender barriers (see illustration 5.1).

Tensions between conflicting gender identities also informed Amy Johnson's account of life at the workshop. She assured her readers that she satisfied popular notions of female stylishness unwillingly. For her, switching between gender roles conventionally associated with masculinity and femininity was, above all, a nuisance. When a journalist appeared at her workshop asking to interview the "'lady engineer' ... everyone looks blank. No lady here. I emerge rubbing dirty hands on the seat of my overall, and join in the search. But when I wash my face and hands for tea, my secret is discovered."[113] Johnson, whose comportment – especially rubbing her "dirty" hands on her backside – suggests masculinity, implied that she did

---

[110] *Ibid.*, 27, 29.  [111] Beinhorn, *Ein Mädchen fliegt um die Welt*, 116.
[112] *Ibid.*, 35, 77, 84, 179.  [113] Dixon, *Amy Johnson*, 28.

Illustration 5.1 From Elly Beinhorn, *Ein Mädchen fliegt um die Welt*. Beinhorn intentionally crossed the gender boundary by donning a captain's uniform and thus signaled that she could defy prevailing expectations.

not conceive of herself in female terms while working as a mechanic. She put on what she considered a disguise for the sake of convention and public image in order to pose for publicity shots that represent her as a cool and detached beauty on the model of a film star (see illustration 5.2):

Oh hurry, where is my heart-shaped helmet, my manicure set and my powder puff? Where can my powder puff be? There is a spanner in my stern pocket, a few loose nuts, screws and bolts in every other pocket, but where, oh where, is my powder puff pocket? I must look nice for the photographer . . . I don my smart flying suit and appear for once as others would have me, but the minute the camera clicked, off it comes, for in roll the machines to be put to bed for the night.[114]

Johnson writes of dressing up as a fashionable "woman" as an exercise in cross-dressing, while stressing that she feels more at ease in dirty, "masculine" flying gear. Yet her narrative disrupts its own inversion of established gender distinctions when she slips into the ultra-feminine language of maternal care to convey her enthusiasm for her duties as a "male" mechanic: like children, the machines have "to be put to bed." Johnson's account hovers between the attractions of the "male" shop floor, notions of female elegance, and an image of the dutiful and loving matron. This description of her identity as a female pilot and engineer combines role models that were gendered in both masculine and feminine ways. In other words, it was impossible for Amy Johnson to position herself within existing gender frameworks in an unambiguous manner.

Marga von Etzdorf provides the most dramatic anecdotes of the gender trouble faced by aviatrixes. When she flew as a co-pilot on commercial airliners – an activity from which German regulations did not ban her – she was sometimes referred to as "*Herr Pilot*" in her flying gear.[115] Never correcting these errors, she considered them to be compliments, which she silently acknowledged. Photographs in her book about her exploits enhanced Etzdorf's androgynous image by depicting her as a short-haired woman in leather jacket and loosely fitting trousers. Etzdorf proudly recounted the effectiveness of her "disguise," which purportedly led on one occasion to an encounter with homosexual undercurrents:

While I was walking down the corridor in the administrative wing of Berlin airport, a girl fell around my neck with a fake stumble and sought steadiness in strong male arms. Should I shatter this illusion? I mumbled 'oops' with my deepest voice, put her back on her feet and marched on in a masculine way [*männlich*].[116]

---

[114] Ibid., 28–9.   [115] Etzdorf, *Kiek in die Welt*, 30.   [116] Ibid., 53.

Illustration 5.2 Star postcard of Amy Johnson as the cool, fashionable, and glamorous diva of the air.

Of course, the humorous tone of Etzdorf's account of machismo in disguise dispelled any doubts about her sexual respectability.[117] Still, her text illustrates the difficulties with which women pilots referred to their activities

[117] On crossdressing and female homosexuality, see Martha Vicinus, "Fin-de-Siècle Theatrics: Male Impersonisation and Lesbian Desire," in Billie Melman (ed.), *Borderlines: Genders and Identities in War and Peace, 1870–1930* (New York, 1998), 163–92.

and described their identities in established languages of femininity. Etzdorf maintained that she felt appreciated as a pilot only when addressed as a man. Like Elly Beinhorn and Amy Johnson, she found that the categories usually employed to describe female occupations were unable to do justice to her achievements.

Given the frequent and conscious invocations of qualities and abilities usually considered to be male preserves, women pilots came dangerously close to the "new" or "modern" woman, one of the most polarizing sociocultural phenomena of the interwar years. A refusal to abide by conventional female role models was often held to characterize the "modern woman." In fact, the parallels between female pilots and the stereotypical "modern woman" extended beyond attempts by aviatrixes to enter the worlds of traditionally male activity. Like the elusive "modern woman," aviatrixes tended to be single, fashionable, metropolitan middle-class women who radiated unprecedented female independence. Because of her removal from maternal roles, this "type" of woman compromised her femininity in the eyes of many contemporaries, often coming to be denounced as a "sexless" being.[118] How, then, did wider British and German public culture accommodate female pilots in the light of the similarities between them and the highly contentious cultural stereotype of the "modern woman"?

Britons responded to Amy Johnson in a variety of ways when she achieved national stardom after her first flight to Australia in 1930. She attracted extensive front-page coverage in the national press, landed an advertising contract for motor oil, and became the subject of popular songs. At the same time, Johnson had to battle against open resistance to her prominence. The *Manchester Guardian* translated its inability to interpret Johnson's exploit clearly as a misogynist attack. Despite its praise for her skill and courage, this newspaper found the "hero worship" for Johnson "pathetic" and "even slightly distasteful," because it remained on the level of superficial "indecent curiosity" rather than initiating a fundamental redefinition

---

[118] For the best discussion see Roberts, *Civilization Without Sexes*, 19–88. On Germany, see Grosmann, "*Girlkultur*"; Lynne Frame, "Gretchen, Girl, Garçonne: Weimar Science and Popular Culture in Search of the Ideal New Woman," in Katharina von Ankum (ed.), *Women in the Metropolis: Gender and Modernity in Weimar Culture* (Berkeley, 1997), 12–40; Cornelie Usborne, "The New Woman and Generational Conflict: Perceptions of Young Women's Sexual Mores in the Weimar Republic," in Mark Roseman (ed.), *Generations in Conflict: Youth Revolt and Generation Formation in Germany, 1770–1968* (Cambridge, 1995), 137–63; Billie Melman, *Women and the Popular Imagination in the Twenties: Flappers and Nymphs* (Houndmills, 1988). Feminist historiography of interwar Britain sometimes questions the prominence of the "New Woman." See Kingsley Kent, *Making Peace*, 114–39.

of "national values."[119] For the *Manchester Guardian*, Johnson's flight seemed pointless.

The culture of empire, however, helped Johnson to overcome the hostility to her public persona because, as an aviatrix who had flown to Australia, she could draw on familiar imagery that depicted women engaged in imperial adventures. In the second half of the 1920s, penny magazines began to publish a flood of adventure stories set in the empire and featuring heroines "who [could] face hardship like any man and who [could] shoot, hunt and ride."[120] Furthermore, a growing number of British women in India learnt to handle firearms in the interwar years, not only to hunt, but also so that they would be ready to defend themselves from potential attacks by militant members of the growing independence movement. Thus there existed a substantial "overlap between femininity and masculinity [that] allowed ... women a broad scope for public activity in the Empire."[121] Britons of Conservative, Liberal and feminist backgrounds could agree on the principle that women could and should play important roles in the empire's "civilizing" mission.[122] Johnson took up this argument and urged her contemporaries "to take advantage of each and every step of progressive transport for more closely linking our scattered empire," not least to "fulfil the great ambition of Cecil Rhodes" in Africa.[123] Casting Johnson in the role of a British woman in the empire thus provided an acceptable way to celebrate her public visibility, since female contributions to the empire were generally perceived as honorable. In addition, the language of empire furnished an opportunity to extol the risk-taking with which her flights were associated. On her return from Australia in 1930, Johnson herself stressed that she "did not believe in 'safety first,' which never got anyone anywhere."[124] The cause of empire justified women's exposing themselves to dangers and shouldering their part of the white man's burden. According to the Secretary of State for Air, Lord Thomson, who greeted Johnson upon her return from Australia in 1930, she "was a young woman ... fired with

---

[119] *Manchester Guardian*, 6 August 1930, 8.
[120] Melman, *Women and the Popular Imagination*, 140.
[121] Mary Procida, "Good Sports and Right Sorts: Guns, Gender, and Imperialism in British India," *Journal of British Studies* 40 (2001), 454–88, here 461.
[122] On "women of Empire," see Antoinette Burton, *Burdens of History: British Feminists, Indian Women, and Imperial Culture* (Chapel Hill, 1994), 33–62; Barbara N. Ramusack, "Cultural Missionaries, Maternal Imperialists, Feminist Allies: British Women Activists in India, 1865–1945," in Nupur Chauduri and Margaret Strobel (eds.), *Western Women and Imperialism: Complicity and Resistance* (Bloomington, 1992), 119–36; Thomas R. Metcalf, *Ideologies of the Raj* (Cambridge, 1995), 109–10.
[123] Amy Johnson, *Sky Roads of the Air* (London, 1939), 31, 61.   [124] *The Times*, 7 August 1930, 8.

that spirit of adventure which has contributed more than any other single factor to the development of the British Commonwealth of Nations."[125] Record flights to South Africa in 1932 and 1936 consolidated her reputation as an imperial heroine, a theme elaborated in the popular film *They Flew Alone* (1942), in which Britain's most famed diva of the silver screen, Anna Neagle, portrayed Johnson as a model of modest, selfless, and committed national service to a country and empire at war.[126]

Reports in the British Socialist press testify to the importance of a positive appraisal of empire for ascribing significance to Amy Johnson's achievement. During the interwar period, British adherents of the Labour party entertained a much more tortuous relationship with imperial ideology than did Conservatives.[127] Although the Left was generally sympathetic to female emancipation, it encountered difficulties in putting forward positive interpretations of Johnson's prominence because it found the open appeals to imperialism to legitimize the aviatrix problematic.[128] In 1930, the Socialist tabloid the *Daily Herald* completely avoided the issue of imperialism and presented Johnson in terms of youthful purity, which, however, did not explain the significance of her achievements. A conversation between Johnson, "the fair haired girl [in] a simple grey coat," and the Secretary of State for Air, Lord Thomson, reminded the journalist "of a schoolgirl – a schoolgirl answering the headmaster who had just presented her with a prize." Invocations of Johnson's juvenile innocence created a stark contrast with her daring enterprise and left the writer unable to account for her heroism. "It is incredible that she had done something which had fired the hearts of the entire world." Although the *Daily Herald* was unable to attach a definitive political meaning to Johnson's star status or to her achievements in imperial terms, the paper was attracted to her informality when, on her return to London, she asked people to "call [her] just Johnny." Because she had until recently been "an almost unknown typist," as the

---

[125] *The Times*, 5 August 1930, 10; *Daily Mail*, 5 August 1930, 10. A biography of Johnson also claimed that her spirit of adventure had motivated her to select Australia as her destination. See Charles Dixon, *Amy Johnson*, 17. An unpublished report to Lord Wakefield, managing director of the Castrol motor-oil company and sponsor of Johnson's flight, also emphasized that she was "a very valuable ambassadress for the British Empire." See Royal Air Force Archive, Hendon (hereafter RAF archive), file AC 77/23/677, Amy Johnson to C. C. Wakefield, 1 July 1930.

[126] Sarah Street, *British National Cinema* (London, 1997), 130–1.

[127] Stephen Howe, *Anticolonialism in British Politics: The Left and the End of Empire, 1918–1964* (Oxford, 1993).

[128] On the relationships between feminism and liberalism or socialism in Britain see Caine, *English Feminism*, 202–9; Pat Thane, "Women, Liberalism and Citizenship, 1918–1930," in Eugenio Biagini (ed.), *Citizenship and Community: Liberals, Radicals, and Collective Identities in the British Isles, 1865–1931* (Cambridge, 1996), 66–92.

paper erroneously pointed out, Johnson's supposed social proximity to the non-privileged sections of British society significantly enhanced her star appeal by lending her public persona the "common touch," which maintained an intangible link with her fans.[129] Johnson and her publicists were aware of this dimension of her appeal. They actively obscured the fact that she could not have launched her flying career without her father's wealth, while they emphasized the hardships and financial deprivations she had allegedly endured during her training period.[130]

Whether or not she was cast in the role of a woman of empire, Amy Johnson's star persona did not, as it emerged, necessarily clash with conventional gender hierarchies. Charles Dixon, whose bestselling biography was published in the year in which its subject shot to fame, did not hide his surprise about her. Although he had "expected to meet one of those modern, masculine women, a woman with a blasé air . . . a firm aggressive step, and a complete disdain of a mere man," he found himself confronted with "a 'feminine' girl," whose "sensitive, shy, animated and often vivacious" ways immediately charmed him.[131] Thus Johnson's non-threatening presence allowed men to regard her as a harmless girl. Later in her career, Johnson transformed herself into "the glamour girl of the machine age" who had outgrown her provincial origins, having toned down her Yorkshire accent in favor of "a West End lull."[132] During the mid-1930s, reports cast Johnson as a society lady, admiring her stylish haircuts, manicured fingernails, and dresses.[133] To emphasize that she considered an elegant appearance of paramount importance even during life-endangering exploits, newspapers claimed that Johnson's survival kit included an ultra-feminine article, a little case of face powder.[134] Thus Johnson's star persona changed from an image of youthful innocence, as articulated through the language of the "flying girl," to that of the society lady who legitimately adopts a public role. British press coverage about Johnson increasingly acknowledged an overlap

---

[129] *Daily Herald*, 5 August 1930, 1.

[130] Amy Johnson's father, a wealthy fish merchant from Hull, supported his daughter's aspirations, paying £100 for her flying lessons and £300 towards her aeroplane. See RAF archive, AC77/23/57, Amy Johnson, "Notes on My Career," 24 April 1930; RAF archive, AC77/23/57, Amy Johnson to Mr. Limb, 19 April 1930.

[131] Dixon, *Amy Johnson*, 10.

[132] *Daily Express*, 7 January 1941, 3. Amy Johnson in fact spoke with an oddly clipped accent that sounded nothing like a "West End lull."

[133] *Daily Mail*, 24 July 1933, 11; 8 May 1936, 12; 16 May 1936, 12; *Daily Express*, 16 May 1936, 12.

[134] *Daily Mail*, 25 July 1933, 9; *Daily Express*, 8 May 1936, 11; *Daily Mail*, 16 May 1936, 12. On make-up and femininity, see Kathy Peiss, "Making up, Making Over: Cosmetics, Consumer Culture, and Women's Identity," in Victoria de Grazia and Ellen Furlough (eds.), *The Sex of Things: Gender and Consumption in Historical Perspective* (Berkeley, 1996), 311–36.

between male and female abilities. In 1936 the *Manchester Guardian*, which had initially viewed her rise to stardom with skepticism, hailed her as a "symbol of feminism, of today's general interest in the women who can equal or surpass men in fields immemorially regarded as masculine."[135] The *Times* and the *Daily Mail* – both politically conservative publications – also paid their respects to Johnson for her success in "a department of activity that requires virtues which are supposed to be the peculiar endowment of the male sex."[136] As a young aviatrix who embraced the conventions of upper-class fashionable glamour, Johnson embodied an acceptable form of female initiative, which could be integrated into British celebrity culture because, like the figures of the woman of empire and of the innocent yet plucky girl, it avoided the impression of an "unfeminine" type of woman aggressively invading spheres traditionally regarded as male preserves. By the mid-1930s, Johnson had secured undisputed star status and was acknowledged as an adventurous female risk-taker.

In comparison, German aviatrixes faced significantly higher cultural and political obstacles as they sought attention, although Germany hosted a female celebrity culture before and after 1933. While movie divas such as Marlene Dietrich were lionized in the Weimar Republic, Lilian Harvey, Zarah Leander, Kristina Söderbaum, and other actresses took up prominent positions in the public spotlight during the Nazi era, as did eminent director Leni Riefenstahl.[137] German airwomen, however, did not enjoy levels of popularity even remotely similar to these of celebrities in the movie world or of Amy Johnson, as the media echo of their expeditions bears out. In August 1931 the liberal *Berliner Tageblatt* reported on the start of Marga von Etzdorf's flight to Japan, but suspended coverage after one day. Although Etzdorf broke world records *en route*, the *Frankfurter Zeitung* remained one of few newspapers to pick up briefly on events once she reached Tokyo.[138] On her return flight, which the Berlin tabloid *Tempo* covered in daily dispatches, she crashed in Bangkok, leaving a blemish on her record.[139] The last time Etzdorf commanded the limelight was after an accident *en route* to Australia in May 1933, which led her to commit suicide,

---

[135] *Manchester Guardian*, 16 May 1936, 12.
[136] *Daily Mail*, 16 May 1936, 12; *The Times*, 16 May 1936, 15.
[137] On the cult of celebrities in the film world, see Antje Ascheid, "Nazi Stardom and the 'Modern Girl': The Case of Lilian Harvey," *New German Critique* 74 (1998), 57–89; Kerstin-Luise Neumann, "Idolfrauen oder Idealfrauen? Kristina Söderbaum und Zarah Leander," in Thomas Koebner (ed.), *Idole des deutschen Films: Eine Gallerie von Schlüsselfiguren* (Munich, 1997), 231–43. On Riefenstahl, see Rainer Rother, *Leni Riefenstahl: Verführung des Talents* (Berlin, 2000).
[138] *Berliner Tageblatt*, 19 August 1931, 8; *Frankfurter Zeitung*, 30 August 1931 (morning edition), 1.
[139] The articles from *Tempo* are collected in Deutsches Museum, Munich (hereafter DMM), archive, Nachlaß Etzdorf, file 001.

an event the Nazi press only partly managed to cover up.[140] Elly Beinhorn, despite sucessfully completing her flight around the world, received only a quiet welcome on her return in 1932, gaining no wider public recognition until the spring of 1933, when a jury of politicians and members of the aviation industry awarded her the prestigious Hindenburg Cup for outstanding aeronautical achievement.[141] During the early years of the National Socialist regime, Beinhorn combined long-distance flying with a career as an author and public speaker on aviation in Germany and abroad. On the whole, newspaper articles about her achievements remained relatively short and colorless, and lacked the enthusiasm generated by Amy Johnson in Britain.[142] An article published in the Catholic *Kölnische Volkszeitung* in 1938 noticed the glaring disparity in the status of female pilots in Germany by comparison with other European countries. Although it conceded that, in the face of successes by Beinhorn and other aviatrixes, flying was no longer labeled as "unfeminine," the paper deplored the lack of "support for female sports pilots."[143]

Several factors prevented German aviatrixes from rising to the levels of stardom achieved by Johnson. Like the British pilots, Beinhorn and, to a smaller extent, Etzdorf sought to draw on the figure of the "imperial woman" to legitimize their public roles. Characterized as a "woman of deep German feeling," Beinhorn devoted herself to supporting those Germans who continued to reside in the former colonies that Germany had lost after World War I. Praising the aeroplane as a prime "propaganda instrument" to invigorate "Germanity abroad" (*Auslandsdeutschtum*), she recalled circling in "a German aeroplane [over] an island [off the African coast] under German administration with a German steamer moored in the harbor" as a climactic moment in her flying career.[144] The emphasis of Beinhorn and Etzdorf on the promotion of "Germanity abroad" through flying in Africa and elsewhere played on the central theme that fueled the revamped activities in German women's colonial associations after the loss

---

[140] See the attempts to deny her suicide in DMM, archive, Nachlaß Etzdorf, file 002. Articles on the accident appeared, for instance, in *Neue Preußische Zeitung* and *Dresdner Nachrichten* on 30 May 1933.

[141] See the articles in *Berliner Tageblatt*, 26 July 1932, 10; *Vorwärts*, 27 July 1932 (evening edition), 6. No mention was made of her in *Frankfurter Zeitung*. On the award, see Beinhorn, *180 Stunden über Afrika*, 8.

[142] For articles on a flight in 1936, see *Völkischer Beobachter*, 8 August 1936, 3; *Motor und Sport*, 16 August 1936, 6.

[143] *Kölnische Volkszeitung*, 25 August 1938. The article can be found in DMM, archive, Nachlaß Reitsch, file 50.

[144] Rolf Italiaander, *Drei deutsche Fliegerinnen: Elly Beinhorn, Thea Rasche, Hanna Reitsch* (Berlin, 1940), 29; Beinhorn, *180 Stunden über Afrika*, 60; Beinhorn, *Ein Mädchen fliegt um die Welt*, 26. For a statement along similar lines by Etzdorf, see Etzdorf, *Kiek in die Welt*, 165.

of formal colonial possessions as a result of the Versailles Peace Treaty. As part of its attempts to strengthen communities of ethnic Germans throughout the world, the Women's League of the German Colonial Association set up girls' schools staffed by German teachers, ran a recruitment scheme for nurses who were seeking work in German hospitals in Africa, and sponsored unmarried women's emigration to suitable settlements outside of Europe. Beinhorn and Etzdorf, however, fit the prevalent mold of a German woman of empire only imperfectly, since their initiatives did not correspond with the "images of maternal solicitude" underlying all of the activities approved by the Women's League.[145] In the context of colonial activities deemed appropriate for German women, the aviatrixes Beinhorn and Etzdorf appeared as unconventional imperial women in Germany during the 1930s.

Another factor added to the difficulties German aviatrixes faced in their search for inclusion in the Valhalla of German aerial heroes. They encountered a much more pronounced gender barrier between notions of femininity and masculinity than pertained in Britain, because German airmen, as we saw above, stressed that unmitigated strength and unwavering toughness were essential prerequisites for a career in aviation. Coverage about Hanna Reitsch, Nazi Germany's best-known aviatrix, illustrates the difficulties the press faced when seeking to praise a female pilot given the insistence that aerial achievement required firm virility. Like Leni Riefenstahl, Reitsch relied on male allies in high places to support her unconventional career, which, after she established her reputation with a world record in gliding at the age of twenty in 1933, included positions as a test pilot for military equipment and experimental helicopters. When Göring promoted her to the rank of a "flight captain" (*Flugkapitän*) in 1937 – a title usually reserved for the male pilots working for the airline Lufthansa – the press did not manage to capture Reitsch's gender status in straightforward terms. Applauding her determination and courage, the *Leipziger Neueste Nachrichten*, for instance, labeled her as a "girl" who "acted like a man." Similarly, after receiving the Second-Class Iron Cross for her bravery in wartime military experiments that resulted in grave injuries, a newspaper characterized her as a "complete man" (*ganzer Kerl*).[146] The phraseology of both texts betrays insecurity about Reitsch's gender status as a woman who

---

[145] Lora Wildenthal, *German Women for Empire, 1884–1945* (Durham, NC, 2001), 172–96, esp. 189.
[146] DMM, archive, Nachlaß Reitsch, file 50, *Leipziger Neueste Nachrichten*, 21 May 1937; DMM, archive, Nachlaß Reitsch, file 51, *Reichskriegerzeitung*, 6 April 1941. In conjunction with the press cuttings in the archive of the Deutsches Museum, Reitsch's unashamedly apologetic memoir provides information on her career. See Hanna Reitsch, *Fliegen – Mein Leben* (Stuttgart, 1956).

performed with the ability of a man. Thus, the celebrity status of Reitsch and other aviatrixes remained precarious because their achievements rested on physical and mental qualities that coexisted uneasily with prevailing German notions of femininity and masculinity. In Germany, there emerged no public language that would have facilitated the celebration of prominent aviatrixes as adventurers handling a high-risk technology.

In Britain and Germany many of these pilots stressed that their exploits were as dangerous as those of their male colleagues, an assertion with considerable implications for conventional understandings of gender distinctions. British aviatrixes avoided public condemnation as masculinized "modern women" by drawing on imageries that mediated between flying and respectable forms of femininity. While the pose of the "lady flier" allowed aristocratic aviatrixes to sidestep associations between flight and danger, the figure of the imperial woman provided the foundation for the outstanding stardom secured by Amy Johnson. As an icon of contemporary "feminism," her glamorous public persona commanded respect across the entire spectrum of the British media by the mid-1930s. By contrast, models of femininity that would have eased the public celebration of daring aviatrixes in similarly straightforward terms did not exist in Germany. Throughout the 1930s, female pilots corresponded only incompletely to the strongly maternal image of the German imperial woman. Furthermore, German male aviators asserted that flying required steadfast virility, thereby erecting a pronounced gender barrier that women pilots found difficult to overcome. Whatever levels of prominence aviatrixes achieved in Britain and Germany during the interwar years, they often differed from their American counterparts in one crucial respect: rather than highlighting the safety of air travel, British and German heroines of the air emphasized perils of flight in their quests for stardom, thereby contributing to the promotion of the aeroplane as a high-risk technology.

After the mid-1930s the star status of individual pilots who had shot to fame in daring flights began to wane. Amy Johnson was not alone in proclaiming the end of "an era" as the appeal of record flights diminished while "huge airliners, with their load of passengers and mail, [were] thronging the aerial roads paved by lonely pioneers."[147] Technical improvements combined with changes in the media landscape to initiate a transformation of aviation's public image, which increasingly highlighted the

---

[147] Johnson, *Sky Roads*, 24–5. A statement to this effect by Mollison can be found as early as 1937. See Mollison, *Playboy of the Air*, 213–14.

safety rather than the dangers of flying. To begin with, the development of powerful yet relatively lightweight radio devices tended to undermine the mystique that had surrounded pilots who launched themselves into protracted silence during their long-distance efforts before the second half of the 1930s. While Johnson, Mollison, and Köhl seemed to reappear from nowhere after their flights, the transatlantic crossings of the late 1930s lacked a similar element of surprise, since the crews were routinely monitored via radio while airborne.[148] In addition, the aero-engines of the 1930s were significantly more reliable than those of the previous decade, reducing an important source of risk. Furthermore, navigation aides such as autopilots facilitated long-distance flying by alleviating the monotony of which many pioneers had complained. Finally, the airline industry supported claims that aviation had "surpassed its experimental stage" with an image of commercial airline pilots as highly trained, responsible professionals who shirked unnecessary risks.[149] Bolstered by larger budgets for public-relations initiatives, Lufthansa and Imperial Airways alerted the public to these technical improvements, not least to generate demand for the expanding passenger services that both companies maintained in the 1930s.[150] The image the industry now projected with increasing efficiency revolved around the notion that air travel was becoming a routine affair. Even in Nazi Germany, where propagandists hardly ever missed an opportunity to resort to hyperbole, the press declared that the non-stop flight by a new Focke-Wulf machine between Berlin and New York in 1938 was a "record flight without any sensation."[151] Similarly, the British journalist who covered the crossing of the Atlantic by an Empire flying boat in the summer of 1937 summed up the event by asserting that "all the world knows [that this] trip was as uneventful as a Channel crossing."[152] Thus press coverage and public-relations efforts emphasized from the mid-1930s that recent

---

[148] For a British report of two simultaneous, non-stop transatlantic crossings by British and American flying boats that were in permanent radio contact in 1937, see *The Aeroplane*, 14 July 1937, 40. For a similar report of a direct flight between Berlin and New York in 1938, see *Völkischer Beobachter*, 12 August 1938, 1.

[149] For characterizations of pilots as thoroughly trustworthy and responsible professionals, see *Die Luftreise*, January 1936, 36–7; *Imperial Airways Gazette*, June 1934, 3–4; *The Aeroplane*, 19 August 1936, 245–6.

[150] For some reports on these issues in public-relations material, see *Imperial Airways Gazette*, May 1935, 5; October 1937, 3; *Luftreise*, January 1936, 36; March 1937, 63–5; September 1938, 209.

[151] *Die Luftreise*, September 1938, 209. The leading Nazi paper took the same line and declared that the flight had gone entirely according to plan. See *Völkischer Beobachter*, 13 August 1938, 1; 14 August 1938, 1.

[152] *The Aeroplane*, 21 July 1937, 87.

technological advances had transformed the character of long-distance flying.

Up to that time, however, the British and German public spheres abounded with stories about aviators that created the popular understanding of aviation and promoted the aeroplane as a high-risk technology. Aircraft came to be seen as unreliable when public coverage concentrated on the heroic qualities pilots required in order to master the vicissitudes of flight. Like public icons of sport, airwomen and airmen were characterized by youth and exceptional physical fitness as well as outstanding mental resilience and stability. In addition, aviators laid claim to unique experiences that, literally and metaphorically, propelled them beyond conventional stardom. Asserting that they found a revitalizing and energizing liberty in the air, pilots described the aeroplane as an instrument to challenge and conquer death. Pilots of both sexes insisted that their feats surpassed other exemplary enactments of masculinity and – in much more spectacular fashions – femininity. While women pilots in both countries struggled to develop autonomous language to describe female heroism in the air, successful flights by aviatrixes attracted more attention in Britain than in Germany. The tropes of the woman of empire, the "flying girl," and the glamorous female celebrity provided acceptable models for the public representation of British women pilots, especially since they did not confront such rigid notions of masculinity as their German counterparts. As many of the attributes ascribed to daring female pilots carried distinctly masculine overtones, the British public sphere succeeded in fashioning language that avoided the tensions so evident in German celebrations of female aerial achievement. As a result, women played a more prominent role in promoting the aeroplane as a high-risk technology in Britain than in Germany. While, in the context of solo aviation, Britain displayed visions of technological modernity that accommodated women with greater ease than was the case in Germany, the dominant ideas about pilots were overwhelmingly similar in both countries. The main difference rested in who embodied those ideas.

Firmly associated with lethal dangers for most of the interwar period, the single-seater aeroplane and its handlers engaged a widespread desire for transcendence. It was the dream of tempting and overcoming death that underpinned visions of the aerial stars who were literally larger than life. The power of these social fantasies reveals itself not just in the intensity of the social veneration for pilots but in the fact that this adulation spread across the political spectrum – and that at a time that suffered from an overabundance of political tensions. Solo pilots, whether male or female, became the icons that highlighted a collective desire for transcendence and fueled a fascination with technology that demanded that individuals embrace lethal risks. Given its emphasis on danger, the star cult surrounding pilots alerts us once more to the crucial roles played by concepts of risk in the creation of cultural climates that drove technological developments ahead. It needs to be stressed that British and German aviators conveyed notions of physical risk that differed profoundly from the exercises in damage limitation that we encountered in debates about aerial and naval accidents. Rather than holding out the prospect of a society that was becoming ever safer, celebrations of pilots extolled the presence of danger in the modern world. Praise for the spectacular dangers of aviation existed side by side with the prospect of its diminishing physical risks, and it was only in the mid-1930s that ideas of aerial safety gained the upper hand in the public sphere and redrew important contours marking Britain and Germany as "risk societies." As technical developments made the outcome of long-distance flights ever more predictable, and as the regular market for passenger travel by aeroplane expanded, the airline industries increasingly managed to sideline the idea that flight was inherently dangerous.

If this development signaled a major transformation in public debate about aviation, there were also clear limits to the success of this public-relations campaign. At no point did the aviation industry strive to convince the public that the aeroplane provided a particularly comfortable, let alone luxurious, mode of transport. Technical constraints that continued to keep the provision of services to a bare minimum were not the only factors responsible for the modesty of most advertising brochures that airlines issued in the 1930s. More ambitious claims would have invited comparisons between airliners and comfortable passenger liners, which had long set the undisputed standard of luxury long-distance travel. Next to the splendor of monumental ocean-going vessels, no large aeroplane could hope to promise more than a functional mode of travel. The idea that passenger liners amounted to "floating palaces" had gained persuasiveness in the 1890s, when shipping lines launched an extensive public-relations campaign to glamorize their vessels. As part of their promotional drives, Cunard, the Hamburg-American Line, and others succeeded in fundamentally redefining popular notions of sea travel by tapping into longings for comfort and a carefree life. Turning to the dreams of material affluence and luxury by which passenger liners were measured, this analysis shifts away from contemporary concepts of risk, but continues to ask how social fantasies shaped public debates about technology. For all their glaring differences, dare-devillish solo pilots in their small, dangerous aeroplanes and luxurious "floating palaces" shared one structural feature: they owed their prominence to widespread dreams and desires that transcended political and social positions and thus rallied support behind new technological devices across a fractured socio-political spectrum.

CHAPTER 6

# *"Floating palaces": passenger liners as objects of pleasure*

"Twentieth-century luxury, comfort, modernity . . . that is what you will enjoy aboard," pledged the prospectus for a Mediterranean cruise on the Cunard liner *Aquitania* in 1931.[1] Proclaiming a symbiosis between modern engineering and pleasure, this booklet espoused a prominent theme in promotional campaigns for passenger liners in the late nineteenth and early twentieth centuries. Visions of on-board luxury engaged social fantasies of a carefree life of material affluence and glamorized passenger vessels. Foregrounding the joys to be found on large boats provided a highly promising advertising strategy for British and German shipping lines because it directed public attention towards the triumphant aspects of engineering while invoking none of the ambivalent shivers that ran down people's spines when they were first exposed to the overpowering sight of a colossal new liner. Public claims that technological breakthroughs filled society with unheard-of delights emerged in a variety of contexts at the time. Automobile manufacturers praised the car for initiating an individualized travel culture that freed drivers from the constraints of the railways with their set routes and timetables. The electrical industry staged spectacular light shows whose dazzling aesthetic effects metaphorically supported pronouncements about the imminence of a bright future. The entertainment sector, meanwhile, nurtured the most intimate link between technology and pleasure. Operators of thrill rides, amusement parks, variety shows, and panoramas prided themselves in adopting advanced equipment to amuse patrons.[2] When film burst on to the scene after the turn of the century, cinematic

---

[1] University of Liverpool, University Archives (hereafter Cunard archives), D42/PR4/21, *Aquitania: Two 30 Day Cruises to Egypt and the Mediterranean* (no place, 1931), 12.
[2] Wolfgang Sachs, *For Love of the Automobile: Looking Back into the History of Our Desires* (Berkeley, 1992), 91–109; Sean O'Connell, *The Car in British Society: Class, Gender and Motoring, 1896–1939* (Manchester, 1998), 77–111; David E. Nye, *American Technological Sublime* (Cambridge, MA, 1994), 143–72; Carolyn Marvin, *When Old Technologies Were New: Thinking About Electric Communication in the Late Nineteenth Century* (New York, 1988), 152–90; Simon During, *Modern Enchantments: The Cultural Power of Secular Magic* (Cambridge, MA, 2002), 107–34.

technology revealed its extraordinarily strong powers to transfix audiences, causing major concern among critics of popular culture. Although, as the case of the film industry demonstrates, the wide range of promotional drives did not guarantee uncontested public acceptance, they hoped to capitalize on technology's potential to usher in a modern age of pleasure.

Operators stressed the delights of novel transport technologies to create public trust and to marginalize suspicions about the physical risks that new forms of travel entailed. As accidents raised questions about contemporary safety arrangements and threatened to erode public confidence, controlling public debate about the physical dangers of technology was a vital concern for companies at the cutting edge of innovation. The aviation sector sought to dispel doubts about air travel by emphasizing the increasing reliability of services, but stopped short of advertising campaigns that hailed flight as a luxurious means of transport. Such assertions would have struck even the keenest enthusiast as unrealistic exaggerations, since weight restrictions ruled out elaborate cabin amenities. During the 1920s and 1930s, manufacturers and airlines primarily promoted aerial services by comparing aeroplane and airship trips with train journeys. Marketing flight as an accelerated train ride emphasized aviation's safety and trustworthiness while simultaneously moderating customers' assumptions about the standards of care they would encounter during aeroplane journeys.[3]

Rather than restricting themselves to praising the reliability of new vessels, shipping lines devised expensive and elaborate promotional campaigns that focused on the unprecedented delights of maritime travel. According to public-relations material, passenger liners embodied technological progress by extending human domination over the exterior world while also enhancing quality standards for travelers. In numerous advertisements, shipping companies ambitiously advanced glamorous visions of vessels as extraordinary places of luxury consumption, their promotions culminating in descriptions of ships as precious works of art. These campaigns provided

---

[3] For an example from 1928 concerning Junkers model *G31*, see Deutsches Museum Munich, Junkers archive, file Propaganda, Firmenschriften, 1928/29, *Junkers-Flugzeuge* (n.p., 1928), 5. Imperial Airways used the same line to promote its *HP42* airliners in 1929. See *Imperial Airways Gazette*, August 1929, 2; November 1930, 2. Junkers compared the interior of its four-engined machine *JU90* with the compartment of an express train as late as 1938. See Deutsches Museum, Munich, archive, Luft- und Raumfahrtdokumentation, LRD 02718, E. Zindel, *JU90: Beginn einer neuen Epoche im Luftverkehr: Vortrag, gehalten anläßlich der Vorführung der JU90 am 19. Juli 1938 in Berlin-Tempelhof* (typescript), 5–6. Press reports about the *JU90* took up this point. See *Der SA-Mann*, 19 August 1938; *Münchner Neueste Nachrichten*, 18 September 1937; 21 July 1938; *Völkischer Beobachter*, 31 August 1937. (All these articles are in file LRD 02718.)

a variation on the wider theme of the pleasures arising from technological innovation and asserted that, as they took to the sea, travelers exchanged the world of everyday existence for a man-made environment that reflected an idealized, prosperous, harmonious, and technologically advanced society. The trope describing passenger liners as "floating palaces" emerged at the center of this protracted international public-relations initiative. From the late nineteenth century on, the shipping industries succeeded in disseminating a sanitized and glamorous public image of sea travel whose appeal by the 1930s was strong enough to prompt the National Socialist regime to launch a program of ocean cruises in keeping with its own ideological premises. Once the idea that passenger vessels amounted to floating palaces had taken root in the 1890s, public-relations campaigns and design initiatives in the interwar years restricted themselves to providing variations on this theme rather than launching fundamental redefinitions. The last decade of the nineteenth century, then, represented the major turning point in the public understanding of passenger liners and harnessed widespread social fantasies of material luxury and comfort in the service of promoting spectacular technological innovations on the water.

## "FLOATING PALACES": IDEALIZING THE GIANTS OF THE OCEAN

The second half of the 1880s witnessed several fundamental transformations in passenger shipbuilding. Naval engineers increasingly resorted to steel as a structural material, a design choice that reduced a hull's weight, enhanced its stability, and permitted the construction of larger and higher decks. As a matter of fact, it was only in the 1890s that, thanks to steel, vessels' tonnage and length began to exceed the dimensions of the ill-fated *Great Western*, which Brunel had conceived in the late 1850s. This development in turn laid the foundation for the extravagant public rooms and private suites that were to grace liners from that point on. Moreover, the liners of the 1880s also dispensed with sails altogether and turned to new propulsion mechanisms in the form of double, and later triple and quadruple, screws.[4] Still, when companies set out to construct larger vessels from the late 1880s on, they not only took on new technological challenges but faced a complex

---

[4] Arnold Kludas, *Geschichte der deutschen Passagierschiffahrt*, vol. 1: *Die Pionierjahre von 1850 bis 1890* (Hamburg, 1986); Eberhard Straub, *Albert Ballin: Der Reeder des Kaisers* (Berlin, 2001), 44; E. C. B. Corlett, "The Screw Propeller and Merchant Shipping, 1840–1865," in Robert Gardiner (ed.), *The Advent of Steam: The Merchant Steamship before 1900* (London, 1993), 83–105; Sidney Pollard and Paul Robertson, *The British Shipbuilding Industry, 1870–1914* (Cambridge, MA, 1979).

public-relations problem. To begin with, there existed a powerful, long-standing cultural tradition that emphasized the dangers of the sea. In the first half of the nineteenth century, problems of orientation and navigation at sea carried religious overtones and aroused widespread interest.[5] Maritime travel frequently served as a metaphor for the precariousness of human existence. Literary images of nutshells struggling in rough seas and pictorial representations of shipwrecks abounded in Western culture, with Turner's extensive oeuvre merely providing the tip of the proverbial iceberg.[6] As late as 1912 and 1914, news of the sinking of the *Titanic* and the *Empress of Ireland*, which together took over 2,500 people to the bottom of the sea, provided powerful reminders of the ongoing risks involved in maritime voyages. In the eyes of many observers, the sea remained treacherous and inhospitable terrain, illustrating man's incomplete control over nature despite numerous triumphant claims to the contrary.

Moreover, the image of shipping lines suffered from well-known reports about scandalous provisions on emigrant ships. Regulations passed by the municipal government of Bremen in 1866 to curb the worst excesses illustrate the sparse conditions under which many embarked on their future across the Atlantic. The authorities required shipping companies to allow 12 square feet per adult steerage passenger, with children under ten entitled to half that space.[7] Along with overcrowding, sanitary arrangements also created hygienic hazards when 200 steerage passengers shared six toilets during voyages lasting several weeks.[8] Even though a regulation ordered passengers to keep their lodgings "spotlessly clean" (*strengste Reinlichkeit*), fears of "infectious diseases" were never conquered.[9] Dietary provisions proved insufficient on many occasions, and authorities repeatedly strove to improve conditions against the resistance of owners.[10] Viewed from this perspective, transatlantic vessels replicated the public-health hazards of the most infamous urban rookeries that concerned social reformers on both sides of the Atlantic in the second half of the nineteenth century.

---

[5] Alison Winter, "'Compasses All Awry': The Iron Ship and the Ambiguities of Cultural Authority in Victorian Britain," *Victorian Studies* 38 (1994), 69–98.

[6] On shipwrecks in general, see Alain Corbin, *The Lure of the Sea: The Discovery of the Seaside, 1750–1840* (Harmondsworth, 1995), 234–49. For a German pictorial representation of a shipwreck, see Regine Falkenberg, "Untergang des Auswandererschiffs 'Austria,'" in *Zeitzeugen: Ausgewählte Objekte aus dem Deutschen Historischen Museum* (Berlin, 1992), 171–4.

[7] Deutsches Schiffahrtsmuseum Bremerhaven, Archive (hereafter DSB archive), file III A 970/94, *Obrigkeitliche Verordnung, die Beförderung von Schiffspassagieren nach außereuropäischen Ländern betreffend: 9. Juli 1866*, 43.

[8] *Ibid.*, 44.   [9] DSB archive, file III A 908/1212, *Schiffsverordnung*, c. 1855, §5.

[10] DSB archive, *Obrigkeitliche Verordnung*, 44–6; DSB archive, file III A 908, *Kreis- und Amtsblatt für die Kreisbezirke Idstein und Langenschwalbach*, 22 January 1856, 1.

While steerage passengers completed their voyages under particularly adverse circumstances, those who traveled first class also told tales of misery; they had to battle with vermin such as lice and other insects, and many succumbed to seasickness for long stretches of their crossings.[11] Undertaking a return voyage between Liverpool and Boston in 1870, Arthur Richard Jones noted in his diary for 6 May: "Very rough today. Rose about 8 o'clock but went to lie down again. Steamer pitched and rolled most horribly. So stayed in berth most day. Most passengers sick." The following day he "managed to sit at table with others but could not eat much, glad when meal was over. Came on deck, but felt squeamish all the time." Upon arrival in Boston ten days later, Jones stayed at his uncle's house, where he was "quite exhausted" and "lay in bed till 5 o'clock" the next afternoon.[12] Increases in the size of ships did not end these discomforts on their own. When Conrad Juchter set out for New York in 1911 on board the *George Washington* – a ship displacing over 25,000 tons and launched only two years earlier – some passengers were immediately struck by seasickness "although seas were quiet." A day later, the ship ran into a gale and Juchter "could not sleep all night"; he "was tossed around in [his] berth," attempting to control his upset stomach in vain.[13] Oral-history interviews of immigrants who entered the United States on the most advanced vessels between 1907 and 1929 also testify to the ubiquity of this malady.[14]

The established imagery of passenger ships as areas of danger and disease created a need for an alternative language if new vessels were to appear in a thoroughly positive light. To remedy this problem, the industry turned the common understanding of ocean travel on its head. In a protracted public-relations effort, it concentrated on novel features that transformed ships into glamorous places of luxury consumption and signaled that crossings had become safe, healthy, and eminently enjoyable. This campaign both capitalized on and amplified the political attention bestowed on maritime technology, which helped liners to become symbols of national prestige around the turn of the century. Advertising copy extolled new ships as unprecedented pleasure-domes, not only because they consigned the former maritime miseries to oblivion, but also because they illustrated the

---

[11] Traveling in 1846, Minna Praetorius wrote to her relatives about bugs and repeated bouts of seasickness. See her account in Uwe Schnall (ed.), *Auf Auswandererseglern: Berichte von Zwischendecks- und Kajüt-Passagieren* (Bremerhaven, 1976), 50–62.
[12] Cunard archives, file D42/PR3/4/6a, *Extracts from A Diary by Arthur Richard Jones (1845–1935)*, 3–4.
[13] DSB archive, file III A 1332/31, *Tagebuch von Conrad Juchter über seine Reise mit der SS "George Washington" im Jahre 1911 von Bremerhaven nach New York*, 3–6.
[14] See Ellis Island Museum, New York City, Ellis Island Oral History Project, interviews with Silvia Jones, Naomi Dorum, Isabel Belarski, Frances Oakley.

competitiveness and dynamism of the national economies capable of creating these maritime wonders. In short, on-board luxury denoted national economic power, which, in turn, provided the material basis for the construction of magnificent prestige objects. As commercial public-relations drives and national enthusiasm reinforced each other, shipping lines exploited their newly gained national fame to install themselves as technological experts that popularized a new generation of vessels as glamorous artifacts.

To achieve this goal, public-relations departments wooed the national and international press in a manner that sometimes matched the hyperbole characteristic of their literary output. When Cunard transferred the *Mauretania* from its shipyard in Newcastle to its port of call Liverpool in 1907, the company invited representatives from fifty-seven newspapers and periodicals from Britain and the United States to a short cruise on the ship, culminating in an exquisite lunch.[15] The Hamburg-American Line issued a luxuriously printed, gilt-edged commemorative album in 1913 that assembled words of praise for its "leviathan," *Imperator*, by journalists who had been treated to a free trip to Southampton on the vessel's maiden voyage, followed by a complimentary two-day stay in London.[16] While advertising departments went to such lengths only on exceptional occasions, they produced a torrent of brochures, pamphlets, press releases, slide shows, and films to disseminate information about the new vessels.[17] In Cunard's case, advertising represented a major investment. In 1913, the company spent over £54,000 on promotions for the Atlantic trade, and by 1928, annual public-relations expenditure had reached £370,000.[18] As we shall see, advertisers followed a careful strategy to transform the public image of ocean liners.

Pamphlets occasionally employed metaphors of urban life to convey the social character of transatlantic ships. In 1914, the Hamburg-American Line labeled its vessel *Vaterland* a "metropolis" (*Weltstadt*) that would make people forget that they were in the middle of the Atlantic.[19] Yet such attempts to link transatlantic liners to the city remained relatively isolated. Even though it was often considered in an enthusiastically favorable

---

[15] Cunard archives, file D42/PR4/11, *R. M. S. Mauretania* (no place, 1907), 11. The same file includes a variety of press releases on different boats.
[16] Deutsches Museum, Munich, archive, brochure collection (hereafter DMBC), *Imperator auf See: Gedenkblätter an die erste Ausfahrt des Dampfers am 11. Juni 1913* (Hamburg, 1913), 9.
[17] For a lantern slide lecture, see Cunard archives, D42/PR4/21, *The Making of Mammoth Liner: Lecture (With a List of Lantern Slides)* (no place, 1914). For a summary of a film, see DSB archive, file II 3 v, 137, *Mit dem Dritte-Klasse-Schiff nach Südamerika: Filmvortrag*, poster (n.d.).
[18] For the figures, see Cunard archives, D42/AC14/31/1, table entitled "Advertising, Atlantic trade 1913," no date; estimate entitled "Advertising General $\frac{1}{2}$ Year Ended 30th June 1928," no date.
[19] DMBC, *Turbinenschnelldampfer Vaterland* (Hamburg, 1914), 5. For another example see DMBC, *Resolute: Reliance* (Hamburg, 1920), 7.

light, metropolitan life was among the most contested of public topics during the late nineteenth and early twentieth centuries.[20] Images of urban volatility not only derived from environmental problems in cities or from social and political conflicts fought out there; they also owed much to prominent disputes about mass consumption and entertainment that often overlapped with controversies about the emergence of new gender roles. As fields of socio-cultural transformation and experimentation, cities signaled a modernity whose disorderly features had a deeply unsettling effect on many contemporary observers.[21] Given that transatlantic liners brimmed with men, women, and children from the most varying social, cultural, political, and ethnic backgrounds, who had boarded the ships to work, emigrate, or simply for leisure, advertising campaigns required a less contentious motif than urban life to highlight the modern qualities of ocean-going vessels while restraining potential concern about their heterosocial aspects.

Shipping companies found a solution in the trope of the "floating palace" or the "luxury hotel."[22] Both images provided public-relations departments with opportunities to extol the luxuries available to first-class passengers. Unlike the city, which contemporaries conceived as a site of consumption *and* production, "hotels" and "palaces" lent themselves more easily to representational modes that prioritized elite consumption and leisure *over* production and promised urbane sophistication. Developments in the hotel sector itself supported this representational choice as establishments on both sides of the Atlantic set new standards in customer service in the late nineteenth century. In particular, the hotels that Charles-Frédéric Méwès designed for César Ritz in London and Paris around the turn of the century gained an outstanding international reputation and provided the ideals aspired to on ocean liners. The introduction of steel into maritime engineering created the vast spaces that allowed architects to transfer internationally acclaimed amenities from land to the sea. In fact, Méwès himself planned the interior designs for the liners *Imperator* and *Vaterland*

---

[20] Donald J. Olsen, *The City as a Work of Art: London, Paris, Vienna* (New Haven, 1986); Andrew Lees, *Cities Perceived: Urban Society in European and American Thought, 1820–1940* (Manchester, 1985).

[21] Judith R. Walkowitz, *City of Dreadful Delight: Narratives of Sexual Danger in Late-Victorian London* (London, 1992). Fears of urban life manifested themselves particularly strongly in narratives of night life. See Joachim Schlör, *Nachts in der großen Stadt: Paris, Berlin, London, 1840 bis 1930* (Munich, 1994).

[22] For some invocations of ships along these lines see Cunard archives, D42/PR3/9/22, *White Star Line: R.M.S. 'Olympic': 46,439 Tons* (n.p., 1921), 4; D42/PR3/9/11, *The Wonders of a Great Atlantic Liner: The "1929" "Aquitania"*, press release, 2; D42/163/4, *Royal Mail Express Turbine Steamer "Lusitania", 32,500 Tons* (no place, c. 1906/7), 17; *Imperator auf See*, 15; DMBC, *Dampfer Kaiserin Auguste Victoria* (Hamburg, 1906), 5.

that entered the service of the Hamburg-American Line in 1912 and 1913, while his British partner Arthur Davis oversaw the fitting out of Cunard's *Aquitania* in 1914.[23] Of course, entrusting a French architect with the interior decorations of British and German national symbols risked inviting charges of aesthetic high treason from nationally minded quarters of the press, and shipping lines consequently remained relatively silent about Méwès's involvement. Like the Ritz and other hotels, ships launched before 1914 embraced an array of historically eclectic period styles and also featured both the "palm courts" (alongside the "Ritz-Carlton" restaurants) and grillrooms that had proved so popular among patrons of luxury hotels in Paris and London. In the early century, "floating palaces" such as the *Lusitania*, *Mauretania*, *Imperator*, and *Vaterland* formed part of an expanding market for luxury goods and services that, together with high-class hotels, spas, and exclusive retail outlets, catered for a clientele of aristocrats, members of the European *haute bourgeoisie*, and American plutocrats.[24] In order to compete successfully in this luxury market, shipping companies strove to meet internationally acknowledged standards of comfort as well as style. This economic consideration gave rise to a highly conventional, homogeneous international style in interior ship design before 1914. Whether it was in Berlin, London, or New York City that prospective travelers picked up brochures of a British or German line, they read assurances that they would find a predictable range of amenities and features while steaming across the sea.

A typical narrative tour of a ship included a description of increasing comfort in cabins. Advertising the spectacular "progress" of its first-class passenger accommodation, Cunard took pride that the *Lusitania* – launched in 1906 – boasted not only multi-bedroom "Regal Suites" but also a number of "commodious, self-contained apartments, replete with every modern convenience and furnished with excellent taste ... All first-class bedrooms are remarkably spacious."[25] Cunard's German competitors from Hamburg

---

[23] See Hugh Montgomery-Massingberd and David Watkin, *The London Ritz: A Social and Architectural History* (London, 1989), esp. 9–78; David Watkin, *Grand Hotel: The Golden Age of Palace Hotels: An Architectural and Social History* (London, 1984), 20–2; Vincent Bouvet, "Méwès: une révolution esthéthique," *Monuments Historiques* 130 (1984), 44–8; Straub, *Albert Ballin*, 102.

[24] On spas, see David Blackbourn, "'Taking the Waters': Meeting Places of the Fashionable World," in Martin H. Geyer and Johannes Paulmann (eds.), *The Mechanics of Internationalism: Culture, Society, and Politics from the 1840s to the First World War* (Oxford, 2001), 435–57. Erika Diane Rappaport, *Shopping for Pleasure: Women in the Making of London's West End* (Princeton, 2000), 142–77. On the development of a plutocracy in Germany, see Dolores Augustine, *Patricians and Parvenus: Wealth and High Society in Wilhelmine Germany* (Oxford, 1994), 191–239.

[25] *Royal Mail Express Turbine Steamer "Lusitania,"* 15. The Hamburg-American Line made a similar claim. See *Dampfer Kaiserin Auguste Victoria*, 15.

still claimed during the interwar years that, on ships, "luxury is space."[26] As shipping companies praised accommodation offering the comforts usually enjoyed on land, they also emphasized improved dining arrangements, to point out that passengers could maintain their usual habits of consumptive individualism on board. "Unlike in former times, guests are no longer forced to take their seats on bolted swivel chairs on long tables – doomed to put up with the same persons sitting opposite and next to them for eight days," the Hamburg-American Line assured customers immediately after the First World War.[27] Rather, the spaciousness of new steel ships allowed for more private seating arrangements at individual tables in dining halls and in *à la carte* restaurants serving gourmet cuisine.[28] A pamphlet for the German liner *Kaiserin Auguste Victoria* went so far as to boast that

> a person in the habit of driving to *Dressel* in Berlin may completely forget that he is at sea and order a carriage. "A carriage is absolutely unnecessary,' the steward will reply, 'My Lord is already there." One step towards the elevator, one second, *et voilà*, the folding doors of the Carlton Restaurant on the Imperial Deck are opening.[29]

Images of individual luxury consumption went hand in hand with evocations of leisurely social intercourse among first-class passengers. Influenced by prevailing models of upper-class domestic and club architecture, shipping companies furnished lavishly decorated halls to which women and men could retire separately, in keeping with gender conventions prevalent in the best circles.[30] Dark, mahogany-clad smoking rooms provided an all-male domestic environment for travelers who wished to sit "around the homelike fire" while enjoying their whiskeys and cigars "as wisely and as well as [they felt] inclined."[31] Correspondingly, ships featured drawing rooms "for the benefit of ladies' social communion," whose designs in brighter colors "followed graceful and tender tastes" and served as "an attractive background for sartorial splendor and charming female elegance."[32] At the same time, shipping lines pledged that they also welcomed the development of more informal sociability between men and women among the upper social echelons, drawing attention to new designated mixed-sex meeting

---

[26] *Resolute, Reliance*, 13.   [27] *Ibid.*, 10.
[28] *Royal Mail Triple-Screw Steamers "Olympic" and "Titanic"* (Liverpool, 1911), 16; *Dampfer Kaiserin Auguste Victoria*, 24.
[29] *Dampfer Kaiserin Auguste Victoria*, 24.   [30] Olsen, *The City as a Work of Art*, 101–12.
[31] *Royal Mail Triple-Screw Steamers*, 41. See also *Die Entwickelung des Norddeutschen Lloyd Bremen* (Bremen, 1912), 73.
[32] *Dampfer Kaiserin Auguste Victoria*, 30. For descriptions in a similar vein, see *Cunard Passenger Logbook*, 48–9; *Royal Mail Triple-Screw Steamers*, 45.

places in response to complaints about "the unwritten law, which . . . reserves the drawing room and music room to ladies only, and thus drives gentlemen to the smoke room, depriving them of society which constitutes not the least pleasure of an Atlantic trip." Cunard therefore provided an "immensely popular" lounge, "which obviates this disability. It is cozily furnished and comfortably seated . . . gentlemen may enjoy their cigar or cigarette without being banished from the society of the ladies."[33] The Hamburg-American Line went one step further and combined motifs of material abundance and female seductiveness to describe in stereotypical sensual terms the atmosphere it sought to create: "When, during dinner, enchanting fragrances ooze from greenhouse flowers on slim stems . . . when conversation in all languages gushes like the champagne in shining glasses and soft, playful tunes blend with the laughter of beautiful, elegant women, the illusion" of not actually traversing the ocean "is perfect."[34] While this passage gave maritime travel an injection of frisson and innuendo, it did so in a highly calculated manner that excluded any hint of social and cultural transgression. In short, it cast the ship as a seductive space that set the scene for conventional forms of sociability devoid of threatening overtones.

Shipping companies resorted to a feminized imagery of elite consumption and leisure for promotional purposes because this iconography was unlikely to spark criticism of excessive "materialism" on board. This advertising strategy profited from a growing social and cultural acceptance of female consumerism during the closing decades of the nineteenth century. Developments in Britain are indicative of wider European trends in this context. As Erika Rappaport points out, in the 1860s and 1870s the first department stores in London's West End triggered vocal moral debates about the habits of consumption of affluent women, but, by the turn of the century, the "image of the deranged and disorderly [female] shopper" struck observers "as more ridiculous than harmful" because of public campaigning by feminist pressure groups and owners of new retail outlets.[35] As the pursuit of leisure and consumption came to be seen as a respectable female activity before World War I, shipping companies gained opportunities to extol liners as settings in which female members of the upper class could ostentatiously enact roles as urbane society ladies. Depicting ships as spaces suitable for upper-class customers invested vessels with social prestige, lent them an aura of exclusivity, and thus counteracted associations between maritime transport, lower-class emigration, discomfort, and

---

[33] Cunard archives, D163/1, *The Cunarder "Campania"* (Liverpool, no place, no date), 6.
[34] *Turbinenschnelldampfer Vaterland*, 21.   [35] Rappaport, *Shopping for Pleasure*, 48–107, 143.

disease. Finally, descriptions of on-board luxury consumption signaled that steamers had transcended their function as a mode of transport and developed into a glittering technology of pleasure. "Floating palaces" represented triumphs of technology not just over nature but over a style of engineering that was dominated by technical considerations of a primarily utilitarian character.

Prospectuses reinforced this last theme by extolling liners as places of recreation that enhanced passengers' health. Although ships still frequently resembled floating hospitals rather than palaces, since scores of passengers fell prey to violent bouts of seasickness, shipping lines nonetheless promised passages devoid of physical inconvenience. For instance, advertising copy insisted that, despite the contrary experiences of many travelers, contemporary vessels were "ships without seasickness," thanks to the anti-rolling tanks that added to their lateral stability on the high seas – though these were ineffective against longitudinal sway.[36] Somewhat improbably, maritime companies also urged customers to consider sea travel "from a medical point of view." Citing scientific authorities, the Hamburg-American Line emphasized the restorative powers of maritime environments and extolled the "pure and genuine" sea air that, unlike polluted, leaden "land air," contained oxygen and minerals that made "man breathe more freely and widely on the ocean."[37] Sea travel also provided other opportunities to reinvigorate the body; first-class passengers could take advantage of sporting facilities on board, including rowing machines, exercise bicycles, squash courts, and "electric horses."[38] In short, ships became analogous to "seaside resorts" that cured rather than caused disease.[39]

While advertisers glamorized ocean travel by underlining the magnificence that awaited customers in first class, the presence of steerage passengers – who often made up the majority of voyagers before the Great War – raised the unwelcome specter of social tensions and disruptions during ocean crossings. Since the population of a ship constituted a social microcosm whose social disparities threatened to poison the sanitized image

---

[36] Deutsches Schiffahrtsmuseum Bremerhaven, Library (hereafter DSB library), *Schiffe ohne Seekrankheit* (Hamburg, 1925), 2. For similar statements, see *Imperator auf See*, 23, 32, 37; DMBC, Norddeutscher Lloyd, *Postdampfer-Dienst Bremen-Nordamerika* (Bremen, 1914), 9; *Royal Mail Express Turbine Steamer "Lusitania,"* 9.

[37] DSB library, *Große Orient- und Indienfahrt 1914* (Magdeburg, 1913), 10.

[38] *Dampfer Kaiserin Auguste Victoria*, 31.

[39] *Resolute, Reliance*, 12. For similar invocations of boats as resorts, see *Berliner Tageblatt*, 8 July 1913 (morning edition), 9; *B. Z. am Mittag*, 8 June 1912, 11; DSB library, *HAPAG Nord- und Polarfahrten 1927* (no place, 1926), 10; DSB library, *To the Land of the Midnight Sun: Spitzbergen and Iceland, England, Ireland and Scotland* (New York, 1905), 3; Hamburg-American Line, *Pleasure and Health Cruises* (no place, 1905).

of the "floating palace," shipping companies devoted considerable attention to measures that enforced social hierarchies and prevented conflict on board. To begin with, promotional material praised maritime architects for creating a tripartite social order in which a "second class" of passengers functioned as a buffer between the very rich and the very poor. This "middle class" enjoyed many first-class amenities, such as cabins for individual families, smoking rooms, drawing rooms, and a wide choice of meals, albeit on a more modest scale.[40] Furthermore, pamphlets stressed that steerage travelers were subject to strict supervision at every stage of the voyage. Prior to boarding, steerage customers had to undergo a medical examination ascertaining their physical health. The industry also emphasized that it fulfilled government regulations about the division of unmarried steerage travelers along sexual lines by "isolating" single women and men "in quarters at either end of the ship" to enforce standards of respectability in the areas housing the "lower orders."[41] Most importantly, such medical and moral initiatives went hand in hand with firm separation between customers booked into different classes. While this seems to have been so self-evident as to go frequently unmentioned, the regime guaranteeing rigid social stratification on board started from the very onset of the voyage. Posters reminded steerage passengers that they "will not be received on board" if they attempted "to embark by Second Cabin or Saloon tender."[42] To demonstrate the uncompromising character of on-board social segregation, pamphlets made it clear that passengers from different classes were isolated from one another both indoors and outdoors, where they used their own promenade decks.[43]

Stable lines of social demarcation and a concomitant sense of security allowed upper-class voyagers to view life in the steerage class – as depicted in promotional material – with a privileged gaze and to cast it in romanticizing narratives that stressed the contentment and happiness of emigrants. During a fictitious tour of a liner, one male saloon passenger "happened upon"

---

[40] See *Cunard Passenger Logbook*, 54–5; *Turbinenschnelldampfer Vaterland*, 29; *Dampfer Kaiserin Auguste Victoria*, 35.

[41] On medical checks, see Cunard archives, D42/PR3/6/11a, John Mortimer, "From London to Chicago," reprint from *Strand Magazine* (1893), 203–4; *Die Auswandererhallen der Hamburg-Amerika-Linie* (Hamburg, 1907). On the sexual separation of steerage passengers, see Mortimer, "London to Chicago," 205.

[42] Cunard archives, D42/PR3/6/14, *Notice to Steerage Passengers* (no place, 1894). For an emphatic assertion of on-board social segregation, see Hamburg-Amerika Linie, *Wiederaufbau der deutschen Überseeschiffahrt* (Berlin, 1923).

[43] *Cunard Passenger Logbook*, 54–5; Mortimer, "London to Chicago," 205; *Royal Mail Triple-Screw Steamers*, 53–4.

a lively scene such as one finds only in the character and surroundings of large emigration ships. Couples were twirling to the captivating sound of an accordion in one of the stirring dances that suit the children from the country of the Magyars so well . . . One could watch the international society of steerage passengers for hours, one could undertake ethnographic studies.[44]

The image of third-class travelers as potential objects of anthropological inquiry underlines the aim of on-board social differentiation. Social divisions should be as marked and as seemingly secure as those imagined to separate people from entirely different cultural orbits.

At the same time, shipping companies prided themselves in improvements of steerage accommodations and drew attention to examples of social progress rooted in conservative paternalism.[45] In 1902, Cunard reprinted the letter of a satisfied steerage traveler whose eloquence indicated that even members of the middle class could confidently contemplate inexpensive ocean travel. Its writer found the social gathering places, hygienic conditions, and beds "remarkable under the circumstances." The diet also deserved mention: "Here we have an abundance of substantial food, well served thrice daily, very good and wholesome."[46] Moreover, after the turn of the century, new ships phased out the traditional steerage accommodation of big dormitories and replaced them with cabins. The Hamburg-American Line housed the "typical" steerage passenger on the *Imperator* and *Vaterland* in "cabins of two, four, or six beds, in which he can live without having to be separated from his family. Excellent artificial and natural ventilation ensure in these rooms a pleasant and constant change of air."[47] Thus, according to promotional material, innovations in marine architecture aimed to spread family-centered domesticity among passengers in third class.

Such reforms in steerage aimed to placate first-class passengers who worried about the strong social contrasts on the floating palaces. "In former times," a German prospectus declared, "the first class suffered much from knowing that scenes of misery took place not far away in what was called the steerage . . . While the beautiful set enjoyed themselves, dirt, want, and misery piled up below." All this had changed because, the pamphlet went on, there existed "no more mass accommodation, where filth and disease take inextinguishable root."[48] After the war, shipping firms removed

---

[44] *Postdampfer-Dienst*, 39. For similar depictions of third-class entertainment, see *Auf Auswandererseglern*, 25; *Illustrated Souvenir of the Cunard Steamship Co., Limited* (Liverpool, 1902), 66.
[45] On the gendered nature of paternalism, see Sonya O. Rose, *Limited Livelihoods: Gender and Class in Nineteenth-Century England* (Berkeley, 1992), 37–45. For German examples, see Christoph Lang, "'Herren im Hause': Die Unternehmer," in Richard van Dülmen (ed.), *Industriekultur an der Saar. Leben und Arbeit einer Industrieregion, 1840 bis 1914* (Munich, 1989), 132–45.
[46] *Illustrated Souvenir*, 66.
[47] *Turbinenschnelldampfer Vaterland*, 29.    [48] *Resolute, Reliance*, 16.

this potential source of social tension altogether by abolishing the steerage class in response to changes in US-immigration regulations, which curbed the number of emigrants. In its place, companies introduced a "third" or "tourist" class that catered for middle-class business travelers and holidaymakers, who enjoyed provisions similar to those of first-class voyagers, albeit on a significantly more moderate scale.[49] Thus shipping propaganda depicted transatlantic liners as stable social worlds under benevolent paternalistic regimes that prevented the eruption of conflict among travelers from the most diverse backgrounds. While strict segregation safely upheld the social distance between rich and poor despite their spatial proximity, material improvements in steerage quarters, it was argued, not only removed potential sources of discontent but also ensured respectable behavior among lower-class passengers.

Shipping companies aimed to extend their conservative vision of a highly stratified ideal social order to include the work force. The "unmitigated coexistence (*hartes Nebeneinander*) of sloth and work," considered as "the sociological wonder of the modern ship," seriously threatened the immaculate image of vessels that the promoters endeavored to project.[50] On board, a multinational and sometimes multiracial labor force carried out some of the toughest and dirtiest jobs that technological innovation had created.[51] Conditions were particularly dangerous and hard for the trimmers, who transported coal from bunkers to furnaces, as well as for the stokers, who were responsible for fueling and cleaning the furnaces. Badly fed and living in crowded quarters when off duty, they had to endure physical violence from their superiors, as well as grime, noise, and heat. It was by no means unusual for trimmers and stokers to commit suicide in mid-ocean by jumping overboard.[52] Others, such as the "engineer's boy" on Cunard's "Round the World Cruise" in 1923, died from exhaustion because, as a stewardess

---

[49] *Wonders of a Great Atlantic Liner*, 4; DSB library, Hamburg-Amerika Linie, *In der dritten Kajüte nach Nordamerika* (Frankfurt, 1930), 2; DMBC, *Überfahrts-Bedingungen für die Dritte Klasse* (Hamburg, 1932), 2; DSB library, *Columbus* (Hanover, no date), 36–42. On German business travelers, see Mary Nolan, *Visions of Modernity: American Business and the Modernization of Germany* (New York, 1994), 18–22.

[50] *Resolute, Reliance*, 22.

[51] On the multiracial composition of the labor force of the British merchant fleet, see Laura Tabili, *We Ask for British Justice: Workers and Racial Difference in Late Imperial Britain* (Ithaca, 1994), 41–57.

[52] On working conditions and suicides on ocean steamers, see Uwe Kiupel, "Selbsttötung auf bremischen Dampfschiffen: Die Arbeits- und Lebensbedingungen der Feuerleute, 1880 bis 1914," in Wiltrud Drechsel and Heide Gerstenberger (eds.), *Arbeitsplätze: Schiffahrt, Hafen, Textilindustrie, 1880 bis 1933* (Bremen, 1983), 15–104; Heide Gerstenberger and Ulrich Welke, *Vom Wind zum Dampf: Sozialgeschichte der deutschen Handelsschiffahrt im Zeitalter der Industrialisierung* (Münster, 1996), 185–212. No mention of these gruesome labor conditions is made in Ronald Hope, *A New History of British Shipping* (London, 1990).

on this trip noted in her diary, "this heat had been too much for him."⁵³ Even though the introduction of oil firing and diesel engines removed the need for trimmers and stokers from transatlantic vessels during the interwar years, service personnel continued to work long hours for low pay. During his tour of England in 1933, J. B. Priestley recorded an encounter with a steward in the first-class smoking room on a Cunard liner who, despite the depressed labor market, had quit his job because "it's a rotten life. Bad quarters. Working all hours. And no proper food and nowhere to eat it . . . Even with the tips, it's no good."⁵⁴

Public-relations departments coped in several ways with the potential image problems that arose from the sharp contrast between the sumptuous surroundings of first-class passengers and the harsh working conditions. As in the context of lower-class passengers, one promotional strategy stressed the separation between workforce and travelers, insisting that walls "decorated by arabesques" concealed "the realm of extensive economic enterprise." In keeping with widespread models of domesticity, servants were to be neither seen nor heard and, as on land, "silent ingeniousness" (*geräuschlose Selbstverständlichkeit*) provided the ideal at sea.⁵⁵ Moreover, detailed descriptions of cleanliness countered notions of dirt and invested ships with an air of purity and, consequently, of security.⁵⁶ Having inspected the *Aquitania* in 1921, Lord Northcliffe, a champion of domestic reform and "efficiency," marveled that "the butcher's shop made me rub my eyes. Never have I seen greater cleanliness; and as for the kitchens . . . one would have thought by the look of the tables and utensils that they had never been used."⁵⁷ Photographs of kitchens further underlined this impression of shipshape tidiness.⁵⁸

Engine and boiler rooms, however, provided the greatest challenge when it came to the question of image. Public-relations departments labored to contain the symbolic dangers of this perilous work environment by extolling the giantism of its machinery while remaining silent about the workforce that built or operated the equipment. Pamphlets provided readers with

---

⁵³ Merseyside Maritime Museum Archive, Liverpool, DX/1166/1991.428, *"Round the World": Diary of Rose Scott, January 6th–June 18th, 1923*, 15.
⁵⁴ J. B. Priestley, *English Journey* (London, 1994 [1933]), 15. See also the account by a steward on the *Queen Mary*, quoted in James Steele, *Queen Mary* (London, 1995), 212–16.
⁵⁵ *Turbinenschnelldampfer Vaterland*, 15.
⁵⁶ Mary Douglas, *Purity and Danger: An Analysis of the Concepts of Pollution and Taboo* (London, 1996).
⁵⁷ Cunard archives, D42/PR3/9/40, *Aquitania: Wonders of the Wonder Ship* (n.p., 1921), 8. On Lord Northcliffe's sponsorship of domestic reform programs, see Deborah Ryan, *The Ideal Home through the Twentieth Century* (London, 1997), 1–32.
⁵⁸ Norddeutscher Lloyd Bremen, *Die Geschichte des Norddeutschen Lloyd* (Bremen, 1901), 105.

seemingly endless enumerations of reciprocating engines, furnaces, steering quadrants, cylinders, electric generators, and more, but hardly ever detailed the occupations of the people employed to work on these mechanisms.[59] Thus a rare and mainly technical description of the duties of trimmers and stokers on the boilers in one company history concluded laconically that a shift in front of the fires was "no easy work" and "require[d] strong, healthy men."[60] Rather than providing information about operational routines or conditions prevalent in the workplace, reports included photographs of workmen, who often adopted artificial poses beside machinery. These pictures served to underline the "breathtaking" dimensions of marine technology. In many photos of mechanical equipment, however, workers were scarcely detectable, if not altogether absent.[61] Depictions of the "Dome of Work," as the Hamburg-American Line called the engine room, may have been filled with technological artifacts, but it generally remained devoid of people.[62] By systematically marginalizing images of workers and foregrounding pictures of impressive machinery, public-relations efforts engaged in "production of ignorance" about labor affairs on vessels.[63]

Such photographs not only marginalized workers' contributions to the creation as well as to the maintenance of innovative technology, and thus directed attention away from the labor force and towards the artifact; they also relied on the "documentary" character of photography to create a purportedly neutral, or objective, image of shipping technology. Because they were widely acknowledged to represent the external world without "distorting" it, photographs transformed "resemblance" into "identity."[64] Thus photography's authority as a supposedly self-effacing medium added weight to promoters' efforts to highlight the huge size of ships while sidelining the presence of workers as well as of their dangerous jobs. These images, which "abolished work" and turned engine rooms into "abstractions devoid of people and actual processes" of operation, conformed to stylistic conventions that also dominated photographic advertising campaigns in the heavy

---

[59] One example is *Royal Mail Triple-Screw Steamers*, 55–70.
[60] Norddeutscher Lloyd, *Geschichte*, 114–15.
[61] *Royal Mail Triple-Screw Steamers*, 68; DSB library, *"Bremen", "Europa": Die kommenden Großbauten des Norddeutschen Lloyd Bremen* (Zwickau, no date), 11.
[62] *Resolute, Reliance*, 22.
[63] For the term "production of ignorance," see Caroline A. Jones and Peter Galison, "Introduction," in idem (eds.), *Picturing Art, Producing Science* (New York, 1998), 1–25, here 11.
[64] Peter Galison, "Judgment against Objectivity," in Jones and Galison (eds.), *Picturing Art*, 327–59, here 333. See also Lorraine Daston and Peter Galison, "The Image of Objectivity," *Representations* 40 (1992), 81–128. The documentary character of photography is addressed in Roland Barthes, *Camera Lucida: Reflections on Photography* (New York, 1981), 4–7.

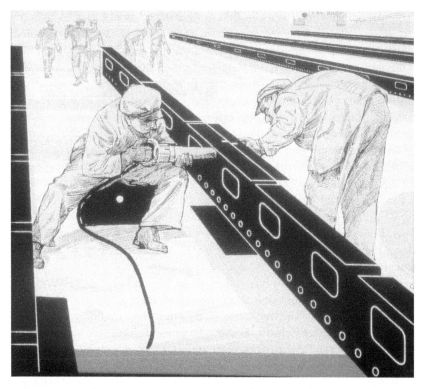

Illustration 6.1  Drawing of riveters in a shipyard. This drawing deprives the riveters of personal traits and excludes any allusion to the dirt in a shipyard, which would have disturbed the glamorous image of the floating palace.

industries in Europe during the early twentieth century.[65] Characteristically, black and white drawings rather than photographs provide some of the few pictorial representations of physical labor in shipping brochures, showing workers with featureless faces absorbed in the process of mechanical riveting in a shipyard (see illustration 6.1) – and significantly not during voyages after construction was complete.[66] Sketches drawn by hand offered a greater degree of representational control over the work process than did photographs, which were more likely to capture the pollution and danger

[65] At Ansaldo, one of Italy's leading industrial concerns engaged in steel production, shipbuilding, steam and electromechanical engineering, the managing director explicitly instructed photographers of public-relations material to select angles that exaggerated the size of castings, boilers, and other large products while avoiding motifs that showed labor processes. Alain Dewerpe, "Miroirs d'usines: Photographie industrielle et organisation du travail à l'Ansaldo, 1900–1920," *Annales ESC* 42 (1987), 1079–114, esp. 1084.

[66] "Bremen", "Europa", 6.

in maritime workplaces. While technical equipment *per se* did not disturb the eulogies of social accord on ocean-going vessels, the presence of working-class operating crews threatened to drag stark disparities and tensions into view. As a consequence, shipping companies went beyond merely maintaining a stubborn silence about working conditions in liners.

The social purification of workplaces on transatlantic liners culminated in a celebration of ships as exceptionally efficient economic environments. Eulogies of naval efficiency reflected prominent concerns about the consequences of conflict-prone social relations, especially in interwar Europe as the United States emerged as a competitor whose industrial strength appeared to derive from new forms of labor organization. Although European debates on "efficiency" and "rationalization" were primarily concerned with what was yet to come, British and German transatlantic liners provided snapshots of a present that was fully analogous to Ford's factories or Taylorist workplaces.[67] As Cunard proclaimed in 1929, the "utmost efficiency" prevailed "in all departments" on board, and Lord Northcliffe saw "that genie, Perfect Organisation," at work in transatlantic liners.[68] German publicity writers joined this chorus of professional admirers and celebrated big ships as manifestations of "the united forces of the intellect (*Geist*) and hands of German engineers, technicians, artists, and workers."[69] These portrayals claimed that ships had found solutions to issues of labor organization, but significantly they rarely detailed specific arrangements. As the sensitive topic of industrial relations was downplayed, with the masculine overtones of production taking a back seat behind feminized images of ships as sites of leisure and consumption, another advertising strategy that became standard practice in the years immediately before the First World War bypassed the naval world of work altogether and singled out ships for their artistic properties.

## MODERNITY, MODERNISM, AND SHIPPING DESIGN

Given their silence on "ugly" labor environments, brochures were in a position to highlight the aesthetic qualities of ships. In 1907, Cunard stressed that the *Lusitania* was designed in accordance with "excellent taste ... and high artistic standards."[70] Similarly, around 1920 the Hamburg-American

---

[67] Geoffrey Searle, *The Quest for National Efficiency* (Oxford, 1971); Anson Rabinbach, *The Human Motor: Energy, Fatigue, and the Origins of Modernity* (Berkeley, 1992); Nolan, *Visions of Modernity*, 30–107, 131–53.
[68] *Wonders of a Great Atlantic Liner*, 2; *Aquitania: Wonders*, 3.   [69] "Bremen", "Europa", 4.
[70] *Royal Mail Express Turbine Steamer*, 15.

Line pointed out that technological "expediency" (*Zweckmäßigkeit*) dictated no "sacrifice of ornament and decoration" on its ships.[71] In fact, promotional texts asserted that their interior design transformed vessels into veritable artworks. In 1914, Cunard, for instance, praised its new liner *Aquitania* as an "epitome of Art in general and Anglo-Saxon Art in particular." With her combination of period styles, ranging from the Renaissance to Louis XVI, evoking a "spirit of antiquity," the *Aquitania* – also labeled the "Ship Beautiful" – had become nothing less than a "Temple of Taste."[72] German advertisers adopted a similar marketing line and described their vessels as both "modern constructions and works of art."[73] Placing the media spotlight on the aesthetic qualities of vessels not only supported other attempts to shift public interest away from labor conditions towards lavish arenas of luxury consumption and leisure; this strategy also provided a justification for the opulence of ocean liners in non-materialistic terms by defining sumptuousness as evidence for refined cultivation. Shipping companies advanced a forceful claim that their ships' aesthetic properties positioned them close to the pinnacles of human creativity, thereby seeking to elevate their products in contemporary British and German cultural hierarchies.

Demands for the inclusion of new ships into the pantheon of outstanding cultural creations were as ambitious as they were problematic, however, because they entangled the industry in conflict-laden debates about contemporary art and design. To begin with, companies had to select individual paintings and sculptures as decorations for the "modern wonders." In the context of the fine arts, the industry had to position itself *vis-à-vis* both aesthetic traditionalists and supporters of modernism, who, since the late nineteenth century, had engaged in increasingly polemical exchanges about whether the advent of the modern age had rendered existing styles obsolete and thereby dictated fundamental artistic innovations. Programmatic and personal conflicts within the modernist camp itself posed further difficulties for shipping companies.[74] As they began to promote liners on the basis

[71] *Resolute, Reliance*, 9.
[72] A. M. Broadley, *The Ship Beautiful: Art and the Aquitania* (Liverpool, 1914), 34, 40.
[73] *Dampfer Kaiserin Auguste Victoria*, 5.
[74] Works on the rise of modernism in Britain and Germany are too numerous to note. For perspectives developed by historians, see Peter Stansky, *On or About December 1910: Early Bloomsbury and Its Intimate World* (Cambridge, MA, 1996); Stella K. Tillyard, *The Impact of Modernism, 1900–1920: Early Modernism and the Arts and Crafts Movement* (London, 1988); Peter Paret, *Berlin Secession: Modernism and Its Enemies in Imperial Germany* (Cambridge, MA, 1980); Thomas Nipperdey, *Wie das Bürgertum die Moderne fand* (Berlin, 1988); Wolfgang J. Mommsen, "Die Kultur der Moderne im Deutschen Kaiserreich," in Wolfgang Hardtwig and Harm-Hinrich Brandt (eds.), *Deutschlands Weg in die Moderne: Politik, Gesellschaft und Kultur im 19. Jahrhundert* (Munich, 1993), 254–74.

of artistic merit, shipping companies faced potentially contentious decisions on which traditional or modernist styles would best correspond with vessels that were widely recognized as symbols of modernity.

Furthermore, the vast spaces in ships provided ample opportunities to bolster claims about liners' outstanding aesthetic qualities through interior design schemes. The question of design touched on a prominent European discussion about the decorative or "useful" arts that overlapped with debates about modernism in fine art but addressed broader aspects of material culture. From the late nineteenth century, a multitude of design reform movements developed the earlier arguments advanced by John Ruskin and William Morris and criticized contemporary material culture on the ground that, as a result of the division of labor and extensive use of machinery, industrial mass production was flooding European societies with shoddy, aesthetically inferior goods. Reformers frequently placed the relationship between culture and technology at the center of their considerations, contending that the ongoing displacement of artisanal by mechanized forms of production presented the decorative arts with unprecedented aesthetic challenges. By the interwar years many followers of Ruskin and Morris "accepted . . . the machine as simply another tool, one capable of great spiritual and economic benefits to society if used properly," and promoted a concept of "industrial art" that disregarded the hierarchical relationship between "useful" and "fine" art.[75]

In principle, shipping companies sided with reformers who believed that industrial products could reach high levels of stylistic sophistication. At the same time, they strove to avoid public controversy by adopting deliberately vague positions on divisive social issues. Given their dubious record as employers, they kept a safe distance from debates that considered which specific modes of industrial production proved most conducive to high-quality designs, a topic that persistently occupied design reformers. Moreover, promotional drives for vessels displayed none of the social and spiritual

---

[75] Michael Saler, *The Avant-Garde in Interwar England: Medieval Modernism and the London Underground* (New York, 1999), 13. Further works on design reform include Tillyard, *The Impact of Modernism*; James Peto and Donna Loveday (eds.), *Modern Britain, 1929–1939* (London, 1999); Paul Greenhalgh, "The English Compromise: Modern Design and National Consciousness," in Wendy Kaplan (ed.), *Designing Modernity: The Arts of Reform and Persuasion* (New York, 1995), 111–39; Adrian Forty, *Objects of Desire: Design and Society since 1750* (London, 1986); Anne Massey, *Interior Design of the Twentieth Century* (London, 2001); Laurie A. Stein, "German Design and National Identity, 1890–1918," in Kaplan (ed.), *Designing Modernity*, 49–77; Joan Campbell, *The German Werkbund: The Politics of Reform in the Applied Arts* (Princeton, 1978); Wolfgang Hardtwig, "Kunst, liberaler Nationalismus und Weltpolitik: Der deutsche Werkbund 1907–1914," in *Nationalismus und Bürgerkultur in Deutschland, 1500–1914* (Göttingen, 1994), 246–73; Magdalena Droste, *Bauhaus, 1919–1933* (Cologne, 1998).

idealism that often underpinned aesthetic reform programs in Britain and Germany. The Bauhaus, for instance, participated in social housing projects during the Weimar Republic, while in interwar Britain, Frank Pick, a leading official of London Underground, aimed to re-spiritualize urban spaces through modern art and innovative decorative schemes. He commissioned a group of like-minded British modernists, including sculptors Eric Gill and Jacob Epstein as well as architect Charles Holden, to design stations for the expanding network. It comes as no surprise that shipping companies lacked comparable high-minded agendas when they adopted specific designs for their products. As they sought to lure well-to-do customers on board, it was primarily for commercial reasons that maritime operators bolstered the status of ships as cultural icons of modernity through artworks and interior decorations.

Debates about modernism began to have an impact on the shipping industry by the interwar period, when the industry abandoned the period styles that had dominated interior design up to 1914. While the Great War intensified the sentiment of living in a new age, the cataclysm on the battlefields alone does not explain the embrace of modernism in the naval sector.[76] The fact that shipping companies shifted from historicizing to modernist styles in maritime architecture also stemmed from stylistic trends in the entertainment and consumer sectors. Whether we turn to film, fashion, or smoking – to name but three examples – advertisers on both sides of the Atlantic increasingly chose modernism to enhance the appeal of their products.[77] Had they stood aside from the international trend that yoked together modernism and consumerism in the interwar period, shipping companies would have given their "floating palaces," which supposedly stood at the apex of all "modern" dream worlds of consumption, an inappropriately anachronistic appearance. Thus operators followed a wider aesthetic trend in consumer industries in order to maintain the cultural status of ships as icons of modernity.

When the liner *Bremen* entered service in the summer of 1929, the North German Lloyd Line issued a richly illustrated catalogue detailing

---

[76] As Jay Winter has persuasively argued in a different context, the exceptional nature of war experiences could also be expressed in conventional, non-modernist artistic styles. See Jay Winter, *Sites of Memory, Sites of Mourning: The Great War and Modern Memory* (Cambridge, 1996).

[77] Matthew Hilton, "Advertising the Modernist Aesthetic of the Marketplace? The Cultural Relationship between Tobacco Manufacturers and the 'Mass' of Consumers in Britain, 1870–1940," in Martin J. Daunton and Bernhard Rieger (eds.), *Meanings of Modernity: Britain from the Late-Victorian Era to World War II* (Oxford, 2001), 45–69; T. Smith, *Making the Modern: Industry, Art and Design in America* (Chicago, 1993); Jeffrey Meikle, *Twentieth-Century Limited: Industrial Design in America, 1925–1939* (Philadelphia, 1979).

Illustration 6.2 Fritz Breuhaus de Groot, *Bauten und Räume*. The modernist design of the shopping area aestheticized the liner *Bremen* and emphasized that this ship was a temple of consumerism.

its modernist design principles. Supervising architect Fritz Breuhaus de Groot explained that his task had entailed bringing interior decorations "to the same level" that the "modern art of engineering" displayed in the "sober, exceptionally beautiful shape of the ship." According to Breuhaus, since "the technical forms of the world of 1930" were "exemplary in artistic respects," the interior designs for the ships had to combine "utility and beauty" in order to offer "spoilt" (*verwöhnt*) passengers the latest conveniences in surroundings reflecting a spirit of modernity. Based on "purity of form, beauty of line, [and] high-quality materials," the *Bremen*'s decorative schemes turned their back on the "bloated luxuries of the past" and took a symbolic step "towards the splendors of our age." Indeed, the *Bremen*'s interior spaces featured no historicizing details, as a view of the shopping arcade for first-class passengers, with its "clean surfaces and curving radii in brown metal, yellow and blue leather," reveals (see illustration 6.2).[78]

---

[78] See Fritz Breuhaus de Groot, *Bauten und Räume* (Berlin, 1935). The illustration is on p. 141. The description of the shopping arcade follows John Heskett, "Design in Inter-War Germany," in Kaplan (ed.), *Designing Modernity*, 263. For another statement about the *Bremen* as a "beauty of technology," see DSB library, *Schnelldampfer Bremen* (Berlin: Otto Elsner, no date), 9.

By the late 1920s, modernist calls attributing "beauty" to machinery were highly conventional in Germany, since diverse German proponents of design reform had celebrated technology's functionality in aesthetic terms since the turn of the century. The Bauhaus was the best-known institution to develop this theme after World War I. While a language of modernism predicated on notions of technological functionalism did not invite controversy, it was, however, unclear how such a design program ought to be implemented. According to some observers, technology's "sobriety" demanded decorative minimalism, rather than splendor that could be read as wastefulness. Critics argued that the sumptuous surroundings for the first-class passengers on the *Bremen*, which aspired to the standards of the "newest hotels and apartment buildings" in the United States, made a mockery of functionalism. Despite rhetorical similarities and overlaps, Breuhaus's designs differed from those hailed by followers of the Bauhaus movement, who favored a sparse modernism.[79] A journalist for the *Frankfurter Zeitung* who inspected the ship prior to her maiden voyage castigated the *Bremen* as an example of crass materialism, because her architects had done "everything" to "out-snob the American snob . . . Despite many styles and architects, the first class has achieved stylistic unity: American taste."[80] In keeping with a widespread theme of anti-American sentiment in interwar Germany, this stylistic critique equated conspicuous affluence with cultural decay. While publicists promoted the *Bremen* as a work of art, then, the ship's modernism also triggered aesthetic opposition that echoed wider conflicts about the supposedly materialistic character of the "modern age."

Cunard was aware of the divisive nature of modernism, and therefore adopted modernist schemes for its new liner the *Queen Mary* only with reluctance. During the first planning stage, in 1930, the company explored a range of French and Italian period styles as well as architectural landmarks such as Versailles and Fontainebleau, which were initially expected to serve as models for decorations in first class. These plans, however, met with an unfavorable reception among staff who were in tune with recent artistic developments.[81] Arguing that "the public expect[s] in design an expression of . . . the Company's policy," the "furnishing department" sent chairman

---

[79] There is no biographical study of Breuhaus that clarifies his position in the German world of design during the interwar years. Still, it is clear that he was not a formal member of the Bauhaus. See Heskett, "Design in Inter-War Germany," 257–62.

[80] *Frankfurter Zeitung*, 18 July 1929 (morning edition), 2. For a similar contemporary criticism, see Heskett, "Design in Inter-War Germany," 262.

[81] Cunard Archive, D4/C3/381, list of motifs, 29 August 1930.

Sir Percy Bates a memorandum warning against historicizing styles because of their traditionalist connotations. The statement pointed out that, since Cunard combined "conservatism in finance with a readiness to exploit the best twentieth-century ideas in administration [and] a firm resistance to the merely fashionable and ephemeral," in-house designers were advised to select "what is good in the modern tendency . . . and merge [this] with traditional design to create a spirit expressive of the twentieth century." Since company directors feared that public controversy over stylistic matters would deter passengers from traveling on the *Queen Mary*, the designers assured the management that "all that is merely startling . . . all that can produce notoriety for a period then to become dated and passé, has been definitely excluded."[82] This last assertion was an unveiled reference to the conflicts among the various modernist avant-garde camps during the 1920s and 1930s. Aversion to aesthetic risk consequently informed the commissions offered to individual artists. Paul Nash, the eminent war artist and surrealist, was ruled out because he was "very modern [and] would require restraint and direction."[83] Duncan Grant, who had filled a canvas 20 feet high with large-limbed females in the style of Matisse, incurred the personal opprobrium of Sir Percy Bates, who had Grant's painting removed from the main lounge. Grant had received a previous warning that his work appealed only "to a limited coterie interested in the development of modern painting."[84]

Commercial factors, however, induced the managers to overcome their aesthetic doubts and adopt modernist designs for the *Queen Mary*. If the new British liner was to compete not only with the *Bremen* but also with the recent French vessels *Ile de France* and *Normandie*, whose Art Deco designs proved popular with American customers, Cunard had to fulfill the aesthetic expectations of prospective travelers from the New World.[85] To this end, the company hired New York-based architect B. W. Morris, who had built Cunard's headquarters on Broadway before the First World War, as a "style advisor" to coordinate contributions by individual artists.[86] To divert attention from Morris's role in the creation of this maritime symbol of national pride, the public-relations department foregrounded the works of British artists that decorated the ship. These covered a broad spectrum

[82] Cunard archives, D4/C3/381, Memorandum to the Chairman, 18 December 1930.
[83] Cunard archives, D4/C3/383, Memorandum, 16 March 1935.
[84] For a detailed account of the conflict surrounding Duncan Grant's commission, see Frances Spalding, *Duncan Grant* (London, 1997), 336–43.
[85] On the designs on the French liners, see Elisabeth Vitou, "Le métal dans le décor et le mobilier des transatlantiques," *Monuments Historiques* 130 (1984), 49–55.
[86] Cunard archives, D4/C3/381, telegram Cunard to B. W. Morris, 25 March 1931.

Illustration 6.3   The *Queen Mary*'s cocktail bar. The *Queen Mary* boasted of the streamlined design of the bar, a nod towards predominant aesthetic trends in the United States during the 1930s.

of non-abstract styles, ranging from a romanticizing circus scene by Dame Laura Knight, through Doris Zinkeisen's colorful depiction of nightclub entertainment, to mildly surrealist works on maritime themes by Edward Wadsworth. Next to these displays of domestic talent, American influences included a cocktail bar for first-class passengers in a "streamlined" design (see illustration 6.3). Thus the *Queen Mary* continued the tradition of aesthetic eclecticism in naval interior design, albeit with modernist premises. This stylistic heterogeneity, oscillating between modernism and traditionalism, not only helped Cunard to appeal to customers with diverse tastes; it also sent out the message that the *Queen Mary* selectively embraced recent styles while striking none of the provocative poses widely associated with the avant-garde. The public-relations campaign that accompanied the vessel's completion successfully alerted the representatives of the press to the company's design ethos. Before the maiden voyage, the *Daily Telegraph* was one of several publications to praise the ship's "extraordinary beauty," which derived from the fact that her decorations avoided the "ultra modern" while "reflect[ing] all that is best in present-day art." Despite the company's initial hesitancy about modernism, the *Queen Mary*'s interior designs ultimately succeeded in winning widespread approval from the traveling public and the media.[87]

## "VOYAGES FOR PURE ENJOYMENT?" THE PROLIFERATION OF OCEAN CRUISING

Thus the shipping sector put forward a relatively coherent vision of its products, stressing the pleasures of ocean travel over its discomforts. Companies succeeded in conveying a sanitized image of liners as engineered spaces that embodied solutions to a variety of social, economic, and aesthetic issues and turned dream-like visions of luxury consumption into material reality. The British and German Conservative and Liberal press wrote about the new passenger liners in exactly the same terms as public-relations departments.[88] In part, this was a direct result of the wooing of the representatives of the press, described earlier. By forging this coalition, the industry managed to overcome the image of ships as places of danger and disease. Nonetheless, large advertising budgets alone did not guarantee positive publicity in the press, as the example of the film industry demonstrates in other contexts. Moreover, it needs to be explained why even liberal papers, which often headed campaigns for social reform in both countries, were reluctant to point out the hard labor conditions on ships. On only one occasion did the British Conservative and Liberal press address this problem directly: when a stoker "went out of his mind" in the heat of the *Lusitania*'s boiler room during a record voyage. Yet these observations led to no overall critique of labor conditions on board.[89]

National pride was central to this hushing-up of exploitation. As we shall see in a later chapter, British and German passenger liners became prominent symbols of national prestige and power in both countries. Respect for these national symbols placed progressive journalists, who were the

---

[87] *Daily Telegraph*, 25 May 1936, Special Supplement, 13. See also *The Shipbuilder and Marine Engine-Builder: Queen Mary Supplement* 44 (1936), 122. Both studies follow closely the explanations in *From the Cradle to the Sea: A Biographical Study of the R.M.S. "Queen Mary"* (Glasgow, 1936), 30. A copy of this pamphlet is preserved in Cunard archive, D42/PR4/25. For an earlier stylistic appraisal of the *Queen Mary*, see *Morning Post*, 19 November 1935.
[88] For some examples, see *Daily Mail*, 21 September 1906, 4; *Manchester Guardian*, 9 September 1907, 7; *The Times*, 20 June 1911, 12; 27 May 1911, 4; *Berliner Tageblatt*, 10 January 1900 (evening edition), 4; *Neue Preußische Zeitung*, 23 May 1913 (evening edition), 1; *Berliner Tageblatt*, 23 May 1913 (evening edition), 5; *Vorwärts*, 15 August 1928, 9.
[89] *Manchester Guardian*, 12 October 1907, 9; *Daily Mail*, 12 October 1907, 7.

most likely to challenge sanitized accounts of steerage-class and work environments, in a difficult position. If they directly criticized ships for their social disparities and labor conditions, they risked being accused of spreading anti-national sentiment. Thus the rhetoric of the floating palace and nationally minded enthusiasm for technology were mutually reinforcing. While, in these circumstances, Liberal newspapers focused on the luxuries of vessels, the Labour and Social Democratic press frequently remained silent about passenger liners. This tactic allowed them to escape accusations of national betrayal and lack of patriotism without compromising their commitment to social-reform programs.[90] In the wake of the constitutional reforms following the Great War, Labour and Social Democratic press organs abandoned this strategy and celebrated ships as evidence of the national importance of the working class. According to this line of reasoning, the fact that laborers dutifully carried out their jobs despite harsh employment conditions revealed the dedication of the working class to shipbuilding as a national cause. As the *Daily Herald* emphasized in relation to the *Queen Mary*, it was the workers' "toil [that] made this triumph."[91] Characteristically, among the radical German Left in the interwar period, a sustained critique of the floating palace developed only from a standpoint of "proletarian internationalism." Communist publications such as the *Rote Fahne* and *Arbeiter-Illustrierte Zeitung* castigated the maltreatment of the crew and their excessive working hours, and branded on-board comfort as luxurious frivolity in times of high unemployment.[92] The majority of press publications, however, either tacitly or actively affirmed the trope of the floating palace.

The steadily increasing popularity of holiday cruises, which operators began to offer on a regular basis from the 1890s, provides the most striking indicator of the public's changed appreciation of vessels. Ocean cruising formed part of the holiday industry that sprang up in the second half of the nineteenth century and, in the context of one-day excursions, soon developed into organized mass tourism. Fueled by increasing levels of affluence, the introduction of paid vacations for salaried employees in particular, and a diverse range of "package" tours offered by professional operators such as Thomas Cook, the holiday industry established an infrastructure consisting

---

[90] For an early criticism of technological nationalism, see *Vorwärts*, 10 January 1900, 1. For silences, see *Daily Herald*, 21 April 1913; 30 May 1914; 11 April 1922; *Vorwärts*, 20 September 1897; 27 September 1897; 23 May 1912; 12 June 1913.
[91] *Daily Herald*, 26 September 1934. For similar evaluations, see *Vorwärts*, 16 August 1928, erste Beilage, 1; *Daily Herald*, 27 May 1936, 1.
[92] *Rote Fahne*, 16 August 1928, 8; 17 August 1928, 7; *Arbeiter-Illustrierte Zeitung*, issue 6, 1928, 4–6; issue 37, 1928, 2; *Arbeiter-Illustrierte Zeitung*, 24 July 1929, 2; *Arbeiter-Illustrierte Zeitung*, issue 32, 1930, 629.

of transport networks as well as a host of spas, seaside resorts, and hotels close to cultural and natural landmarks – all extensively described in a burgeoning travel literature.[93] Vacations on the sea brought together the central aspects of early-twentieth-century holiday culture by taking travelers to unfamiliar yet acknowledged places of note by a "modern" means of transport that blended the familiar features of the hotel and the seaside resort.

Cruise ships epitomized contemporary tourism and radiated precisely the aura of unmitigated pleasure and leisure that shipping companies had labored to convey in promotional drives for the floating palace. Like other ocean tour organizers, the Hamburg-American Line, which gained an internationally leading position in the business before World War I, could declare by 1914 that "half a generation ago" only a "dreamer" would have predicted the advent of a time when people undertook "long-distance voyages for pure enjoyment."[94] Initially directing their offers exclusively at a clientele able to pay fares of up to 10,000 Marks (£500), shipping operators devised an expanding array of trips to Norway, the Mediterranean, the Caribbean, India, and, eventually, around the world.[95] During the interwar years, as companies sought to stimulate customer demand at a time of over-capacity in maritime transport, advertisements increasingly drew attention to the "cheapness" of this form of vacationing and introduced trips that cost between 300 and 1,000 Reichsmarks, within the reach of solid middle-class budgets. Although holiday-makers could book tours for as little as 140 Reichsmarks by the late 1920s, this still meant that vacations on the sea remained beyond the means of most skilled workers, who, if entitled to a week-long annual break, usually set aside about 100 Marks for this purpose.[96] While naval operators converted a few smaller vessels to serve as luxury yachts for the top end of the market, most cruise ships were

---

[93] On Thomas Cook, see Piers Brendon, *Thomas Cook: 150 Years of Popular Tourism* (London, 1991). On guidebooks, see Rudy Koshar, "'What Ought to Be Seen': Tourist Guidebooks and National Identities in Modern Germany and Europe," *Journal of Contemporary History* 33 (1998), 323–40. A general overview of tourism is Lynne Withey, *Grand Tours and Cook's Tours: A History of Leisure Travel, 1750–1915* (London, 1997). The seaside resort is covered in John K. Walton, *The English Seaside Resort: A Social History 1750–1914* (New York, 1983). An excellent overview is Orvar Löfgren, *On Holiday: A History of Vacationing* (Berkeley, 1999).

[94] DSB library, Hamburg-Amerika Linie, *Große Orient- und Indienfahrt 1914* (Magdeburg, c. 1913), 3. On Hapag's leading role in the cruise business, see Straub, *Albert Ballin*, 107–9.

[95] For these prices, see DMBC, *Polarfahrt 1911 mit dem Doppelschraubendampfer "Großer Kurfürst"* (Bremen, 1910); DSB library, Hamburg-American Line, *Cruises by the New Twin Screw Pleasure Yacht "Prinzessin Victoria Luise", 1900–1901* (Magdeburg, 1900), 8–11.

[96] DSB library, Norddeutscher Lloyd Bremen, *Vier billige Lloyd-Mittelmeerfahrten 1932* (Bremen, 1931), 5; Norddeutscher Lloyd Bremen, *Volkstümliche Reisen zur See* (Bremen, 1925), 5. The lowest figures are from Christine Keitz, *Reisen als Leitbild: Die Entstehung des modernen Massentourismus in Deutschland* (Munich, 1997), 116; on working-class holiday budgets, see *ibid.*, 112.

regular liners that earnt more on holiday routes than during slack periods in regular service. Customer care varied considerably across the price range. Unlike Thomas Mann, who thoroughly enjoyed his Mediterranean voyage on a purpose-built "capitalist pleasure barge" (*kapitalistische Lustbarke*) in 1925, Victor Klemperer, a professor of romance literature, found in the same year that his "study trip" to Latin America sometimes stood in clear "contrast" to the promises he had read in "glossy prospectuses."[97] Reports in the popular national press also illustrate the growing appeal of leisure travel by boat. In 1922, the *Daily Mail* recommended "a summer trip on an Atlantic liner between her home port and her 'first call,'" because of its "moderate fare" and because a "liner's population is always interesting . . . and the shortest experience on board these great ships is a flash of British sea-history." Ten years later, the same paper maintained that "cruises to Madeira and Mediterranean ports . . . have captured the imagination of the nation," since "eight of Britain's greatest ships" had left Southampton "fully booked" with 11,000 passengers on a single day in late July.[98] By the early 1930s, maritime holiday travel had become both a widespread practice among the middle class and a national news item.

In Germany, the general recognition of ships as sites of enjoyment, which had displaced older notions of maritime discomfort, laid the cultural foundation for an appropriation of the sea holiday by National Socialism after 1933. Given the widespread appeal of sea travel, the Nazi dictatorship could expect to reap considerable national and international political prestige when it set out to organize subsidized ocean cruises after 1933. Under the auspices of the party's leisure organization Strength through Joy (*Kraft durch Freude*, or *KdF*), which became Germany's largest tourism operator and arranged trips for a total of 54 million Germans before war broke out, about 690,000 tourists boarded cruise ships bound for the Mediterranean, Norway, and Madeira at prices between 56 and 140 Marks.[99] Although sea holidays thus accounted for only a small fraction of the travelers who

---

[97] Thomas Mann, "Unterwegs," in *Bemühungen* (Berlin, 1925), 257–67, here 266; Victor Klemperer, *Leben sammeln, nicht fragen wozu und warum: Tagebücher 1925–1932*, ed. Walter Nojowski (Berlin, 1996), 96.

[98] *Daily Mail*, 11 April 1922, 3; 30 July 1932, 7.

[99] This is the figure provided in Hasso Spode, "Arbeiterurlaub im Dritten Reich," in Carola Sachse et al. (eds.), *Angst, Belohnung, Zucht und Ordnung: Herrschaftsmechanismen im Nationalsozialismus* (Opladen, 1982), 275–328, here 298. The total figure of *KdF* tourists is from Shelley Baranowski, "Strength through Joy: Tourism and National Integration in the Third Reich," in Shelley Baranowski and Ellen Furlough (eds.), *Being Elsewhere: Tourism, Consumer Cultures, and Identity in Modern Europe and North America* (Ann Arbor, 2001), 213–36, 216. On the prices, see N. S. Gemeinschaft Kraft durch Freude, *Jahresprogramm 1935 Gau Franken* (Nuremberg, 1934); idem, *Hochsee- und Landreisen im Urlaubsjahr 1938* (Bayreuth, 1937).

availed themselves of *KdF*'s services, they played an important propagandistic role. Featuring prominently in the party press, Strength through Joy cruises, it was argued, illustrated the material achievements of the regime and testified to its commitment to constructing a "national community" (*Volksgemeinschaft*) devoid of the class privileges that had formerly reserved this form of recreation mainly for affluent social sectors.[100] This agenda of social inclusion, which was, of course, restricted to so-called Aryans, sat uneasily with the elitist overtones that adhered to the dominant image of ocean holidays in general and to large vessels, as floating palaces, in particular. Nazi propaganda, therefore, reconfigured the pleasure cruise in accordance with its ideological premises.

To begin with, Strength through Joy tempted holiday-makers with the promise of light-hearted excursions whose "unforgettable experiences" left behind the "ugliness" and "eternal uniformity of everyday life."[101] Dances, concerts, costume balls, sunbathing, and deck games featured among the diversions on board, while, at ports of call, tourists participated in sightseeing tours, spent their modest foreign currency allowance on inexpensive souvenirs, and marveled at the "strange" tastes, sounds, and smells of Madeira and Italy.[102] The organizers aspired to an informal atmosphere, and participants, once embarked, were usually on first-name terms.[103] Furthermore, prospectuses emphasized that travelers did not have to bring along elaborate evening attire for social occasions, another hint at casual forms of sociability. If the rich society lady provided the archetypal consumer on the "floating palace," the *KdF* ships overtly catered for "ordinary" women and men who enjoyed simple pleasures. In fact, several photos display vacationers in positively childish poses as they competed in eating contests and became entangled in hopeless piles of bodies during group games. To prevent exuberance from spiraling out of control (parties sometimes went a "little wild"), brochures reminded participants that they were obliged always to behave in a manner that did "not endanger the honor and reputation of the National Socialist community."[104]

---

[100] For reports in the main party paper covering the beginning and end of the cruise season, see *Völkischer Beobachter*, 16 March 1936, 2; *Völkischer Beobachter*, 19 September 1935, 2.

[101] N.-S. Gemeinschaft Kraft durch Freude, *Jahresprogramm 1936: Gau Franken* (Nuremberg, 1935), 4; *Arbeitertum*, 15 March 1936, 5; 1 April 1936, 7.

[102] Photos depicting these entertainments accompanied the tour programs cited in the previous and ensuing footnotes. The experience of foreign lands is covered in Karl Busch (ed.), *Nach den glücklichen Inseln: Mit KdF-Flaggschiff "Robert Ley" nach der farbenprächtigen Welt von Madeira und Teneriffa* (Berlin, 1940), 22–4, 38–40, 54.

[103] Jakob Schaffner, *Volk zu Schiff: Zwei Seefahrten mit der 'KdF'-Hochseeflotte* (Hamburg, 1936), 21.

[104] The "wild" dance is reported in Schaffner, *Volk zu Schiff*, 115. Calls to order are included in *Jahresprogramm 1936*, 32; *Jahresprogramm 1935*.

Of course, these party-led pleasure excursions served purposes beyond pure enjoyment and recreation. Motivating the labor force for the challenges of their occupations provided one rationale for sea holidays. "Everybody returns relaxed and happy because they know that much work awaits them . . . in Germany" concluded one richly illustrated booklet about a cruise to Madeira.[105] The productivist agenda that underpinned the *KdF* program went hand in hand with political indoctrination on the boats. Travelers routinely attended morning calls to the flag, witnessed military exercises at sea, staged memorial services for the fallen soldiers of World War I, and were treated to readings from *Mein Kampf*.[106] This intermingling of politics, productivism, and pleasure was constitutive of the National Socialist ideology of leisure. According to the party press, Hitler himself had allegedly ordered the foundation of Strength through Joy in 1933 in order to grant "the worker sufficient holidays," with the intention of creating a "people (*Volk*) with strong nerves because one can only pursue truly great policies with a people that keeps its nerves."[107]

If *KdF* ambitiously identified the entire "people" as its clientele, the organization had to demonstrate its dedication to make its program attractive and accessible to a wide range of Germans with limited means. Consequently, advertisements for cruises emphasized that "active *Volk* comrades (*schaffende Volksgenossen*) who are unable to afford a normal holiday" with a commercial tourist agency enjoyed priority in booking.[108] Articles and books about ocean trips drew attention to the presence of, among others, female sales assistants, unmarried secretaries, and manual workers, all happy Nazis who were dutifully "grateful" for their experience of an unexpected "dream."[109] Propaganda material credited female and male travelers from modest yet differing social backgrounds with a sense of "comradeship" that confirmed the development of the "national community" in Germany.[110] Significantly, this "national community" also comprised the workforce on board. Pleasure seekers praised the "good stewards" for their efforts to render trips enjoyable.[111] Having initially chartered and bought

---

[105] *Nach den glücklichen Inseln*, 64. A similar point is made in *Jahresprogramm 1936*, 4.
[106] *Arbeitertum*, 15 May 1935, 12–13; 15 October 1935, 13; 1 August 1938, 12.
[107] The quotation is from an article on a cruise to Norway. See *Arbeitertum*, 15 February 1936, 6. The origin of this "order" is unclear. See Hermann Weiß, "Die Ideologie der Freizeit im Dritten Reich: Die NS-Gemeinschaft 'Kraft durch Freude,'" *Archiv für Sozialgeschichte* 33 (1993), 289–303, esp. 293.
[108] *Jahresprogramm 1936*, 32; see also *Jahresprogramm 1935*.
[109] *Arbeitertum*, 15 March 1936, 8; 1 April 1936, 6–9; 15 April 1936, 10–13; Schaffner, *Volk zu Schiff*, 13, 16.
[110] *Arbeitertum*, 15 March 1935, 11.   [111] *Ibid.*, 15 May 1935, 12; 15 March 1936, 8–9.

existing liners from private operators, *KdF* launched a construction program for purpose-built cruise ships with standardized accommodation for all voyagers, to convey its vision of a "classless" community at sea more effectively. The manufacture and operation of the *Wilhelm Gustloff* and *Robert Ley*, the first two of twenty projected vessels, generated the most comprehensive propagandistic statements about the *Volksgemeinschaft* in maritime contexts. Party journalists did not restrict themselves to celebrating these vessels for offering "the most modern of modern" provision for customers and crew; they also detailed in word and photographs the contributions of a "thousand" workers who "built this ship for workers" and who, serendipitously, were also among its first customers.[112] The *Wilhelm Gustloff*, for instance, the "symbol of a people that knows how to create its own presents," was proclaimed a "masterpiece of socialism" on National Socialist premises.[113]

While *KdF* liners, like previous ships, embodied idealized social microcosms, they thus diverged from the traditional model of the floating palace in several respects. Strength through Joy propagandists explicitly rejected conventional elitist social models that focused on clear on-board stratification, and instead envisioned their vessels as prototypes of a "classless" national community. Furthermore, they broke the widespread silence surrounding maritime work arenas, placed a spotlight on service personnel during cruises, and extolled manual laborers as the creators of new ships. Strength through Joy also claimed to target a mixed-sex clientele from lower parts of the social spectrum, including manual workers, than had previously been the case, thereby allegedly democratizing the ocean cruise. Lastly, unlike commercial tour operators, who built on the notion of the floating palace to promote ocean cruises as a form of leisure for its own sake, the Nazis conceived of sea holidays as a means towards the eminently political end of fostering a motivated and productive labor force in tune with the party's principles.

Propaganda material does not permit evaluations of the extent to which these journeys, combining discipline and diversion, lived up to their objective of giving lower-class travelers "unforgettable experiences," but several factors indicate that *KdF* cruises were a considerable popular success. Voyages sold out quickly and, on postcards, holiday-makers bragged about "marvelous trips" with their "fabulous atmosphere" on "wonderful steamers," claiming to regret that these voyages were not within reach of "more

---

[112] *Ibid.*, 1 February 1938, 20–2; 15 March 1938, 8–11; 15 April 1938, 4–5; 15 August 1938, 16–17.
[113] *Ibid.*, 1 February 1938, 22; 1 July 1939, 8–10.

Germans."[114] At the same time, the notion of liner populations as miniature "national communities" remained largely a propagandistic fiction. Lower-middle-class and middle-class salaried employees and craftsmen, rather than manual workers, made up the bulk of voyagers to Italy between November 1938 and March 1939, as one study has shown.[115] Neither the ideological celebration of manual labor nor "low-cost travel as a substitute for higher wages" drew a sufficient number of workers for *KdF* to achieve its stated aim of populating its ships with a majority of working-class vacationers.[116] In fact, frustration about the working class found its way into prospectuses that lashed out against the "eternal doubters" who "simply don't belong to this time" because they did not appreciate Strength through Joy's achievements.[117] The leisure organization developed its strongest appeal among the parts of German society that traditionally provided the core support for National Socialism and who gradually transformed Strength through Joy into a "travel agency for party members" in the late 1930s.[118] *KdF*'s overtly political orientation may explain its limited appeal, not only among workers, but also among many middle- and upper-class vacationers who continued to book themselves on to privately operated luxurious "floating palaces" at prices as high as 1450 Reichsmarks in 1938.[119] Although Strength through Joy established a powerful alternative public meaning of liners as symbols of an elusive "national community," it did not altogether displace the older, socially exclusive notion of the floating palace that had transformed the public image of sea travel in Britain and Germany since the 1890s.

Shipping companies achieved extraordinary success when they began to promote their products as glamorous "floating palaces" and swept aside long-standing notions of ships as places of disease and danger. Ships were

[114] These quotations are from postcards in the author's collection. They are dated 29 April 1938, 20 April 1938, 26 April 1938, 10 May 1939. Hasso Spode states that many cruises sold out quickly; Spode, "Arbeiterurlaub im Dritten Reich," 297.
[115] Salaried employees (*Angestellte*) accounted for roughly 40 percent, craftsmen for 10 percent, and workers for 17 percent. See Keitz, *Reisen als Leitbild*, 250.
[116] Baranowski, "Strength through Joy," 215. On the celebration of labor for ideological purposes see Alf Lüdtke, "'Ehre der Arbeit': Industriearbeiter und die Macht der Symbole: Zur Reichweite symbolischer Orientierungen im Nationalsozialismus," in Klaus Tenfelde (ed.), *Arbeiter im 20. Jahrhundert* (Stuttgart, 1991), 292–343.
[117] *Jahresprogramm 1935*; *Jahresprogramm 1936*, 55.
[118] Spode, "Arbeiterurlaub im Dritten Reich," 304.
[119] DMBC, Hamburg-Amerika Linie, *Hapag-Herbstfahrten 1938 mit MS. "Milwaukee" ins Mittelmeer und nach Westafrika* (Hamburg, 1938), 14; DSB library, Norddeutscher Lloyd Bremen, *Lloyd Mittelmeer-Fahrten 1934* (no place), North German Lloyd, *Round the World in Ninety Days in the Bremen* (Boston, 1937).

extolled as sites of luxury consumption and leisure, with a strictly enforced social hierarchy between travelers in different classes. The systematic downplaying of exploitative labor conditions in advertisements further stabilized images of social harmony on vessels and put companies in a position to celebrate ships as exemplary in terms of their social organization, their economic efficiency, and their aesthetics. Advertising campaigns, although they could become embroiled in conflicts, urged contemporaries to view vessels as works of art, and the proliferation of ocean cruises provides striking evidence that ships came to be considered as objects whose technological features transcended primarily utilitarian styles of engineering. Indeed, the popularity of holidays on the sea led the National Socialists to incorporate leisure cruises into their subsidized holiday program in ways that combined recreational, political, and productivist motifs to demonstrate the regime's ostensible commitment to bringing previously elitist consumer practices within reach of the lower-middle and working classes. By the 1930s, ocean travel exerted an irresistible appeal across profound social and ideological divides.

Several factors explain the success of the shipping industry's public-relations campaigns. The technical properties of maritime artifacts influenced advertising drives from the time when the introduction of steel into naval construction allowed shipping companies to transform the architecture of ocean vessels fundamentally by incorporating lavish provision for first-class passengers. Technological developments, therefore, shaped the material characteristics on which promoters based their descriptions of ships as luxurious. Yet it would be simplistic to suggest that technological features determined the success of advertising campaigns. After all, in the light of widespread traditional views of vessels as unhealthy and perilous, the naval industry set itself an ambitious promotional aim that required a tremendous leap of faith when it began to celebrate ships as sumptuous consumer environments. Extensive public-relations budgets enabled shipping companies to flood the public sphere with advertising material that combined to tell similar stories. Unsurprisingly, neither prospectuses nor pamphlets nor press releases mentioned that passengers fell prey to seasickness, and instead praised the restorative qualities of ocean travel. Moreover, promotional drives constantly directed attention away from the social disparities and working conditions that went against the image of ships as icons of pleasure. National fervor also supported propagandistic fictions of ships as exemplary social microcosms, since Liberal and Left-leaning opinion makers exposed themselves to accusations of unpatriotic behavior if they criticized labor conditions on board the prominent vessels that figured as national symbols. Only National Socialist ideology, with its emphasis on a "national community" supposedly devoid of social tensions, permitted mention of the roles played by service personnel and maritime workers while leaving intact the status of large ships as places of enjoyment. Thus technical characteristics, the financial resources available for public-relations campaigns, nationally minded enthusiasm, and the semantic consistency of the stories with which promoters wooed customers and the wider public all worked to make promotional efforts a success. In instrumentalizing social fantasies of a materially secure and carefree life, sustained advertising drives demonstrated how technological advances transformed travel from an uncomfortable and dangerous chore into a safe source of pleasure by turning large passenger ships into icons of luxury, comfort, and modernity.

The major transformation in the public understanding of ocean liners had already occurred in the last decade of the nineteenth century. To be sure, some modifications were required to adapt the idea of leisured ocean travel to the ideological preoccupations of National Socialism, but, on the whole, visions of material affluence and comfort provided the dominant social fantasy that created a widespread fascination with large passenger vessels. Prior to the advent of the Nazi regime, transnational similarities outweighed national differences in the public meanings of ocean liners in both countries. A comparable pattern of Anglo-German similarities disrupted by National Socialism also emerges in another context that illustrates how social fantasies motivated contemporaries to embrace new technological objects. When Britons and Germans, mostly from middle-class backgrounds, took up film making as a new hobby from the mid-1920s on, they frequently did so because amateur cinematography allowed hobby directors to carve out a sphere of individual independence. Rather than pointing towards technology's material benefits, as was the case with passenger liners, amateur film, it was claimed, demonstrated how new devices could enrich one's private life in distinctly non-material ways. In their home movies, film amateurs staged their lives as they wanted them to be remembered. It was the social fantasy of individual autonomy, often revolving around idealized notions of family, that prompted amateurs to become cinematographers in their spare time, in the process investing film with a set of meanings that could hardly have been more different from those that cultural critics ascribed to commercial cinema culture.

CHAPTER 7

# *Fantasy as social practice: the rise of amateur film*

Before joining the armed forces in 1942, Charles Chislett, a bank manager from Rotherham in South Yorkshire, took his wife and children on a family trip to the scenic Yorkshire Dales. As an avid amateur film maker, who had already documented the *Queen Mary*'s maiden voyage as a passenger in 1936, the *pater familias* brought along his cine camera to commemorate what, given the war, might well have turned out to be the Chisletts' last family vacation. In addition to capturing on celluloid the barren, craggy rock formations of the Dales, *Dale Days*, as the film came to be entitled, featured, among many other scenes of family bliss, Chislett fishing, his son's engagingly unskilled experiments with a cricket bat, and his daughter chasing balls. The film's emphasis on fun and games culminated in a lengthy passage about local cheese production that concluded with a comical attempt on Chislett's part to feed fresh cheese to a cow who stubbornly ignored this offering that was derived, as the narrative implied, from her own milk. *Dale Days* is a testimony to the personal value of amateur film – and not just because it was shot on expensive, rare color stock and thereby represented a significant financial investment for Chislett. More to the point, the amateur and his family dedicated considerable parts of their wartime holiday to time-consuming activities carried out exclusively in the service of film making. The Chisletts probably did not find this obsession unusual, since Charles was, after all, a highly committed amateur who produced more than 100 reels between the 1930s and 1960s. In Chislett's case, amateur film textured, if not dominated, his private life.[1]

Charles Chislett was one of a steadily growing number of enthusiasts who, since the mid-1920s, had discovered film making as a hobby. Amateur directors and manufacturers of amateur equipment extolled the

---

[1] Yorkshire Film Archive, York (hereafter YFA), film no. 328, Charles Chislett, *Dale Days* (1940). Biographical information on Chislett was provided by the director of the Yorkshire Film Archive, Sue Howard. On Chislett, see also the informative article by Heather Norris Nicholson, "Seeing It How It Was? Childhood Geographies and Memories in Home Movies," *Area* 33 (2001), 128–40, esp. 131–3.

creative possibilities of new apparatuses suitable for use by non-professional cinematographers, ignoring the manifold charges against cinema entertainments that could have tainted amateur film. Instead, a large number of books, journals, and prospectuses stressed that the movie camera was uniquely suited to advance individual creativity by enabling amateurs to produce visual records of their lives. Here was an entirely beneficial application of film technology that put amateurs in a position to enact a wide range of social fantasies in their private domain and under their exclusive control. The link between private film and social fantasies of individual autonomy found its most eloquent expression in films about personal relationships in the family circle. Most importantly, cameras could be used to produce records of private life that expanded the material stock of individual memories. Thus, in contradistinction to representations of commercial films as morally dubious, evaluations of films that were privately produced and consumed concluded that they strengthened the private sphere, and attributed salutary effects to film technology. In short, such reasoning credited amateur film with establishing a wholesome bond between individuals and technological artifacts.

To sustain a beneficial connection with their cameras, aspiring hobby directors had to master the medium of film even if they commanded only limited technical expertise. Lack of technical knowledge could wreck any beginner's enthusiasm as long as the equipment was frustratingly complicated. Enlightening amateurs about the technical principles of cameras and projectors was one strategy for confronting these problems, and most hobby film makers acquired at least a rudimentary understanding of how their equipment worked. Another promising response to the issue of technical expertise rested in the development of user-friendly – or, to put it differently, "fool-proof" – product design that allowed non-professionals to handle complicated apparatuses confidently. Rather than expecting amateurs to become technical wizards, Kodak, AGFA, and other companies developed devices that technical laypersons with little previous experience could operate successfully. The rise of amateur film thus illustrates one solution to the epistemic problems that, as we have seen earlier, could trigger ambivalent reactions when innovations appeared on the scene. User-friendly designs took into account laypersons' limited technical horizons and thus allowed them to familiarize themselves with new objects. While not altogether removing the knowledge problems that technological innovation generated, design solutions enabled technical laypersons to establish close and at times intimate relationships with numerous devices whose workings they understood only incompletely. In contrast to aviation and passenger shipping, where non-experts had to content themselves with the passive role of observer, amateur film allowed laypersons to become actively engaged with innovation. Practical rather than theoretical knowledge, then, overcame their initial insecurity.

Since home movies frequently captured scenes of private life, the practice of producing amateur reels domesticated film technology. Highlighting the benefits of camera work for the individual, promotional and prescriptive literature about amateur movie making assisted the public acceptance of film technology in both Germany and Britain. Nonetheless, promoting amateur film as a wholesome hobby for individual self-expression through a distinctly modern medium was by no means unproblematic, even if amateurs mastered the medium's technical challenges. Since a thoroughly conventional male gender bias provided a main ingredient of many a standard recipe for home movie making, camera work could give rise to friction as men struggled to assert their authority as technological "experts" in the home while women defended their position as household managers. More importantly, by strengthening the private realm as a sphere of individual autonomy, the proliferation of amateur film came to be regarded with suspicion after 1933 by the Nazi rulers who only reluctantly tolerated social arenas that evaded direct state control. Thus "public discourse on amateur films" was by no means a straightforward "form of social control," as Patricia Zimmermann writes, because its normative prescriptions could unintentionally import gender frictions into the private sphere and because, by marking out an arena of individual independence, it indirectly erected a barrier to state authority.[2] Before turning to these issues in detail, however, we must first recount the developments that boosted the popularity of film as a hobby in the mid-1920s.

## "ONE, TWO, THREE, AND YOU CAN FILM": MAKING FILM AMATEUR-FRIENDLY

Around 1925, producers of photographic and film equipment first began to market a variety of movie cameras and projectors specifically aimed at amateurs. Non-professional film makers could choose from a wide range of new

---

[2] Patricia R. Zimmermann, *Reel Families: A Social History of Amateur Film* (Bloomington, 1995), 5. Zimmermann's book relies on somewhat dated Marxist approaches to cultural studies and concerns itself with the amateur scene in the United States.

This chapter focuses on promotional literature as well as advice manuals penned by amateurs for amateurs and primarily employs the topics and narrative sequences in amateur films to test to what extent practitioners followed printed advice. This cautious use of amateur film as a "source" acknowledges the difficulties of "reading" these movies as texts. The fact that many amateurs had a limited technical command of their medium renders semiotic interpretations of these productions hazardous.

devices manufactured by firms such as Kodak, AGFA, Zeiss-Ikon, Pathé, and Niezoldi und Krämer, which differed from earlier models in several respects.[3] They all used film considerably smaller than the standard 35mm format of professional film productions. The size of cameras also decreased because they dispensed with technical refinements that were considered unnecessary for home use, typically featuring rudimentary viewfinders and providing only single, fixed-focus lenses rather than offering different options for close-ups and panoramic views. These designs were motivated in part by the manufacturers' desire to supply potential customers with equipment at a much lower price than previous models. Both the purchase of cameras as well as filming itself became considerably cheaper during the 1920s as companies aimed to expand the market.

Before the advent of cameras specifically manufactured for private leisure use, high costs had barred numerous enthusiasts from taking up filming as a pastime, because amateurs had to use the same equipment as professionals. In 1919, the German company Ernemann sold a camera at 1,450 Marks, a price that only very wealthy members of the upper-middle class could afford.[4] Early amateurs lamented not only the prohibitive start-up costs of home movie making but also the outlay required to maintain their hobby. The new amateur equipment admitted a much wider group of middle-class people to the circle of film makers. Although amateur cinema was still anything but a cheap pastime, given that prices of cameras produced by Kodak, AGFA, and Zeiss-Ikon ranged from £15 15s 0d to £20, and from 210 to 320 Marks, the cost of basic equipment had dropped significantly by 1930.[5] A handbook from 1931 pointed out that while a minute's filming required an outlay of around £10 when shot on 35mm standard film, the same project done in an amateur format cut the price to 36s. Hobby film making, however, remained a pricey pleasure if we consider that an annual income at the lower end of the German middle class amounted to no more than about 1,500 Marks at the beginning of the 1930s. In Britain, a solid middle-class lifestyle required a minimum of £500 a year during the interwar years.[6] Given the considerable costs of their hobby, amateur film makers

---

[3] For an overview of models available at the time, see Michael Kuball, *Familienkino: Geschichte des Amateurfilms in Deutschland*, vol. 1: *1900–1930* (Reinbeck, 1980), 101–12.

[4] Deutsches Museum, Munich, archive, brochure collection (hereafter DMBC), *Ernemann Normal-Aufnahme-Kinos* (Dresden, 1919), 2.

[5] See the advertisement in Edward W. Hobbs, *Cinematography for Amateurs: A Simple Guide to Motion Picture Taking, Making and Showing* (London, 1930), 1; DMBC, *Die große Überraschung: Das AGFA Heimkino* (Berlin, 1932), 5; *The Kodak Catalogue, 1932* (London, 1932), 27.

[6] The figures are from Hans-Ulrich Wehler, *Deutsche Gesellschaftsgeschichte*, vol. IV: *Vom Beginn des Ersten Weltkriegs bis zur Gründung der beiden deutschen Staaten, 1914–1949* (Munich, 2003), 288;

continued to pay close attention to technical developments that promised further savings. When Kodak launched 8 mm film equipment in 1932, a leading German amateur periodical welcomed the new film format because it roughly halved running costs for non-professional cinematographers. The journal anticipated that this innovation would make film making accessible to "a new wide circle of people, who possess the necessary and comparably modest means" to afford the innovative design.[7]

Along with initiating a series of significant price reductions, producers aimed to make non-professional equipment more user-friendly to attract more customers. Before the mid-1920s, film making could be a cumbersome, if not frustrating, hobby. It was not simply that amateurs had to handle heavy cameras mounted on tripods, which allowed for few spontaneous snapshots; developing films also presented a complex technical challenge that only very skilled enthusiasts could face with glee. Early handbooks abounded with pages of instructions about chemical solutions and blueprints for assembling darkroom accessories.[8] Even loading a camera with film could be a troublesome exercise during a family holiday: "It is best to insert the film under a thick blanket at night. If one is careful, no light will mar the fresh film," warned one early enthusiast.[9] Finally, exhibiting one's cinematic treasures presented a fire hazard, because highly flammable celluloid film reels easily burst into flames that could not be extinguished with water. Manuals therefore admonished amateurs never to smoke in the proximity of film material, and to throw burning projectors out of the window, even if that required smashing the glass pane.[10] Thus financial expense, complicated equipment, and safety concerns were major deterrents that limited the appeal of film making as a hobby.

After 1925, manufacturers insisted that the cameras, projectors, and film materials that entered the amateur market should conform to new standards of safety as well as ease of use. The development of user-friendly

---

Francois Bédarida, *A Social History of England, 1815–1990* (London, 1991), 206. According to Eric Hobsbawm, a male elementary-school teacher earned £334 in 1928. See Eric J. Hobsbawm, *Industry and Empire* (Harmondsworth, 1986), 277.

[7] *Film für Alle* 6 (1932), 227. For a similar statement in Britain, see *Parliamentary Debates: House of Commons*, fifth series, vol. 204 (London, 1927), 247.

[8] Thomas F. Langlands, *Popular Cinematography: A Book for the Camera* (London, 1926), 31–50, 58–78; Ewald Thielmann, *Amateur-Kinematographie* (Leipzig, 1925), 40–8. For a late and particularly elaborate example, see Paul Herrnkind, *Die Schmalfilm-Kinematographie* (Vienna and Leipzig, 1929).

[9] Helmut Lange, *Wie entsteht ein Amateurfilm? Ein Amateur plaudert von seinen Filmarbeiten* (Berlin, 1930), 8.

[10] Friedrich Willy Frerk, *Der Kino-Amateur: Ein Lehr- und Nachschlagebuch* (Berlin, 1926), 181–3. Fears of fire also lingered on later. See A. Stüler, *Filmen leicht gemacht*, 2nd edition (Stuttgart, 1931), 13.

equipment for home cinematography formed part of a wider proliferation of consumer technologies in the domestic sphere. In the late 1920s, the radio industry, for instance, energetically moved beyond the stage when "listening in" involved donning headsets attached to mechanisms that openly displayed their electrical components such as tubes, condensers, and resistors, which required constant manipulation by skilful operators. The new wireless apparatuses featured wooden casings designed to blend in with living-room furniture, were more reliable, and could be operated by listeners with little technical training.[11] The success of the wireless industry in domesticating the radio, however, paled beside the earlier expansion of still photography. At the turn of the century, Kodak had been the first company to provide both the hardware and a service infrastructure to transform the taking of pictures from an exclusive hobby pursued by technically accomplished, affluent amateurs into a widespread activity. Kodak developed simple, inexpensive, hand-held cameras whose film rolls customers sent to processing centers to be developed. "You press the button and we do the rest" – Kodak's famous advertising slogan encapsulated the company's system, which rendered a hitherto esoteric technology accessible to a large clientele and set the industry standard. Since many manufacturers of photo equipment branched out into the hobby film market in the 1920s, central aspects of the public-relations campaigns for amateur cinematography drew on the earlier promotion of spare-time photography.[12]

Manuals began by emphasizing improvements in safety, whereby the fire-resistant qualities of newer rolls of film reduced the risk of fire. Hobby cinematographers no longer had to forgo smoking during home-movie viewings and, as an AGFA prospectus assured potential practitioners in 1932, "everybody can store films in their houses as well as display pictures in a carefree manner."[13] Most importantly, however, brochures stressed that film making had relinquished its status as a "secret science" (*Geheimwissenschaft*), since "there are simple and small apparatuses for amateurs, which are easy to handle."[14] In other words, making film pictures required little

---

[11] The best study of the domestication of wireless technology during the 1920s is Susan J. Douglas, *Radio and the American Imagination: From Amos 'n' Andy and Edward R. Murrow to Wolfman Jack and Howard Stern* (New York, 1999), 55–82. On Germany, see Karl Christian Führer, "A Medium of Modernity? Broadcasting in Weimar Germany, 1923–1932," *Journal of Modern History* 69 (1997), 722–53, esp. 732–6.

[12] The literature on Kodak is immense. See Elizabeth Brayer, *George Eastman: A Biography* (Baltimore, 1996); Nancy Martha West, *Kodak and the Lens of Nostalgia* (Charlottesville, 2000); Colin Harding, "A Transatlantic Emanation: The Kodak Comes to Britain," in David E. Nye and Mick Gidley (eds.), *American Photographs in Europe* (Amsterdam, 1994), 109–29.

[13] *Die große Überraschung*, 3.

[14] *Ibid.*, 2. A similar line is taken in *The Kodak Magazine*, January 1931, 10; March 1931, 42.

technical training and could therefore be learnt quickly. To lend authority to such claims, Kodak enlisted the help of celebrities, who described how effortlessly they had learnt to make amateur films thanks to new equipment. In 1932, the bi-monthly *Ciné-Kodak-Magazin* featured an article in which the German transatlantic pilot Hermann Köhl recounted his first successful endeavors as a film amateur. Although he possessed no previous experience in film work and lacked the time to acquaint himself thoroughly with his new acquisition, he shot away at an air show. To his surprise, his initial "skepticism" about the "mysterious box" proved entirely unfounded. As he watched his own "sharp and life-like" film a few days later, Köhl could not believe that he had actually produced the reel himself. After all, he had performed only "a few meaningless actions" with a mechanism whose workings he admitted he did not understand.[15] Köhl owed his successful start as an amateur film maker not only to a camera that could be operated with little training, but also to the increasingly common processing service for enthusiasts who had neither the time nor the skill to develop their films at home. More and more camera dealers began to accept orders for processing reels, thus removing another deterrent that had previously dampened interest in home-movie making.[16] In short, it was in the mid-1920s that the industry designed amateur film equipment in ways that compensated for the limited expertise of many customers.

To demonstrate the simplicity of amateur film making further, advertising drew on gendered notions of technological competence that had already commanded prominence in campaigns for still cameras. In 1926, a German promotion featured an illustration of a fashionably dressed woman holding a camera, with the caption: "One, two, three, and you can film" (see illustration 7.1).[17] Apart from conveying the sense that filming represented a trend-setting pastime, the image stressed that using a camera had become so easy that *even* a dainty society lady – hardly a typical personification of technological prowess – could acquire the requisite skills in no time. As other advertisements mobilized similar motifs that depicted women effortlessly mastering amateur equipment, the fashionable lady behind the camera became as common a visual stereotype in campaigns for amateur film equipment during the late 1920s as the "Kodak Girl" had been in publicity drives for still cameras earlier.[18]

---

[15] *Ciné-Kodak-Magazin*, August 1932, 4.
[16] Stüler, *Filmen leicht gemacht*, 18; Lange, *Wie entsteht*, 26.
[17] Deutsches Museum, Munich, illustrations archive, file Schmalfilm, 570.3.04.
[18] One example can be found in *The Kodak Magazine*, March 1931, inside front cover. On the Kodak girl, see West, *Kodak and the Lens of Nostalgia*, 53–60.

Illustration 7.1  Cine Nizo advertisement, emphasizing the simplicity and fashionable nature of amateur film during the 1920s.

Manufacturers achieved considerable success in generating interest in amateur film through marketing that linked aggressive promotion, lower prices, and simplified use. Leading firms such as AGFA and Kodak aimed to extend and supplement existing markets by making amateur film attractive to those who were already captivated by photography. In Britain, Kodak employed *The Kodak Magazine* – a journal at first exclusively published for amateur photographers – to disseminate information about its emerging venture into film by introducing a regular "movie page" in 1931.[19] Connections between photography and film making as a pastime surfaced when photographers encouraged one another to turn to cinematography because "filming is definitely simpler than taking photos."[20] Although photography supported the spread of amateur film, no reliable figures are available of the total number of film amateurs in Germany and Britain.[21] Evidence, however, suggests that the popularity of amateur film increased significantly during the second half of the 1920s. Submissions to a national competition for the best amateur reel doubled in Germany between 1929 and 1931.[22] Moreover, reports about amateur film began to reach the national press. In 1928 an article in the German daily *B. Z. am Mittag* informed readers about "the world's first film premiere on the high seas," which had taken place during a cruise whose passengers had been shown an amateur reel shot on board.[23] In Britain, where film amateurs founded clubs all over the country, by 1932 enthusiasts had organized no fewer than twenty associations in the capital city, including a "Civil Service Amateur Ciné Society" and a "Jewish Amateur Ciné Club."[24] By the mid-1930s, a German observer estimated that the British scene comprised around 250,000 hobby film makers, about 3,000 to 4,000 of whom held a club membership. It is unclear how the correspondent compiled his figures, but he admitted that these numbers were "overwhelming by Continental standards."[25] By comparison, the League of German Film Amateurs (*Bund Deutscher Filmamateure*) counted a total of 16,000 owners of film equipment in 1935, with membership of amateur societies running at a level of around 1,000.[26]

---

[19] *The Kodak Magazine*, January 1931, 10.
[20] Erna Engel, "Meine erste Filmaufnahme," *Kino-Amateur* 2 (1929), 30.
[21] Kodak estimated that there were 50,000 amateurs in Britain and Ireland. See *The Kodak Magazine*, January 1931, 10.
[22] See *Kino-Amateur* 2 (1929), 337; 4 (1931), 85.     [23] *B. Z. am Mittag*, 17 April 1928, 9.
[24] Andrew Buchanan, *Films: The Way of the Cinema* (London, 1932), 228–32; Marjorie A. Lovell Burgess, *A Popular Account of the Development of the Amateur Ciné Movement in Great Britain* (London, 1932), 189–95.
[25] *Film für Alle* 9 (1935), 134.     [26] *Ibid.*, 9 (1935), 77.

Thus film making became a more widespread hobby after 1925, when manufacturers brought out more affordable, safe, and easy-to-handle equipment for non-professionals. With the help of dynamic advertising, the British and German camera industries stressed that these innovative products differed significantly from earlier devices whose high costs, fire hazard, and complicated use had deterred potential customers. Once the new designs reached the market, they quickly attracted a wider following and spread amateur film as a leisure activity in Britain and Germany. Nonetheless, new equipment on its own does not explain the motivations prompting increasing numbers of people to take to the camera. Why, then, did Britons and Germans consider movie making a rewarding pursuit, to which they devoted considerable amounts of money and spare time?

### "INDEPENDENT FROM ALL DICTATES": AMATEUR FILM AND INDIVIDUAL AUTONOMY

Like hobby photographers, film amateurs captured a variety of topics with their cameras. Local events, such as visits by prominent public figures, spurred spare-time directors into action, as did festivities such as pageants, carnivals, and folk-dance performances, which celebrated local customs and traditions. While travel and holiday films had proved a popular genre from the very beginning, some entrepreneurs and engineers during the 1930s took to documenting their work environments.[27] Despite such forays into topics found in places far and near, it was the family circle that provided the most common film set, and relatives – wife or husband, children or siblings, aunts or uncles – hardly ever escaped the enthusiast's attention. All these types of films, however, signaled the more general motivations that drove devotees to invest time, money, and energy in movie making. Directing the viewfinder at their subjects, amateurs tried to capture on the reel something more than the immediately visible subject matter. While amateur films themselves yield little unambiguous information on the desires that fueled amateur cinematography, written accounts reveal the aspirations pursued by amateurs through film making.

Many enthusiasts found that filming as a leisure pursuit gave them an opportunity for self-realization that "everyday life" frequently denied them.

---

[27] YFA, film no. 37, *Kirby* (c. 1930–2); film no. 58, *Jubilee Celebrations at Clayton West* (1936); film no. 22, *Visit of HRH the Princess Mary and Viscount Laselle to Halifax* (1925). On work environments as a film topic, Michael Kuball, *Familienkino: Geschichte des Amateurfilms in Deutschland*, vol. II: *1931–1960* (Reinbeck, 1980), 10.

To be sure, amateur film makers did not seek to escape from harsh labor conditions in a factory or in the outdoors, as most of them belonged to solidly middle-class sections of British and German society, pursuing careers as, among others, businessmen, doctors, lawyers, salaried sales personnel, engineers, teachers, and civil servants.[28] Unlike commercial film, amateur film remained devoid of associations with working-class and, by extension, "mass" culture. At the same time, the incomes of many amateurs were sufficiently restricted for them to welcome technical innovations that cut the costs of their hobby, as we have seen. In their self-characterizations, they revealed themselves as typical members of the middle class. One author melancholically imagined his reader as "a poor, exhausted office person (*armer, abgehetzter Büromensch*) who goes to, and returns from, his job" in the darkness of an October morning and evening, looking for relief from the burdens of everyday routine through camera work.[29] Discontent with civilization also left its imprint on scripts such as the one for *City Dweller's Weekend* (*Großstädters Wochenende*), which lamented the way crowds clogged up trains and beaches during a Sunday trip to the outskirts of Berlin.[30] Thus some amateurs used film as a medium for addressing dissatisfaction with contemporary life in a playful manner. On other occasions, filming offered a temporary flight from the mundane troubles of workaday tedium. In these cases, the movie camera served as an "escape tool" that transposed its users into a realm where they claimed to find individual autonomy.[31] Amateur film, then, provided an active rather than a passive form of escapism. While the discontent with the historical present, which fueled at least parts of the British and German amateur film movements, was not strong enough to entice amateurs into joining one of the numerous

---

[28] According to a survey from 1935, the members of the Association of German Film Amateurs belonged to the following socio-economic categories: 41 percent were salaried employees (*kaufmännische Angestellte*) and self-employed businessmen; 18 percent had qualified as engineers, while 22 percent worked as doctors, lawyers, civil servants, and teachers. The survey did not list any workers. See *Film für Alle* 9 (1935), 79. Amateurs' wealth was demonstrated by the presence of nannies and cars, which claimed center stage in several more ambitious scripts. See Helmut Lange, *Filmmanuskripte und Film-Ideen: 121 Ideen für den Kino-Amateur* (Berlin: Photokino Verlag, 1931), 53–6; Alex Strasser, *Kind und Kegel vor der Kamera* (Halle, 1932), 87, 100. On socio-economic aspects of the British and German middle classes, see Jürgen Kocka, "The Middle Classes in Europe," *Journal of Modern History* 67 (1995), 783–806; Harold Perkin, *The Rise of Professional Society: England since 1880* (London, 1989), 78–101, 266–73. On middle-class culture, or "*Bürgerlichkeit*," see Hermann Bausinger, "Bürgerlichkeit und Kultur," in Jürgen Kocka (ed.), *Bürger und Bürgerlichkeit im 19. Jahrhundert* (Göttingen, 1987), 121–42.
[29] Lange, *Wie entsteht ein Amateurfilm*, 22.
[30] Lange, *Filmmanuskripte*, 46–51. A melancholy script for a film to be made in the rain is in *Film für Alle* 3 (1929), 66–9.
[31] Kuball, *Familienkino*, vol. 1, 40.

life-reform organizations, they still strove to create a sphere in which to act out a sense of personal independence.[32]

This search for individual liberty within "modern" society through camera work sometimes went hand in hand with mildly transgressive behavior that was too harmless to provoke legal prosecution but still provided the pleasures of subversion. One amateur, shooting a documentary on a train, unscrewed a surface panel that connected two carriages to gain a better view of the bumpers he wanted to film in motion. When railway staff detected the loose part at the terminus, the film maker smiled at "the great fuss" he had caused and walked away "with twenty-five meters of successful film."[33] Another enjoyed defying authority by disregarding a German ban against filming railway installations during the Weimar Republic, writing in 1932 that "when I had finished my take, a voice boomed behind me. 'The taking of films is forbidden in Germany.' But that did not worry me too much, neither then nor now."[34] Filming could thus be as much about picture taking as about symbolic acts that demarcated and extended the amateur's personal domain.

Home-movie making also provided a source of pleasure because it disregarded the supposedly stifling requirements that tainted the worlds of work and business. By pursuing film as a pastime, amateurs often hoped to create a private sphere that would remain isolated from the intrusive demands of everyday life. Filming in one's spare time represented a "hobby" or a "sport" – terms that derived their meanings from an opposition to the semantics of professionalism. Since these enthusiasts did not have to take third parties' interests into account, they could freely film whatever they wished.[35] As an "antithesis to the concept of work," as one German manual from 1929 pointed out, hobbies were "free and independent from all dictates (*Notwendigkeit*)" and subject only to an amateur's turn of fancy.[36] In keeping with the Latin roots of the term "amateur," these film makers insisted that, first and foremost, they sought "enjoyment" through an occupation they cherished.[37] Spare-time directors appreciated the movie camera

---

[32] See Ulrich Linse (ed.), *Zurück, o Mensch, zur Mutter Erde: Landkommunen in Deutschland, 1890–1933* (Munich, 1983); Frank Trentmann, "Civilization and its Discontents: English Neo-Romanticism and the Transformation of Anti-Modernism in Twentieth-Century Western Culture," *Journal of Contemporary History* 29 (1994), 583–625.

[33] *Film für Alle* 3, no. 2 (1929), 37.   [34] Strasser, *Kind und Kegel*, 35.

[35] *Film für Alle* 1 (1927), cxxxiii.

[36] Andor Kraszna Krausz (ed.), *Kurble! Ein Lehrbuch des Filmsports* (Halle, 1929), 57.

[37] Kraszna Krausz (ed.), *Kurble!*, 57; *Film für Alle* 1, no. 1 (1927), 1; Robert Dykes, *The Amateur Cinematographer's Handbook on Movie-Making* (London, 1931), 45; Ernst Rüst, *Der praktische Kinoamateur* (Stuttgart, 1925), 7.

as an instrument of private pleasure that provided them with a demarcated individual sphere subject to minimal disturbance from outside.

Given the earlier proliferation of photography, the taking of amateur films was not the only hobby that enabled amateurs to document their lives by visual means. Many ciné enthusiasts, however, preferred the moving to the still image for more than one reason. To begin with, amateur film making remained the pursuit of a propertied minority before World War II because, despite the fall in prices mentioned earlier, new cameras in the mid-1930s still cost at least £10 or 270 Marks. This level of initial expenditure ensured that hobby cinematography did not become a mass phenomenon to the same extent as photography, which, according to a contemporary estimate from 1925, commanded around three million followers of varying ambition in mid-1920s Britain.[38] The virtual exclusion of the working class by dint of expense not only homogenized the camp of sparetime cinematographers along social lines;[39] it also saved middle-class film amateurs from developing fears of losing control of their medium to masses of casual film makers, who allegedly lowered the aesthetic standards of nonprofessional cinematography. Although the denigration of the "snapper" or mere "film user," who purportedly never moved "beyond the 'press-the-button' stage" (as one incensed commentator wrote with an unveiled reference to Kodak's hallmark marketing slogan), became widespread in photographic circles during the interwar years, similar complaints did not surface within the expanding ranks of amateur film makers.[40] Since embarking on a career as an amateur cinematographer set film enthusiasts apart from the broad "mass" of still photographers, film making in one's spare time could – somewhat ironically in the light of the central place of commercial film in interwar popular culture – amount to an expression of cultural elitism.

Indeed, hobby directors insisted that the very nature of their medium demanded considerable dedication in order to achieve encouraging results and, therefore, that it precluded superficial treatment. For all their effusion about their hobby's potentially pleasurable self-fulfillment, amateurs

---

[38] The prices are from representative advertisements in *The Amateur Photographer and Cinematographer*, supplement, 31 May 1933, 1, and *Film für Alle* 8 (1934), 89. It is unclear how many amateurs actually pursued still photography in interwar Britain. The contemporary estimate is from John Taylor, "Kodak and the 'English' Market Between the Wars," *Journal of Design History* 7 (1994), 29–43, here 37.

[39] Workers managed to save up to buy a film camera only with great difficulty. For a description of a rare working-class film amateur, see Kuball, *Familienkino*, vol. II, 16.

[40] *The Amateur Photographer and Cinematographer*, 21 May 1930, 450; 14 May 1930, supplement, 5; 7 May 1930, 411; 21 May 1930, 457. For an early complaint along these lines, see *The Amateur Photographer*, 11 January 1901, 23.

claimed that film making was a serious business that required considerable skill and experience.[41] Although film cameras became easier to handle after 1925, manuals emphasized that improved user-friendliness ought not to be mistaken for an invitation to start shooting random subjects in an incoherent fashion. Better amateur equipment, the argument ran, put enthusiasts into a position to focus more exclusively on story-telling and perspective instead of fretting over technical intricacies.[42] Since producing an interesting amateur film took effort, planning, and training, publications sponsored by manufacturers, and textbooks, bombarded novices with advice. Those who had mastered cinematography sufficiently to screen their films for fellow enthusiasts could expect to encounter stiff criticism that was purportedly offered in a constructive spirit. In both countries amateur film societies held meetings in which, as one participant from the 1920s recalls, members "thoroughly ripped apart" one another's productions, and periodicals such as *Amateur Ciné World* made a habit of unsparingly singling out "offending" films in their review sections.[43] If pursued in strict accordance with the advice advanced in instructional literature and clubs, hobby film making presupposed an almost masochistic willingness to embark on a sustained, open-ended, and often frustrating learning process.

Visual education took a prominent place in German and British instructional literature in order to help amateurs to identify rewarding perspectives as well as effective set arrangements. Most frequently, books of advice urged beginners to concentrate on the "essential" features in a particular image rather than obscuring motifs through an abundance of detail. Manuals also emphasized that many themes could be found in everyday contexts and therefore encouraged amateurs to train their vision by observing their surroundings.[44] Moreover, pamphlets alerted enthusiasts to the importance of sustaining a plot that increased viewers' interest in a film. Rather than a haphazard succession of individual scenes, a good amateur movie required narrative organization around a particular theme. Film makers received tips on effective film scripting and rehearsing individual scenes, and learnt about post-production techniques such as editing and titling.[45] At times, recommendations threatened to stifle spontaneity altogether when authors

---

[41] On the seriousness of hobbies, see Robert A. Stebbins, *Amateurs, Professionals, and Serious Leisure* (Montreal, 1992), esp. 1–19.
[42] Strasser, *Kind und Kegel*, v.
[43] Kuball, *Familienkino*, vol. II, 47; *Amateur Ciné World*, September 1934, 266–8; October 1934, 317.
[44] For a few such examples, see *Ciné-Kodak-Magazin*, April 1932, 5–7; Burgess, *A Popular Account*, 53; *Film für Alle* 1, no. 5 (1927), lxv; Friedrich Willy Frerk (ed.), *Jahrbuch des Kino-Amateurs, 1931–32* (Berlin, 1931), 28–9.
[45] For several examples, see Lange, *Filmmanuskripte*, 5–20; Stüler, *Filmen leicht gemacht*, 78–80; *The Kodak Magazine*, September 1931, 160; Burgess, *A Popular Account*, 23–34.

codified elaborate instructions as the "Ten Commandments" of amateur cinematography.[46] Of course, the extent to which hobby film makers actually followed suggestions in manuals varied widely, and a great number of films amounted to little more than a succession of unrelated individual shots.[47] Many writers accepted the reality that individual film amateurs pursued their pastime with differing degrees of intensity and ambition.[48] Despite making these allowances, practically all brochures, journals, and books agreed that visual training, basic script writing skills, and a passing acquaintance with editing constituted the minimum preparation that a film maker needed to have mastered in order to produce a film whose content provided more than an erratic array of single takes. In short, home movie making provided an intellectual challenge.

Individual amateurs, whether they were seeking to escape the alienation of the workplace, to carve out a niche of individual autonomy, or to demonstrate their cultural elitism by taking up a demanding spare-time activity, would have agreed that, first and foremost, hobby cinematography facilitated personal retrospection because it enabled them to record their personal histories. Instruction books and promotional material unanimously praised the camera's potential to create personal memories by capturing fleeting impressions and "preserving for To-morrow the precious moments and happy scenes of To-day."[49] One observer argued that "to look back occasionally" was particularly important "in the whirl of modern life," in which the present tended to crowd out the past.[50] From this perspective, the camera became an instrument that endowed their life stories with coherence in a world in rapid transformation. Taking up film as a hobby allowed individuals to construct personal histories that rooted them within individualized time frames. Amateur cinematography might even counteract the disruptions of "modernity" by rescuing events from oblivion, since the projector could bring back past occurrences at will. Advertising copy pressed this point: "whenever you desire they [the events] can happen again – on the screen in your home."[51] Amateur film promised to furnish people's personal

---

[46] Frerk, *Der Kino-Amateur*, 132–3. For a warning against overly rigid scripts, see Lange, *Filmmanuskripte*, 15.

[47] YFA, film no. 58, *Jubilee Celebration at Clayton West* (1936); film no. 93, *Normanton Gala* (1926); Centre of South Asian Studies, University of Cambridge, film archive (hereafter CSAS film archive), *Films by Judge Kendall* (no date).

[48] Burgess, *A Popular Account*, 7; Lange, *Filmmanuskripte*, 16–17.

[49] The quotation can be found in *The Kodak Catalogue, 1930* (London, 1930), 45. See also *Ciné-Kodak-Magazin*, October 1931, 3. Scholars have only recently taken the study of personal memories beyond oral history. On photography in this context, see Raphael Samuel, *Theatres of Memory*, vol. 1: *The Past and Present in Contemporary Culture* (London, 1994), 315–77; Eva Jantzen and Merith Niehuss, *Das Klassenbuch: Geschichte einer Frauengeneration* (Reinbeck, 1997).

[50] Hobbs, *Cinematography for Amateurs*, 14.     [51] *The Kodak Catalogue, 1930*, 45.

lives with a sense of continuity by creating *and* re-enacting their individual memories. Thus amateurs as well as suppliers agreed on cinematography's potential to assist individuals to write personal "chronicles" of their lives by building up "archives."[52]

Amateurs found that film was particularly well suited for this task, since the cinematographic medium enhanced the capacity of memory by preserving recollections and by subsequently putting them on display in a manner that surpassed other technologies. Unlike photography, which presented single moments frozen in time in static pictures, cinematography exhibited its subjects "in living images" and depicted them in all their "naturalness."[53] Rather than being restricted to the display of single, isolated poses, film, as one manual explained, captured "characteristic expressions, postures, or positions [and documented] how these singular personal features blended into one another in completely unique ways" in an individual's movements.[54] Manufacturers also praised film for its superiority over other mnemonic technologies, singling out the medium's ability to bring the flow of past events back to "life," thereby stimulating particularly vivid memories in the beholder.[55]

In addition to recommending film for its "liveliness," amateurs found cinematography attractive because manifold creative interventions allowed them to enhance the recording of their personal memories. Time and again, manuals, advice books, and periodicals impressed on aspiring amateurs that a film required a storyline to organize its visual material, or, in the words of the *Kodak Magazine*, that it had to "tell a story in an interesting and attractive way."[56] After all, a badly scripted film – one, for instance, that lacked carefully set scenes or smooth transitions between individual takes – was bound to bore viewers and frustrate the director. Unlike people looking through a snapshot album, who could turn the pages at will and focus on the shots that captured their imagination while passing over unimpressive material, film audiences were forced to sit through an amateur director's production no matter how fascinating or tedious they found its individual parts. If amateur-movie audiences were to be properly entertained, planning, scripting, rehearsing, editing, and titling formed essential stages in the production process, in addition to the actual shooting. Amateur movies

---

[52] Frerk, *Der Kino-Amateur*, 10; Lange, *Filmmanuskripte*, 21–2.
[53] *Film für Alle* 1 (1927), vi; Burgess, *A Popular Account*, 62.
[54] Rüst, *Der praktische Kinoamateur*, 7–8.
[55] DMBC, *Movector A. S.: Das AGFA-Heimkino* (Berlin, c. 1930), 1.
[56] *The Kodak Magazine*, September 1931, 160; See also *Ciné-Kodak-Magazin*, October 1932, 14; Stüler, *Filmen leicht gemacht*, 40.

*Fantasy as social practice* 209

were expected to represent tailor-made personal memories that sparked the curiosity of their audiences. At the same time, no reflection arose in amateur circles on the question whether hobby directors compromised the authenticity of their works through the various narrative and visual manipulations they carried out. On the whole, the factual subject matter and the documentary intentions of the film makers ensured that home movies were widely regarded as true visual records.

Some amateurs took this advice to heart and devoted substantial energy and imagination to rendering their productions attractive and accessible to viewers who may not have been familiar with the topics they filmed. Charles Hunter, an engineer in colonial India, began his home movie entitled *A Glimpse of India* (1928) with a particularly elaborate sequence to establish a narrative frame. The film opened with two shots of black paper cut-outs on a white background, the first depicting the silhouette of a mosque, followed by the movie title with the shapes of palm-trees on the left and a camel with its driver on the right. The next scene briefly showed off the lush vegetation of Hunter's garden before giving way to a title that informed viewers that they were about to watch a film about "HOME LIFE." Dressed in a light suit and tie, Charles Hunter then identified himself as the director of this oeuvre by talking into the camera in a casual manner, holding a cigarette in his left hand while his right rested in the pocket of his jacket. Since the viewers could not hear what Hunter was saying, the next take consisted of a banner that bade them "Good evening" and asked the audience to "imagine that we have invited you out to India." The shot of a handwritten invitation note signed by "Charlie," his wife Mary, and daughter Frances completed the opening sequence and provided the transition to the main part of the movie, which detailed the domestic environment of the Hunter family in Meerut.[57] The fact that Hunter and whoever helped him – most probably his wife – chose a labor-intensive introduction to their film, involving time-consuming titles and artistic designs, illustrates the personal importance that he attached to amateur cinematography. Hunter was by no means alone in his dedication, and memoirs abound with anecdotes about hobby directors who insisted on repeated rehearsals and shots until they were satisfied with the quality of a take.[58] Of course, all this effort did not preclude mixed results: even Hunter's opening sequence contains titles that are virtually unreadable because they flash across the screen at high speed.

---

[57] CSAS film archive, Charles Hunter, *Home Life* 1 (1928).
[58] This recollection on the part of Charles Chislett's daughter was passed on to me by Michael Howard, who screened several of Chislett's films for me at the Yorkshire Film Archive. For a similar anecdote, see Kuball, *Familienkino*, vol. II, 32.

Virtually any topic lent itself to memorialization by the home movie. The reels produced by Percy Berridge, an engineer for the North Western Railway in colonial India, were not only devoted to subjects he encountered as a tourist to the Taj Mahal and the Sikh Golden Temple in Amritsar; Berridge also captured local "curiosities" such as a "snake-charmer," and "fanatics beating their breasts till they bleed" in a religious procession, and recorded high points of the colonial calendar that were marked by viceregal visits, military parades, and horse races. Although books on technique always encouraged amateurs to restrict their activities to the production of pleasant memories, Berridge documented a devastating earthquake that cost 20,000 lives in Quetta in 1935; he covered the destroyed city, the search for bodies in the rubble, and the "piles of ashes from cremated Hindus" that awaited "despatch . . . to the sacred waters of the Ganges." Berridge did not rank among the amateurs who saw in their hobby an antidote to discontent with the workplace. On the contrary, he celebrated his job as a civil engineer in a series of heroicizing films about railway stations, tunnel inspections, and track repairs that, on one occasion, required the installation of new bridge girders 230 feet above a stream that flowed through "awesome" mountainous areas.[59] If Berridge's work provides an example of the breadth of topics that film amateurs treated in their spare time, it is atypical of the genre in one crucial respect: not once did the director capture himself or his family on celluloid in a private setting.

The majority of hobby cinematographers concentrated on films that foregrounded aspects of family life, with one manual estimating that 90 percent of movies covered domestic topics.[60] Amateurs chose their families as cinematographic subjects partly because they could effectively control the contents of their films in this setting. In addition to this largely practical consideration, the middle-class domestic realm offered an ideal backdrop for the stories emphasizing personal autonomy and self-determination that hobby film makers often strove to enact. After all, the private sphere was conventionally conceived of as detached from the professional and political concerns of the "outside" world. Male breadwinners, especially, dropped their protective guard of professional demeanor in the domestic circle and enjoyed expressing themselves in more relaxed and, by implication, more

---

[59] CSAS film archive, Percy Berridge, *Films 1–7*. The quotations are from the unpaginated shot list that is held in the archive. A copy is in the author's possession.

[60] Lange, *Wie entsteht ein Amateurfilm*, 10. See also Paul Gross-Talmon, *Filmrezepte für den Hausgebrauch: Kleine Drehbücher für jedermann* (Halle, 1939), 16.

authentic ways.⁶¹ Books of advice favored the family film as a sub-genre because it permitted explorations of the self as well as of the people closest to the amateur.⁶² To be sure, some enthusiasts found the dominance of domestic themes stiflingly repetitive, and urged their peers to expand their range of motifs once they "knew Aunt Clare's bow legs well enough."⁶³ Since, in other words, purely documentary reels did not fully satisfy amateurs' ambitions, film makers assumed the role of directors who staged family life in ways that rendered their topic worthy of the camera's attention. Rather than representing fathers, mothers, and children in mundane contexts, amateurs often foregrounded dimensions of family life that invoked idealizing notions of peaceful happiness and undisturbed bliss. Allowing relatives to play themselves, amateur cinematography attempted to create and simultaneously to capture scenes of the notional, role-defining families that people "lived by," which John Gillis has distinguished from the "actual" families they "live with" on a daily basis.⁶⁴

The topics that manuals and promotional material recommended for family films, as well as the movies themselves, bear out this tendency. Model scripts in manuals tended to focus on the relaxing and recreational atmosphere of weekends or public holidays, including scenes in which family members slept, ate cake, played games, and teased one another harmlessly. In his film about British domestic life in India, Charles Hunter not only included scenes of his family reading the daily and weekly papers on their verandah and taking tea in the garden; he also staged a sequence in which, pretending to return from his office, he hands his bicycle to a servant, dashingly jumps over a row of rosebushes, and joins his wife, with whom he takes a few steps towards the camera before kissing her.⁶⁵ On these occasions, hobby directors highlighted the family as a refuge from the pressures of daily existence. Holiday films took this theme one

---

⁶¹ For an exploration of the sanctity of family life, see John R. Gillis, *A World of Their Own Making: Myth, Ritual, and the Quest for Family Values* (New York, 1996). On men's roles in the family, see John Tosh, *A Man's Place: Masculinity and the Middle-Class Home in Victorian England* (New Haven, 1999).
⁶² Helmut Lange urged amateurs to "research" (*durchforschen*) their own lives. See Lange, *Filmmanuskripte*, 6.
⁶³ *Film für Alle* 2, no. 13 (1928), 45. See also *Film für Alle* 1, no. 12 (1927), 12; *Jahrbuch des Kino-Amateurs, 1931–32*, 28.
⁶⁴ Gillis, *A World of Their Own Making*, 7. See also John R. Gillis, "Making Time for Family: The Invention of Family Time(s) and the Reinvention of Family History," *Journal of Family History* 21 (1996), 4–21.
⁶⁵ Lange, *Filmmanuskripte*, 24–7, 41–5; Strasser, *Kind und Kegel*, 7–8; CSAS film archive, Hunter, *Home Life*.

step further, springing from what Orvar Löfgren has termed the "urge to narrate, to depict, to memorize, and communicate" the private joys the family found upon leaving the home to get away from their everyday concerns.[66] As ocean cruises came within financial reach of the middle classes in the interwar years, hobby cinematographers eagerly published reports of their film adventures on voyages to Scandinavia, the United States, the Pacific, and the Mediterranean.[67]

In both vacation and family films, children often took center stage. A host of films displayed their little subjects proudly showing off toys, frenetically jumping, running, dancing, and pursuing sports at varying levels of accomplishment. The impulse to capture one's offspring in elaborately "cute" poses did not just derive from a romanticizing childhood ideal that nostalgically equated this stage in life with innocence and unalienated selfhood.[68] Filming children also offered adults legitimate pretexts for regression, and many a father only too willingly shed his air of respectability when he played to the camera with his daughters and sons. In *Rachel Discovers the Sea* (1939) by Charles Chislett, the movie's main character, a girl of pre-school age, manages to drag her father into the waves despite the latter's ostentatious arm-waving protestations. Later on in the same reel, Chislett joins his offspring in what can only be described as a comically inept dance routine.[69] Furthermore, by visually preserving how they grew up and developed over the years, films of children narrated a family's private history. In a fictionalized account, a wife pressed her husband to buy a film camera because in the future they would want to "re-visit our past; we could see the kids again, when they were six, seven, eight or ten years old."[70] An equipment manufacturer hoped to tempt potential customers with the same line of reasoning and pointed out in a prospectus how its products would "preserve the growth of children in life-like images for days to come."[71] Family films often zoomed in on rites of passage in a child's

---

[66] Orvar Löfgren, *On Holiday: A History of Vacationing* (Berkeley, 1999), 73. See also Ingrid Thurner, "'Grauenhaft. Ich muß ein Foto machen': Tourismus und Fotografie," *Fotogeschichte* 44 (1992), 23–42.

[67] See *Ciné-Kodak-Magazin*, June 1931, 5–7; October 1931, 1–3; April 1932, 1–3; *Kino-Amateur* 5 (1932), 40–84, 133–5; *The Kodak Magazine*, August 1931, 138. Kodak even ran a tourist agency to help customers go on holiday. See *The Kodak Magazine*, May 1931, 83.

[68] On the development of this idea, see Anne Higonnet, *Pictures of Innocence: The History and Crisis of Ideal Childhood* (London, 1998), 7–14, 87–96; Carolyn Steedman, *Strange Dislocations: Childhood and the Idea of Human Interiority, 1780–1930* (London, 1995), esp. 1–20; Gunilla Budde, *Auf dem Weg ins Bürgerleben: Kindheit und Erziehung in englischen und deutschen Bürgerfamilien, 1840–1914* (Göttingen, 1994). Particular consideration to amateur films is given in Norris Nicholson, "Seeing It How It Was?"

[69] YFA, film no. 354, Charles Chislett, *Rachel Discovers the Sea* (1939).

[70] Strasser, *Kind und Kegel*, 3. [71] DMBC, *Ernemann* (Dresden: Ernemann, 1925), 29.

life, such as the "graduation" from tricycle to bicycle and the first day in school.[72] Thus casting children as symbolic embodiments of a family's history enhanced the value of home movies as documents cementing domestic intimacy and unity.

The appeal of amateur film as a tool for creating an idealized version of family life extended beyond the immediate subject matter of a production because movie making could act as a rallying point for communal family activities. To begin with, all family members were expected to be involved in shooting films, as advertisements and practitioners pointed out. As "motion picture making [was] a hobby which [gave] pleasure to all the family [and] everyone [could] join in," the family became a group of performers staging plays of domesticity with themselves in the lead roles.[73] Furthermore, families had the chance to show themselves as tightly knit social units, since films were shown in front of domestic audiences. Advertisements displayed core families gathering around a projector, harmoniously watching themselves in their latest movie (see illustration 7.2) – a scene that actually became a take in one of Charles Chislett's productions.[74] Thus camera and projector added to the set of rituals through which families enacted their unity and expressed a sense of togetherness. One manual went a step further and anchored home movie equipment so deeply in the domestic sphere that the technological artifact came to be narrated as veritable family members. This text, which provided advice in a series of fictional episodes about one family's experiments with amateur film making, cast the camera as an anthropomorphized technological object from the outset. After unpacking the instrument the husband had placed under the Christmas tree, his wife Suse "embraced the apparatus lovingly: 'Your name shall be *Kurbulus*!' she exclaimed. 'And you will always be our good and faithful friend! Cheers!' Suse and I drank to our new family member."[75] For all the initial elation which welcomed the new addition, however, it could also give rise to disturbances in the established domestic relations, as a small number of articles in amateur manuals and periodicals indicates.

While acquiring a film camera created new self-representational opportunities to document the joys of the home, introducing this technological artifact into the family could confuse the conventional domestic gender

---

[72] The change from tricycle to bicycle is captured in Hunter, *Home Life*. A still showing a child's first day at school is in Kuball, *Familienkino*, vol. II, 32.
[73] *The Kodak Catalogue, 1930*, 45. See also Stüler, *Kind und Kegel*, 87.
[74] *The Kodak Magazine*, December 1926, ii. See also Kraszna Krausz (ed.), *Kurble!*, 1; *Die große Überraschung*, 2; *Ciné-Kodak-Magazin*, October 1931, 4–5; YFA, film no. 328, *Dale Days*.
[75] The nickname *Kurbulus* alludes to the German noun *Kurbel*, meaning the "crank" of an engine. Strasser, *Kind und Kegel*, 4.

Illustration 7.2 Advertisement for a Kodak film projector. This advert cast amateur cinematography as an activity that rallied the entire family.

order. Although middle-class fathers began to assume increasingly active roles in the family during the wars, they had traditionally occupied a place "at the threshold of family life" rather than at its center, as John Gillis has written.[76] On the whole, contemporaries continued to conceive of the middle-class domestic sphere as a primarily female realm under the wife's management. By taking up home-movie making, men tended to expand their involvement in domestic affairs, and this could lead to misgivings on their wives' part. Moreover, not all wives were content to act parts in family films under their spouses' direction and, consequently, became film makers themselves to shoot their own versions of domestic life. These female initiatives, in turn, potentially challenged the position of husbands as the main technological experts around the house. In Britain and Germany, where mechanical appliances entered the domestic sphere more slowly than in the United States, amateur film equipment was among the first devices to prompt both spouses to position themselves in relation to domestic technology.[77]

Despite a humorous tone that glossed over these tensions, gender friction permeated various accounts in which women handled film cameras. Some women willingly sought male guidance while familiarizing themselves with their novel technological possession, but others set out to explore the camera's potential independently.[78] As she enthusiastically embarked on her experiment "as director of the giant film entitled *Life*," a character named Lotte began shooting scenes with her fiancé Erwin as *Life*'s main protagonist. Lotte initially defied Erwin's suggestions on several occasions, which led to a predictable outcome. After her cinematic experiment had failed miserably, she pledged to heed her lover's recommendations because "Lotte knew that she would not only win Erwin's appreciation; she herself . . .

---

[76] Gillis, *A World of Their Own Making*, 179. For an argument emphasizing male participation in domestic life more strongly before the First World War, see Tosh, *A Man's Place*, 53–142. On gender relations in the family in the interwar period, see Leslie Hall, "Impotent Ghosts from No Man's Land, Flappers' Boyfriends, or Cryptopatriarchs? Men, Sex and Social Change in 1920s Britain," *Social History* 21 (1996), 54–70; Marcus Collins, *Modern Love: An Intimate History of Men and Women in Twentieth Century Britain* (London, 2002).

[77] On these issues, see Sue Bowden and Avner Offer, "The Technological Revolution That Never Was: Gender, Class, and the Diffusion of Household Appliances in Interwar England," in Victoria de Grazia (ed.), *The Sex of Things: Gender and Consumption in Historical Perspective* (Berkeley, 1996), 244–74; Carroll Pursell, "Domesticating Modernity: The Electrical Association for Women, 1924–1986," *British Journal for the History of Science* 32 (1999), 47–67; Joachim Radkau, *Technik in Deutschland: Vom 18. Jahrhundert bis zur Gegenwart* (Frankfurt, 1989), 232–4; Ruth Schwartz Cowan, *More Work for Mother: The Ironies of Household Technology from the Open Hearth to the Microwave* (New York, 1983), 151–91.

[78] A story of male advice willingly sought is in *Kino-Amateur* 2 (1929), 29.

would also enjoy her films more thoroughly."⁷⁹ In this account, a brief challenge to male expertise quickly resulted in the reaffirmation of dominant gender hierarchies, placing the female amateur in a subordinate position as the recipient of male advice. A few other articles, however, figured more assertive female film makers, who proved more resistant to men's objections. These narratives featured women amateurs who departed from scripts that cast the domestic sphere in a purely recreational light and developed an alternative view of the home.

"Suse," for instance, the woman whom we encountered a few pages ago, who gave the family camera a nickname, employed her "*Kurbulus*" to explore the household as a site of work rather than of recreation. Suse, depicted as a "modern" woman with short hair, who smoked and wore neckties, made an amateur film that covered the entire house from the basement to the attic and documented activities such as cleaning and cooking. When she had to lengthen an extension cord to light up the family laundry room for one scene, she brushed aside the protestations of her husband, who feared she would electrocute herself. While Suse happily participated in family films taken during birthday parties and skiing holidays, she overrode her spouse's resistance and pursued her own cinematic project, which showed the domestic sphere as a world of female labor and management.⁸⁰ Another text published in a leading German amateur journal similarly encouraged women to develop their own perspectives in hobby films. In this article, a female author recounted the content of her movie about daytime life in a Berlin backyard, in which she aimed to capture the boredom as well as the toil of seamstresses working in a sweatshop.⁸¹

Even if texts about women as active amateur film makers remained few and far between in the interwar years, they serve as a reminder that the topics suggested to readers by a host of manuals and advertisements depicted leisure-time cinematography as a deeply gendered enterprise that tacitly assumed the presence of a man behind the viewfinder. After all, promotional texts, manuals, and films were mostly authored by men who imagined the family circle as a self-contained repository of domestic bliss rather than as a site of household labor and parental responsibilities.⁸² On the whole, for various reasons, this gender bias did not give rise to conflicts that would have escalated beyond the level of humorous banter. For one thing, affirming elements of traditional gender hierarchies did not

---

⁷⁹ *Ciné-Kodak-Magazin*, April 1932, 5–7.    ⁸⁰ Strasser, *Kind und Kegel*, 85–7, 111.
⁸¹ *Film für Alle* 2, no. 13 (1928), 44–5.
⁸² John Tosh has noted that English middle-class husbands tended to harbor higher expectations of the domestic space as a refuge than their wives. See Tosh, *A Man's Place*, 55–6.

necessarily stand in the way of explicit celebrations of family members of both sexes. In the opening sequence to *An Englishman's Home*, Noel Beardsell, having declared his residence a "castle," introduced himself as the "King," his wife as the "Queen," and their son as the "Prince," with each of these royals ruling over their own domestic realms: the garage, the kitchen, and the playroom.[83] Family films primarily served to create joyful memories, thereby affirming, rather than critically analyzing, conventional domestic relations. Furthermore, since harmony on the domestic "set" was a precondition for successful takes, directors of home movies usually began their films after they had secured their actors' consent to individual scenes, a step that required them to take into consideration their relatives' suggestions. Put differently, male directors had to gain the good will of their spouses to ensure that productions proceeded smoothly. Thus, at a time when middle-class "marriage . . . was being reconstructed as a partnership, a 'companionate' institution, with a conscious rebellion against 'Victorian patriarchy,'" as Lesley Hall has written, home movies allowed wives and husbands to act out a new spirit of mutuality that began to erode the traditional family ideology of separate spheres.[84] Moreover, many home-movie productions called for active cooperation between husband and wife on practical grounds: if the *pater familias* wanted to make a personal screen appearance in his work, his spouse had to take over behind the camera for several takes.[85] Women were therefore actively involved in the shooting of family movies whether these productions affirmed conventional gender notions or not. Thus a combination of wider cultural trends, practical prerequisites, and the aim of fashioning agreeable memories through amateur cinematography ensured that gender bias only rarely translated into overtly detectable conflicts between the sexes in home movies.

As movie making gained popularity as a spare-time activity after the mid-1920s, British and German amateurs, by concentrating on family films, employed their cameras mostly for similar purposes. The most striking contrast between the amateur film scenes in both countries arose from the fact that a far larger proportion of the British middle class could afford to pursue this hobby than was the case in Germany, where the First World War and the hyperinflation of 1923 had severely diminished this group's prosperity. From the mid-1930s on, however, further differences between the German and British amateur scenes emerged as the National Socialist

---

[83] YFA, film no. 73, Noel T. Beardsell, *An Englishman's Home* (c. 1935).
[84] Hall, "Impotent Ghosts from No Man's Land," 67.   [85] Kuball, *Familienkino*, vol. II, 28.

government identified new uses for spare-time cinematography in the Third Reich. What significance, then, did the Nazis attribute to amateur film?

For all their enthusiasm for the cinema, the National Socialists initially regarded the amateur film scene with considerable suspicion. The anti-individualism that permeated Nazi ideology, with its emphasis on collective organization, was bound to create mistrust towards home-movie making as a cultural practice that celebrated the autonomy of the individual. Furthermore, gleeful accounts of mildly subversive behavior hardly ingratiated amateur directors with a political movement that fetishized a public imagery of rigid discipline. Finally, some prominent members of the German amateur scene did little to inspire confidence in party circles. *Film für Alle* – the monthly publication of the League of Film Amateurs and, as such, the country's leading periodical in the field – remained under Andor Kraszna Krausz's editorship until the middle of 1935, when he was deposed because of either his Jewish ancestry or his sympathies for progressive aesthetics such as the Bauhaus movement, or both. Up to this point, Krausz had used his position not only to publish contributions by Jewish Left-leaning film critic Rudolf Arnheim as late as the summer of 1933, but also to give column space to émigré Alex Strasser, who reported on developments in Great Britain for two years after the "seizure of power."[86] Amateur film circles thus contained elements of nonconformity which, while, of course, never mounting a threat to the authorities, nonetheless signaled a will to elude the dictates of the regime.

Most importantly, the practice of amateur cinematography raised the question of who controlled the nation's visual culture, and the new rulers did not hesitate to defend claims to exclusive authority, as one German couple learnt in a rather draconian manner upon their return from a two-year work-related stay in France in 1934. Unaware of a ban that prohibited the production of private films featuring military and police units, husband and wife took turns recording a parade without a permit, an act resulting in their arrest, the confiscation of their film, and a police search of their premises.[87] This forceful assertion of state power to discipline a harmless transgression provides a prime example of how, in Detlev Peukert's words, the "all-pervasive interventionism of the Nazi regime" imported "political claims into domains that had previously been private."[88] In fact, the private nature of the medium complicated the control of hobby cinematography and, therefore, stood in fundamental tension with the government's demand for ubiquitous, unrestrained power.

[86] *Film für Alle* 7 (1933), 159; 8 (1934), 16; 9 (1935), 134–7.   [87] Kuball, *Familienkino, Band 2*, 52.
[88] Detlev J. K. Peukert, *Inside Nazi Germany: Conformity, Opposition, and Racism in Everyday Life* (New Haven, 1987), 84.

## Fantasy as social practice 219

The administration attempted to establish unequivocal authority over the hobby film scene by several measures. In February 1934, new guidelines for the screening of amateur films significantly extended the government's reach over hobby projectionists. Unlike in Britain, where amateurs required official permission to show films to a wider audience only when they tried to make a financial profit from ticket fees, in Nazi Germany all non-commercial screenings, including free events at sports clubs or charitable gatherings, were now subject to official oversight.[89] Only when arranging programs for their families or small circles of friends at their private residences did German amateurs escape the need to secure state approval for projections. To gain exhibition certificates for their films, amateurs had to demonstrate "Aryan" ancestry for two generations, subject their work to an examination by censors at the propaganda ministry, and prove membership in the Reich Film Chamber (*Reichsfilmkammer*) through one of Germany's hobby film societies.[90] These extensive censorship regulations illustrate the Nazis' wariness of amateur film as a spare-time activity that in many cases eluded the regime's direct control. While the administration, in effect, tolerated limited visual autonomy in a narrowly circumscribed private sphere, it simultaneously ensured that none of the products that might have been deemed suitable for private consumption gained unauthorized wider circulation.

In addition to placing tight reins on screenings, the National Socialists wielded direct influence over hobby cinematography's associational landscape. Since 1935, Karl Melzer, a party official who rose to the vice-presidency of the Reich Film Chamber in 1939, employed his position at the helm of the League of German Film Amateurs to restructure the amateur scene along party lines by, for instance, incorporating local clubs into regional organizations grandiloquently termed "Gaue."[91] In keeping with his own ideological preoccupations, Melzer also intensified earlier attempts to endow hobby film making with a political mission, which he outlined in a speech on the tenth anniversary of the League in 1937: "It is the purpose of amateur cinematography to enable individuals to contribute to the best of their abilities to the cultural work of National Socialist reconstruction. For this reason, it is our aim to go beyond the narrow-minded ideas that have previously characterized this organization [i.e. the League] and embrace the all-encompassing idea of the national community."[92] Under Melzer's aegis the League of German Film Amateurs worked to integrate hobby circles

---

[89] On the British regulations, see *Amateur Ciné World*, May 1934, 19.
[90] *Film für Alle* 8 (1934), 23; Helmut Lange, *Der neue Schmalfilmer* (Berlin, 1941), 367–9.
[91] *Film für Alle* 11 (1937), 242.   [92] Ibid. 11 (1937), 200.

into the Third Reich's apparatus for political indoctrination. Having introduced a separate honor for films on "German tradition and customs" in its annual competition for best home movie in 1935, the national organization for film clubs added a prize for the best self-made propaganda reel two years later.[93] This trend reached a climax early in the war when, in 1940, the League bestowed its most prestigious award upon a movie glorifying the Carinthian capital Graz as the "City of the People's Rising" during Austria's incorporation into the *Reich*. In the same year, *Film für Alle* also praised amateur film for its potential to enhance public morale in air-raid shelters.[94]

The politicization of amateur societies triggered a mixed reaction among German spare-time directors. Two years after the Nazis had gained power, the author of a manual reveled in the prospect of documenting "the difficult struggle of our people in all these years" by interweaving records of high political events with private episodes that, at times, could add a lighter touch by reminding viewers that "even that time had its bright sides."[95] As this amateur recommended blending the personal and the political in hobby reels, the proliferation of political subjects in pastime cinematography did not meet with universal approval among amateurs, as the high membership turnover in amateur associations during the regime's early years indicates. According to a survey, the League of German Film Amateurs featured a membership approaching 1,000 individuals in 1935. A year later, *Film für Alle* congratulated the national association for having registered 383 new enrolments but this recruitment drive had not increased overall membership: it remained at no more than 914 hobby cinematographers by the end of 1936.[96] Established amateurs who had been active in hobby film clubs prior to 1935 must have left their associations in droves throughout 1935 and 1936, while a regular influx of new, ideologically "reliable" enthusiasts stabilized overall enrollments. Put differently, party official Karl Melzer's rise to prominence in the League coincided with a transformation of the German amateur scene in which active support of National Socialist ideology began to play a much larger role than during the regime's first two years. Amateurs lacking either a substantial commitment to the party line or the will to tolerate the increasing ideological saturation of club proceedings left to pursue their hobby on their own from the mid-1930s. The fact was that, by confronting amateurs with ideological concerns, the politicization of hobby film societies stood in direct tension with many a spare-time

---

[93] *Ibid.* 9 (1935), 249–50; 11 (1937), 210–11.    [94] *Ibid.* 14 (1940), 114–17, 159.
[95] Heinz Lummerzheim, *Das AGFA-Schmalfilm-Handbuch* (Harzburg, 1935), 16.
[96] *Film für Alle* 9 (1935), 78; 11 (1937), 211.

director's motivation for taking up camera work in the first place. After all, they employed their hobby to carve out a niche of private autonomy, which, in turn, required them to dissociate themselves from the concerns of public life – including political affairs. By the mid-1930s, then, amateur cinematography was a medium whose intimate associations with social fantasies of individual autonomy facilitated the domestication of film technology while at the same creating, to a limited extent, a barrier against the medium's comprehensive appropriation by the National Socialist regime.[97]

When manufacturers started to market relatively inexpensive, safe, and user-friendly amateur equipment around 1925, film making became an increasingly popular leisure activity in Britain and Germany. Contemporaries took up filming as a "hobby" for various reasons. As a spare-time activity, film making served as an antidote to the pressures of everyday life under the conditions of modernity, offering a welcome release, especially from the professional and economic dictates of the workplace. Amateur cinematographers also revealed a certain social elitism in their frequent assurances that fellow enthusiasts could hope to master their medium only if they possessed an unusual degree of dedication. Beyond the pleasures of the act of picture taking itself, shooting one's own movies provided opportunities for demarcating and extending an amateur's individual domain. While amateurs produced reels on a broad range of topics, family films were particularly suited for the purpose of defining a private realm of autonomy. These small films served not only to create personal memories that documented a family's history, thereby inscribing continuity into individual life stories; they also provided rallying points for communal family activities because all family members could become involved in the production of home movies. Yet, while they encouraged amateurs to focus on the relaxing and recreational atmosphere of family life during weekends and on holidays, the idealizing and idyllic storylines that informed manuals, promotional material, and many films reveal a distinctly male view of domesticity.

Although amateur film making and its associated literature sometimes illuminated and enhanced strained gender relations within the family, promoting the camera as a tool to improve private life cast film technology in a thoroughly positive light. Rather than exhibiting the dubious and deceptive qualities that were ascribed to film as a commercial entertainment, in the context of amateurism camera work was seen as a salutary pastime with

---

[97] Numerous amateurs continued to shoot family films independent of club organizations after 1935. See Kuball, *Familienkino*, vol. II, 92–3, 97–8.

altogether beneficial effects. In several respects, public debate conceived of amateur film as a reverse mirror image of popular cinematic entertainments. Where rampant commercialism allegedly corrupted many products of the cinema industry, home-movie making preserved its moral purity because amateurs sought self-fulfillment instead of financial benefits from their hobbies; where the cinema created a worrying example of "mass" culture, amateur film promoted individualism; and where watching cinema films was considered a process of merely passive reception, producing amateur movies required active personal initiative and effort that established an intimate bond between individuals and technology. Public evaluations of amateur film indirectly threw into relief some of the antipathies to commercial film and simultaneously drew attention to film's potential for enhancing private life in seemingly non-material ways – a claim conveniently overlooking the financial costs of home-movie production. Despite its seemingly harmless and apolitical nature, hobby cinematography aroused suspicions in Germany after 1933, and it is after this date that the meanings of amateur cinematography begin to diverge in Britain and Germany. The National Socialists strove to reshape Germany's amateur film scene to ensure the regime's control over an aspect of visual culture whose emphasis on individual autonomy stood in tension with the regime's claims to all-encompassing power. By redefining amateur cinematography primarily as a propaganda instrument, however, the National Socialists drove away those amateurs who pursued camera work to escape from the pressures of everyday life into a world of private fantasies. Thus amateur cinematography's public status as a technology of private pleasure passively stood in the way of the medium's straightforward ideological appropriation by the Nazis.

Amateur film brings into view a pattern that has already emerged in earlier chapters: before National Socialism's ascent to power, British and German public attitudes towards innovative technologies exhibited far more similarities than differences. Whether in the context of the knowledge problems that arise in the wake of technological changes, the debates about physical risks, celebrations of passenger liners as luxurious "floating palaces," or amateur film – on all these occasions considerable transnational similarities existed before the rise of the Nazi government. It would, however, be erroneous to assume that noteworthy differences were absent from British and German public debates about technological change before Hitler's "seizure of power." These come into view when one accounts for a crucial factor that drove technological innovation forward: Britons and Germans were convinced that technological progress was an indispensable prerequisite for maintaining, expanding, or restoring their respective nations' power and international status. It is by turning to contemporary evaluations of technology's national importance that it becomes clear that, long before the ascent of National Socialism, political differences deriving from each country's varying political culture as well as from their positions as competitors on the international scene overarched many cultural Anglo-German similarities. The following chapter thus provides several political counterpoints to the manifold cultural similarities identified previously, allowing a comprehensive assessment of the origins of similarities and differences between British and German public meanings of technology.

CHAPTER 8

# *Technology and the nation in Britain and Germany*

It would have been hard to find any British or German commentators between the late nineteenth century and World War II who would deny the importance of technology for their nations. Technological innovations not only underpinned the competitiveness of national economies as well as both countries' military might; a large range of artifacts also became national symbols and prestige objects that signaled international leadership in a variety of engineering disciplines. Given technology's widely acknowledged national significance, drawing attention to the national benefits of an innovation provided a promising strategy to generate enthusiasm for changes in engineering. Although, as we saw chapter 4, film did not altogether shed the stigma stemming from its association with mass culture, and continued to be regarded suspiciously as a deceptive tool, in the cases of aviation and passenger liners public emphasis on the intimate connections between technology and the nation offered a rich arsenal of arguments in favor of innovation from the beginning of our period. This chapter examines the political, military, and economic arguments that were brought into play in contemporary debate to demonstrate how both countries sought to achieve and retain technological leadership in the air and on the water. Whichever specific aspect of an engineering development public eulogies chose to single out, praising the advantages that technological leadership conferred upon each country provided drives for new technologies with a sense of collective purpose, showing how innovation served the nation as a whole rather than merely furthering the interests of specific industrial sectors and certain individual pursuits. Playing up technology's national significance, therefore, engendered understandings that overcame public resistance to new artifacts and instead highlighted their promise and led technological laypersons to embrace advances, thereby contributing to a culture supportive of technological change in general. Seen from this angle, debates in both Britain and Germany have much in common: widespread agreement that technology was of national significance worked to bias the public atmosphere in favor of innovations in both countries. When we move from the results of these assessments to the meanings they ascribed to technology in Britain and Germany, however, a different picture comes to light. Since the international position and political culture of Germany and Britain varied significantly, the public in both countries arrived at divergent evaluations of the specific meanings of technology for each nation. The political climate in which public debate unfolded left a deep imprint on the national semantics of passenger shipping and aviation.

For much of our period, Britain viewed itself as a nation facing novel competition and challenges after having enjoyed the unparalleled economic and political supremacy of mid-Victorian times, when the country prided itself in being the "workshop of the world" at the heart of a worldwide empire. While warnings about national decline had occasionally surfaced during Britain's industrial and imperial heyday in mid-century, the late nineteenth and early twentieth centuries witnessed a spread of public concern about the future of the country as a global leader, given the rise of prominent rivals such as Germany and the United States.[1] Military victory in World War I displaced fears about national decline to the sidelines of public discussion, and the question how technology could assist in maintaining an empire that had reached its largest territorial expansion attracted considerable attention from the 1920s. Budgetary constraints, increased domestic social tensions arising from economic problems, and the development of vociferous independence movements, especially in India, lent further urgency to this debate.[2] The build-up of a sizeable air force in Nazi Germany compelled British politicians and the wider public to revisit an issue in the 1930s that had already exercised them before World War I: how to deal with the danger of aerial attacks, which posed an unprecedented existential threat to the island nation.[3] At times portraying the country as a

---

[1] An early warning about decline could be found even during the Great Exhibition of 1851. See Jeffrey Auerbach, *The Great Exhibition of 1851: A Nation on Display* (New Haven, 1999), 124–5. On Anglo-German rivalry before the Great War, see Paul Kennedy, *The Rise of Anglo-German Antagonism, 1860–1914* (London, 1980); Aaron L. Friedberg, *The Weary Titan: Britain and the Experience of Relative Decline, 1895–1905* (Princeton, 1988); Geoffrey Searle, *The Quest for National Efficiency: A Study in British Politics and Political Thought, 1899–1914* (Oxford, 1971). On political tensions arising in the context of technological initiatives, see Daniel R. Headrick, *The Invisible Weapon: Telecommunications and International Politics, 1851–1945* (New York, 1991), 50–116; Gregor Schöllgen, *Imperialismus und Gleichgewicht: Deutschland, England und die orientalische Frage* (Munich, 1992).

[2] David E. Omissi, *Air Power and Colonial Control: The Royal Air Force, 1919–1939* (Manchester, 1990); David Edgerton, *England and the Aeroplane: An Essay on a Militant and Technological Nation* (Houndmills, 1991), 31–2, 39–43.

[3] On pre-war fears, see Alfred Gollin, *No Longer an Island: Britain and the Wright Brothers, 1902–1909* (Stanford, 1984). The interwar period is covered in Malcolm Smith, *British Air Strategy Between the Wars* (Oxford, 1984).

beleaguered global player, British debate about technology and the nation tended to adopt defensive motifs that differed sharply from the public voices that dominated public considerations about technology's national importance in Germany.

If engineering professor Franz Reuleaux had caused considerable public soul searching in 1876 by proclaiming Germany's technological entries at the world's fair in Philadelphia as overwhelmingly "cheap and shoddy," an altogether different tone prevailed in assessments of the work of German engineers by the turn of the century.[4] In the two decades preceding World War I, the Wilhelmine public celebrated maritime and aerial products as symbols of the country's new role as a prominent international player with global aspirations. Although military defeat reduced Germany's international standing after 1918, its passenger vessels, aeroplanes, and airships succeeded in attracting worldwide attention during the interwar period.[5] Time and again, political commentators noted a discrepancy between the country's actual political and economic position in the wider world and its leadership potential as symbolized by its technological achievements. Simultaneously, public debate fueled considerable political conflict in the Weimar Republic as the Left attacked the aggressive nationalism that accompanied technological successes. After 1933, the National Socialists took up the public rhetoric that cast technology as an embodiment of frustrated national ambitions and transformed it into a language of open aggression as their rearmament effort proceeded.

On the whole, German assessments of the national significance of technology displayed more pugnacious motifs than comparable British evaluations. At no time did this contrast become more marked than during both World Wars, when propagandists employed divergent strategies to legitimate the use of advanced technology in war, a divergence that highlights the forms of militarism at work in each country.[6] As symbols of technologically

---

[4] Joachim Radkau, *Technik in Deutschland: Vom 18. Jahrhundert bis zur Gegenwart* (Frankfurt, 1989), 152–4.

[5] On the fleet construction program, see Michael Epkenhans, *Die wilhelminische Flottenrüstung 1908 bis 1914. Weltmachtstreben, industrieller Fortschritt, soziale Integration* (Munich, 1991). On aviation, see Peter Fritzsche, *A Nation of Fliers: German Aviation and the Popular Imagination* (Cambridge, MA, 1992); Guillaume de Syon, *Zeppelin: Germany and the Airship, 1900–1939* (Baltimore, 2002).

[6] On militarism in the British context, see Michael Paris, *Warrior Nation: Images of War in British Popular Culture, 1850–2000* (London, 2000); Anne Summers, "Edwardian Militarism," in Raphael Samuel (ed.), *Patriotism: The Making and Unmaking of British National Identity*, vol. 1: *History and Politics* (London, 1989), 236–56, esp. 254–6; Martin Dedman, "Baden-Powell, Militarism, and the 'Invisible Contributors'," *Twentieth-Century British History* 4:3 (1993), 201–23. On Germany, see Geoff Eley, "Army, State, and Civil Society: Revisiting the Problem of German Militarism," in *From Unification to Nazism: Re-thinking the German Past* (Boston, 1986), 85–109; Dieter Düding, "Die Kriegervereine

modern nations at war, military pilots became public icons that brought out the contrasts between British and German enthusiasm in the starkest manner.[7] In the pages that follow, an exploration of transnational similarities sets the stage for a subsequent discussion of the ways British and German views of technology took on nationally specific meanings.

## ANGLO-GERMAN SIMILARITIES IN ASSESSMENTS OF TECHNOLOGY'S NATIONAL SIGNIFICANCE

Around the turn of the century, the British and German publics began to display a heightened awareness of the international prestige that technological leadership conferred upon their nations. While the engineering feats of railway engineers Isambard Kingdom Brunel and Thomas Telford – to name but two examples – had generated national pride in Britain throughout the entire nineteenth century, fundamental shifts in the international economic and political scene underlay the stronger sense of national urgency that greeted technological innovations after 1900. In this period, by contrast with the mid-Victorian boom, Britain could no longer claim to be the only "workshop of the world," for Japan, the United States, and, closer to home, Germany developed into rivals able to match and even outperform the United Kingdom. After decades of unchallenged economic leadership, competition from these countries came as a shock to the world's foremost imperial power and immediately conjured up the specter of "decline." In the European political arena, the German government compounded British concerns by abandoning Bismarck's policies designed to maintain an international equilibrium through a series of diplomatic and military initiatives after the mid-1890s. In its grandiloquent quest for "a place in the sun," the Wilhelmine Reich sought to extend its influence in the colonial world and, in 1898, launched a battle-fleet construction program that provided the most striking example of Germany's intention of challenging British global supremacy.

im Wilhelminischen Reich und ihr Beitrag zur Militarisierung der deutschen Gesellschaft," in Jost Dülffer and Karl Holl (eds.), *Bereit zum Krieg: Kriegsmentalität im Wilhelminischen Deutschland, 1890–1914* (Göttingen, 1986), 99–121; Thomas Rohkrämer, *Der Militarismus der "kleinen Leute": Die Kriegervereine im Deutschen Kaiserreich 1871–1914* (Munich, 1990); Thomas Nipperdey, *Deutsche Geschichte, 1866–1918*, vol. II: *Machtstaat vor der Demokratie* (Munich, 1992), 230–8.

[7] In addition to the works already cited, see Michael Paris, *From the Wright Brothers to Top Gun: Aviation, Nationalism and Popular Cinema* (Manchester, 1995); George Mosse, "The Knights of the Sky and the Myth of the War Experience," in Robert A. Hinde and Helen E. Watson (eds.), *War: A Cruel Necessity? The Bases of Institutional Violence* (London, 1995), 133–42; John H. Morrow, "Knights of the Sky: The Rise of Military Aviation," in Frans Coetzee and Marilyn Shevin-Coetzee (eds.), *Authority, Identity and the Social History of the Great War* (Oxford, 1995), 305–24.

Before the First World War, Anglo-German political and economic rivalry encouraged contemporary observers to view prominent innovative artifacts as indicators of their countries' international status. The scramble for the Blue Riband, awarded for the fastest transatlantic crossing, illustrates the way highly publicized speed competitions invested passenger liners with international prestige. Previously an exclusively British affair, between 1897 and 1907 new vessels operated by the North German Lloyd and the Hamburg-American Line turned the transatlantic race into a contest between two German companies. With the German empire only beginning to transform itself into a major military and civilian naval player, these feats initially received little attention in the German national newspapers.[8] The British press, however, displayed a greater sensitivity to the issue and immediately detected an assault on Britain's leading economic position.[9] After the *Lusitania*, of the Cunard Line, had eventually regained the record for Britain in 1907, British journalists assumed an air of ostentatious self-confidence that compensated for the insecurity of previous years. While *The Times* breathed a sigh of relief that no German ship "could in any way compare" with the *Lusitania*, the Liberal *Manchester Guardian* considered British maritime pre-eminence secure for years to come:

> The *Lusitania* . . . has made a clean sweep of all the Atlantic speed records, and has proved (what, of course, we all knew) that she is the fastest liner on the ocean. She may, and almost certainly will, break her own records from time to time, but until her sister ship, the *Mauretania*, joins her in the service, she can have no possible rival . . . We shall thus have a domestic instead of an international competition, and one of much more genuine interest; for, after all, there is no particular glory in beating a German steamer of considerably inferior size and power.[10]

Against all British hopes, German shipping lines did not abandon their quest for leadership. Since technological constraints prevented the design of ever-faster boats before World War I, the Hamburg-American Line opted for increases in size and luxury when it decided to build a trio of liners between 1912 and 1914 that outstripped all previous constructions.[11] By now, passenger ships had become eminent objects of national pride in Germany because, in stark contrast to initiatives in the arena of military

---

[8] Devoting only a few lines to the crossing in which the liner *Wilhelm der Große* gained the Blue Riband, the *Berliner Tageblatt* was one of the few national newspapers to take note of the event. See *Berliner Tageblatt*, 27 September 1897 (evening edition), 4; 7 October 1897 (evening edition), 4.
[9] *The Times*, 28 September 1897, 4; 29 September 1897, 9; 7 October 1897, 7.
[10] *The Times*, 9 September 1907, 10; *Manchester Guardian*, 12 October 1907, 8.
[11] Arnold Kludas, *Die Geschichte der deutschen Passagierschiffahrt*, vol. III: *Sprunghaftes Wachstum 1900 bis 1914* (Hamburg, 1988). On the role of the North American Line in Anglo-German rivalry, see Eberhard Straub, *Albert Ballin: Der Reeder des Kaisers* (Berlin, 2001), 180–214.

fleet construction, where Britain maintained a frustrating and continuing supremacy, civilian vessels successfully demonstrated German leadership on the seas. As a consequence, the Conservative and Liberal press enthusiastically welcomed the first of the new "giants," named the *Imperator*, hailing it as both the most comfortable and "the biggest ship of the world."[12] Thus German shipping companies renewed their earlier challenge to British dominance by pursuing technological superlatives with respect to size rather than speed. Despite Germany's defeat in the Great War, Anglo-German maritime competition continued to thrive during the interwar years. When the German ship *Bremen* captured the Blue Riband in 1929, the British government agreed within a year to subsidize the construction of two new Cunard liners to counter the German initiative.[13] Of course, British and German attempts to secure national and international recognition for technological leadership were not restricted to maritime engineering. In both countries, records set by airships and aeroplanes also attracted immense attention because they commanded acclaim across national borders.[14] As the quest for superlatives provided frameworks to measure achievements in technological competitions, it also engendered permanent insecurity about a country's international status, as technological leadership could be gained and defended only through continuing exertions, thereby reinforcing and energizing the rivalries between Britain and Germany.

The status of new ships, airships, and aeroplanes as objects of national prestige rested on the premise that individual technologies demonstrated a country's creative potential. German and British shipping lines, for instance, emphasized that their vessels had been built on shipyards in their own countries and, therefore, underlined the nation's "productive power" (*Leistungsfähigkeit*).[15] The same argument surfaced in the context

---

[12] *B. Z. am Mittag*, 23 May 1912, 1; *Berliner Tageblatt*, 23 May 1912 (evening edition), 5; *Neue Preußische Zeitung*, 23 May 1912 (evening edition), 1.
[13] For a triumphalist report on the *Bremen*, see *Berliner Tageblatt*, 23 July 1929 (evening edition), 4. The decision to subsidize the new boats for Cunard was widely reported with approval. See *Daily Mail*, 2 August 1930, 9; *Daily Herald*, 2 August 1930, 1.
[14] For descriptions of airships as indicators of national leadership, see *Zeppelin fährt um die Welt: Ein Gedenkbuch der "Woche"* (Berlin, 1929), 5; Max Geisenheyner, *Mit "Graf Zeppelin" um die Welt: Ein Bild-Buch* (Frankfurt, 1929), 6, 97; *Völkischer Beobachter*, 10 May 1936, 1; *Daily Mail*, 1 August 1930, 8; *The Times*, 29 July 1930, 14; *Manchester Guardian*, 1 August 1930, 11.
[15] Deutsches Museum, Munich, archive, brochure collection (hereafter DMBC), *Dampfer Kaiserin Auguste Victoria* (Hamburg, 1906), 5. See also Deutsches Schiffahrtsmuseum Bremerhaven, library (hereafter DSB library), *"Bremen", "Europa": Die kommenden Großbauten des Norddeutschen Lloyd Bremen* (Zwickau, no date), 6; *Die Entwickelung des Norddeutschen Lloyd Bremen* (Bremen, 1912), 5. British praise for the steam turbine on the *Lusitania* as evidence of national creativity can be found in *The Times*, 12 October 1907, 10; *Daily Mail*, 21 September 1906, 4.

of aviation. One of the few agreements at which German Social Democrats, Liberals, Conservatives and, later, the National Socialists arrived after World War I concerned the appreciation of airships as quintessentially German inventions. Although the Liberal *Frankfurter Zeitung* expressly admonished its readers to refrain from arrogance after a German airship had successfully crossed the Atlantic to international acclaim in 1924, the paper hardly lived up to its own advice when it went on to comment that "the facts demonstrate [this] clearly to the entire world: efficient airships can only be produced either in Germany or according to German guidelines."[16] British aerial technology was subject to similar evaluations. When Imperial Airways presented its new *HP 42* long-distance airliners and the Empire flying boats in the 1930s, both the company and Fleet Street emphasized that "these giant machines should be, as they are, entirely of British design and construction."[17]

By generating international prestige and serving as indicators of national creativity, technological innovations came to be viewed as manifestations of national power around which Britons and Germans gathered. Providing national rallying points – or, as Peter Fritzsche put it in a different context, bringing "the idea of the nation to life" in the shape of cheering crowds – firmly anchored innovations among the material stock defining Britain and Germany as technological nations.[18] Events that celebrated individual technological artifacts and achievements commonly attracted several hundred thousand or occasionally over a million spectators. Like other festivals of national pride, such as patriotic holidays and the dedication of monuments, nationally minded celebrations of technology tended to be highly stage-managed affairs planned by members of the political and economic elites, who usually reserved the most prominent parts in the festivities for themselves.[19] With cheering multitudes signaling widespread enthusiasm, local notables, national

---

[16] *Frankfurter Zeitung*, 16 October 1924 (morning edition), 1. See also *Neue Preußische Zeitung*, 16 October 1924 (morning edition), 1; *Vorwärts*, 15 October 1924 (evening edition), 1.

[17] On *HP 42*, see *Imperial Airways Gazette*, November 1930, 1. Similar descriptions are in *The Aeroplane*, 19 November 1930, 1138; *The Times*, 18 November 1930, 11; *Daily Mail*, 18 November 1930, 12. On the Empire flying boats, see *Imperial Airways Gazette*, March 1936, 2–3; *The Aeroplane*, 30 December 1936, 819.

[18] Peter Fritzsche, *Germans into Nazis* (Cambridge, MA, 1998), 3.

[19] Charlotte Tacke, *Denkmal im sozialen Raum: Nationale Symbole in Deutschland und Frankreich im 19. Jahrhundert* (Göttingen, 1995); Stefan Ludwig Hoffmann, "Sakraler Monumentalismus um 1900: Das Leipziger Völkerschlachtsdenkmal," in Reinhart Koselleck and Michael Jeismann (eds.), *Der politische Totenkult: Kriegerdenkmäler in der Moderne* (Munich, 1994), 249–80; Alon Confino, *The Nation as a Local Metaphor: Württemberg, Imperial Germany, and National Memory, 1871–1918* (Chapel Hill, 1997); Jay Winter, *Sites of Memory, Sites of Mourning: The Great War in Cultural History* (Cambridge, 1996), 78–116.

politicians, and captains of industry portrayed themselves as the energetic leaders who introduced new technologies into their countries' material inventory. The print media and, after the Great War, newsreels and the radio instantly turned new mechanisms into national icons. In the case of shipping, launches provided occasions for a host of magnificent ceremonies, while first flights of airships and aeroplanes as well as spectacular airshows featuring "aerobatic" demonstrations, air races, and aerial joyrides, regularly attracted six-figure crowds throughout our time period.[20]

Although British and German public rituals that celebrated engineering leadership conveyed the differing political messages that we shall examine subsequently, in both countries they emphasized links between governments, the wider population, and technology. Nonetheless, the control exerted by the political authorities over patriotically minded displays of enthusiasm for technology had its limits in Britain and, before the Nazi regime, in Germany. To begin with, popular enthusiasm for technological feats sometimes swept aside the stages the political establishments carefully prepared for orderly festive events. During the 1920s and 1930s, detailed government arrangements could not prevent the tumultuous scenes that accompanied the return of British solo pilots such as Alan Cobham and Amy Johnson from successful flights across the empire, bringing London to a standstill.[21] While these disruptions were relatively harmless short-term disturbances, other occasions laid bare the significant political rifts that could exist between promoters of individual artifacts and the state authorities. In Germany, the airship owed its survival to public collections in 1908 and 1924, rather than to state support, because officials proved reluctant to fund this accident-prone technology that possessed only limited military and commercial use.[22] A similar instance of innovation against official resistance occurred in Britain in August 1935, when press baron Lord Rothermere trumpeted in his *Daily Mail* the arrival of the twin-engined *Bristol 142*, an aeroplane whose development he had bankrolled partly in protest against the government's supposedly sluggish rearmament efforts in the face of growing German aerial threats.[23]

[20] Public festivities surrounding aerial technologies have attracted considerable scholarly attention. See Fritzsche, *A Nation of Fliers*, 11–22, 136–7; Robert Wohl, *A Passion for Wings: Aviation and the Western Imagination, 1908–1918* (New Haven, 1994). Somewhat uncritical descriptions of launches of passenger liners can be found in Straub, *Albert Ballin*, 7–10; James Steele, *Queen Mary* (London, 1995), 47–52.
[21] On Cobham, who was welcomed by a million people, see *Daily Mail*, 2 October 1926, 11; *The Times*, 2 October 1926, 11; Alan Cobham, *Australia and Back* (London, 1926), 122–4. On Johnson, see *Daily Mail*, 16 May 1936, 13–14; *Daily Express*, 16 May 1936, 1.
[22] See Fritzsche, *A Nation of Fliers*, 14–17; Syon, *Zeppelin*, 124–8.
[23] S. J. Taylor, *The Great Outsiders: Northcliffe, Rothermere and the Daily Mail* (London, 1996), 308–10.

While patriotic enthusiasm generated by celebrations of and demands for technological excellence was particularly strong among Liberal and right-wing sections of the German and British public, it remained open to challenges from the British and German Left, which displayed a wariness of the links between the rhetorics of national unity, imperialism, and militarism.[24] In 1908, *Vorwärts* openly opposed the public collection for Zeppelin's airship, claiming that it served militaristic ends.[25] On other occasions before the Great War, Socialist and Labour newspapers such as *Vorwärts* and the *Daily Herald* either entirely ignored the launch of large liners such as the *Imperator*, or dealt with them in a cursory manner.[26] During the 1920s, the German Communist and Social Democratic press continued to warn of the military implications of technological progress as much as newspapers sympathetic to the Labour party.[27] In Britain the *Daily Herald* also drew attention to the dangerous military potential of aviation when it hailed pilot Alan Cobham as a national hero on his return from Australia in 1926:

> What is really deplorable is the fact that such gallant feats as those of Mr. Cobham are being checked and "graphed" by man for man's destruction; that . . . the loud-mouthed "defenders of civilisation" are working blindly for the destruction of that same Civilisation . . . while people are applauding Alan Cobham . . . their rulers are busy collating the "lessons" of his trip, with a view to making the dealing of disease, death and desolation from the air more effective. The madness of it.[28]

The Socialist end of the political spectrum thus criticized the military uses of technology while welcoming the benefits of innovations for civilians. Despite the ubiquitous celebrations of technology as national symbols among Liberals and Conservatives, political tensions lingered beneath the surface because of rival interpretations about the appropriate uses of technology both before and after the First World War. Although British and German public debate about the national significance of technology generated conflict, the intensity of political acrimony was different in each country. To a large extent, the diverging forms of political strife about

---

[24] Paul Ward, *Red Flag and Union Jack: Englishness, Patriotism and the British Left, 1881–1924* (Woodbridge, 1998); Dieter Groh, *"Vaterlandslose Gesellen": Sozialdemokratie und Nation, 1860–1990* (Munich, 1992); Nicholas Stargardt, *The German Idea of Militarism: Radical and Socialist Critics, 1866–1914* (Cambridge, 1994).

[25] *Vorwärts*, 6 August 1908, 1.

[26] See *Daily Herald*, 21 April 1913, 3. *Vorwärts* did not report on the launch of the *Imperator*, which dominated the headlines of the Liberal and Conservative press. See *Vorwärts*, 23 May 1912.

[27] For a Social Democratic example, see *Vorwärts*, 19 June 1928, 5. For Communist statements to this effect, see *Rote Fahne*, 14 April 1928, 1; 16 August 1928, 8; 12 October 1928, 7; *Arbeiter-Illustrierte Zeitung*, 47 (1930), 924.

[28] *Daily Herald*, 2 October 1926, 1.

technologies arose because public debate ascribed differing purposes to aviation and shipping in Britain and in Germany.

## EMPIRE AND DEFENSE IN BRITISH DEBATE ABOUT THE NATIONAL IMPORTANCE OF TECHNOLOGY

The imperial *leitmotif* figured as an important, long-standing theme in British discussions about technology. In the nineteenth century, telegraph cable networks proved crucial for communication between the center and the periphery, while British mail ships ensured the movement of correspondence, goods, and people. The technological leadership that had underpinned imperial leadership was most overt in the case of the Royal Navy and few, therefore, doubted that improvements in shipping and aviation were of major significance for the empire subsequently. Some proponents of flight even went so far as to state, in the words of Alan Cobham in 1926, "that the progress of flying within the Empire is of more importance to us than any other modern development." With an eye to the subsidies required by Imperial Airways, he declared categorically that "any money spent in furthering the cause of aviation is a sound investment for Britain."[29]

Mechanized flight enhanced British colonial power because it secured imperial rule over extended areas. Aircraft, as "airminded" imperialists pointed out, put the British military in a position to react to insurgent activities in remote regions without having to engage in direct combat. The pilot Jim Mollison recalled a campaign against an insurrection on India's northwest frontier to convey how the machine maintained the colonizer's superiority: "We would fly over the deserted villages, drop bombs on the mud houses, machine gun the cattle and burn the crops by means of incendiary bombs."[30] Public displays such as the annual Hendon air pageant, which often culminated in displays of bomb attacks on model "Arab" or "African" villages, drove home the same message to hundreds of thousands of spectators in the metropolis.[31] Because it could reach even the most distant areas of imperial possessions, aviation fueled hopes of "control without occupation" at exactly the time when the British empire had achieved its largest expansion and was, consequently, faced with novel

---

[29] *Daily Mail*, 1 October 1926, 9.
[30] Jim Mollison, *Death Cometh Soon or Late* (London, 1932), 68. On the extent and limitations of such policies, see Omissi, *Air Power and Colonial Control.*
[31] See David E. Omissi, "The Hendon Air Pageant, 1920–37," in John MacKenzie (ed.), *Popular Imperialism and the Military* (Manchester, 1992), 198–220.

administrative and political challenges.[32] Amid public concerns about the strategic and budgetary implications of potential imperial overstretch, flight promised a comparatively inexpensive means of policing and securing overseas territories.

In addition, the British public had long identified technological innovations as instrumental in linking the center and the periphery more efficiently, thereby facilitating imperial communications. While in the nineteenth century such considerations had concentrated on the invention of the telegraph and the opening up of new shipping lanes such as the Suez Canal, aviation occupied the focus of attention during the interwar years. Imperial Airways, of course, employed this argument at great length to justify the "mail contracts" that kept the company operational, and its public-relations efforts met with considerable interest in Britain and the empire.[33] In the summer of 1934, the company staged an exhibition of aerial landscape photographs entitled "Flying Over the Empire" in a gallery on Bond Street; it subsequently traveled to eight English provincial cities as well as to Australia and Canada, where it attracted over 100,000 spectators in Ottawa and Montreal.[34] A year later, the "Empire's Airway Exhibition" detailed the airline's global operations in museums in London and Canada, attracting even bigger audiences. The appeal of the exhibitions led the company to produce a set of 100 slides that interested individuals could hire, and to design an "Empire's Airway Exhibition Train," which toured Britain from August 1937, charging a 2d admission fee.[35] While these promotional efforts could not divert attention from mismanagement in the company in the late 1930s, when a commission of inquiry appointed by the government criticized its acrimonious industrial relations and excessive dividends, they nonetheless did much to familiarize the public with the role of aviation in imperial commerce.[36]

Finally, the language of imperial responsibility on the part of the metropolis called for active state support of flying in the interwar period. In 1924 the military establishment lobbied for the development of a sizeable peacetime air force despite the absence of an obvious European threat.

---

[32] *Parliamentary Debates: House of Commons* (hereafter *Parliamentary Debates*), fifth series, vol. 161 (London, 1923), 1608; vol. 169, 1683; vol. 180, 2201.

[33] Daniel R. Headrick, *The Tools of Empire: Technology and European Imperialism in the Nineteenth Century* (Oxford, 1981). On the postal service in the empire, see Martin J. Daunton, *Royal Mail: The Post Office Since 1840* (London, 1985), 146–90. For contemporary statements stressing the communicatory function of aviation in the press and Parliament, see Alan Cobham, *Skyways* (London, 1925), vi; *Parliamentary Debates*, fifth series, vol. 161, 1607.

[34] *Imperial Airways Gazette*, August 1934, 4; September 1934, 5; March 1935, 3; June 1936, 5.

[35] Ibid., October 1935, 4–5, 4; *The Aeroplane*, 30 September 1936, 426; August 1937, 3.

[36] Edgerton, *England and the Aeroplane*, 31–2.

Conservative MP Major-General John Seely made his "appeal in the interest of Imperial safety. It is a fact that the British Empire cannot survive if this country is destroyed. It is an Imperial question." Although adherents of Labour and Liberalism voiced concerns that such demands would initiate a new arms race and eventually lead to another world war, they did not publicly challenge the correctness of the argument's imperial dimension.[37] Emphasizing the intimate link between aviation and imperialism, politicians, pilots, and the aviation industry produced an irresistible argument for the promotion of aerial technology, reining in even pacifist critics. Furthermore, imperialists drew attention to the comparative cheapness of flight as a benefit in the maintenance of the empire, an argument in part directed at the "little Englanders" who lobbied for a withdrawal from international responsibilities out of fear of political and financial overstretch. In fact, air-minded imperialists pointed towards the economic benefits of a strong, internationally competitive airline industry, which was required to maintain imperial communications. Rather than sap precious funds from the center, an aerial service of imperial proportions would generate income, or so the public argument ran.[38]

If there was widespread agreement that state-of-the-art technology was indispensable for the empire, debates about national defense made a similar point. Seely's portrayal of Britain as a country under threat of attack and destruction was no coincidence. British promoters of aviation and shipping aimed to demonstrate that defensive rather than offensive motives underlay their demands for technological innovation. Perceptions of threats to the foundation of Britain's outstanding international position often informed public pleas to achieve or regain technological leadership. Essentially a warning against the potential military consequences of technological decline, this argument emerged around the turn of the century as maritime challenges affected the core of Britain's identity as the world's premier trading and naval power. The rise of German shipping lines alone does not account for the extensive public alarm at the time. In 1902 an American shipping trust led by financier Stanley Morgan seized control of the British

---

[37] *Parliamentary Debates*, fifth series, vol. 167, 103; vol. 169, 1669–70. See also the assessment of Alan Cobham's return flight to Australia in 1926 in the Liberal and Labour press. Critiques of militarism went hand in hand with silence about the imperial implications of the flight. See *Daily Herald*, 2 October 1926, 1; *Manchester Guardian*, 2 October 1926, 14.

[38] These argument repeatedly emerged during parliamentary debates in the 1920s. See *Parliamentary Debates*, fifth series, vol. 161, 1620; vol. 180, 2201. For a critique of this argument by Ramsey MacDonald, see vol. 167, 75–89. This debate has been explored in a wider context about British armaments policies by David Edgerton, "Liberal Militarism and the British State," *New Left Review* 185 (1991), 138–69, esp. 150–3.

White Star Line, turning public concern turned into "panic" because the change in ownership rendered it unclear whether the British Admiralty would be able to use White Star liners as troop transports in case of war.[39] Many feared that the world's foremost maritime power was about to lose economic control of its commercial fleet. In the light of the recent Boer War, during which the military had employed civilian vessels as troop carriers, the Secretary to the Admiralty presented his view of the problem to the House of Commons as a rhetorical question: "What would be the position of the Admiralty and of the country, if, in a naval war, no vessel carrying the British flag could cope with merchant cruisers as those we might find employed against us? The Admiralty had to consider the cheapest and most efficient method of meeting this menace – for menace it must be considered."[40] The need to avert a perceived threat to a central foundation of British power motivated the resolution to defend the country's technological leadership through a highly visible project by subsidizing the construction of the *Lusitania* and *Mauretania* and by stipulating that these ships had to remain under British ownership.[41]

The advent of aviation in the first decade of the twentieth century lent a particular sense of urgency to public demands to defend the nation's position through technological innovation. The nation that had deemed itself immune to military intrusion since the Napoleonic era realized with trepidation that, as German airships extended their range and French aeroplanes dominated European contests, neither its geographical character as an island nor its navy would secure Britain from external aggression in the future. Frenchman Louis Blériot's cross-Channel flight in July 1909 galvanized diffuse British anxieties and established a persistent tradition of publicly invoking the dangers of an "aerial invasion" of the British Isles that lasted through World War II.[42] A month after Blériot's feat, Conservative MP Arthur Lee charged the Government with irresponsible negligence because it allegedly disregarded the fact that "men [were] actually flying about in other countries, and Frenchmen [were] landing like migratory birds upon our shores." He came to the conclusion that Britain could "not

---

[39] For the word "panic" used in this context in 1903, see *Parliamentary Debates*, fourth series, vol. 127, 1083. For anxious debates about the wartime status of White Star liners in 1902, see vol. 106, 1181–2, 1322–4; vol. 107, 29–31.

[40] *Parliamentary Debates*, fourth series, vol. 127, 1104.

[41] Similar considerations informed the debates that led to the construction of the *Queen Mary* in the early 1930s. See *Daily Mail*, 2 August 1930, 9; *Daily Herald*, 2 August 1930, 1; *Manchester Guardian*, 2 August 1930, 11.

[42] On Blériot's flight, see Wohl, *A Passion for Wings*, 44–66. Public debate about the potential threats of aviation has been covered in extensive detail in Gollin, *No Longer an Island*.

afford to be left behind" in aviation because of the industry's importance "from the standpoint of national defence," and urged the Government to launch a British aerial program as a counter-measure.[43] The first decade of the twentieth century thus witnessed several prominent instances when technological initiatives abroad came to be viewed as existential challenges and contributed to a wider debate about potential national decline. Despite their limited military significance, German airship and bomber raids on London and other cities during the Great War confirmed Britain's prewar fears of new territorial vulnerability.[44]

Although British anxieties about the country's international decline did not reach prewar levels after Germany had been defeated, public apprehension did not entirely disappear during the 1920s.[45] After 1918, fiscal restraint kept defense budgets at low levels to help to service war debts. Furthermore, by dismantling significant portions of its armed services in general and of its air force in particular, Britain hoped to demonstrate her commitment to a foreign policy aimed at preventing a new arms race similar to the one that had exacerbated international tensions before 1914. British debate about international security often stressed the nation's commitment to the avoidance of armed conflict, not least to take into account the arguments advanced by a peace movement that had gained a strong public presence as a result of the horrors of the First World War.[46] Eyeing the implications of these policies with suspicion, Conservative opinion demanded that comprehensive technological and military leadership should underpin Britain's status as the world's prime colonial player, irrespective of financial cost. Given the strong anti-war sentiment, characterizing Britain as a nation in peril from abroad provided one of the few promising strategies to bolster calls for military innovation. In debates about aviation, claims that stressed the need to meet international challenges persisted to such an extent that some contemporaries lost patience and questioned their factual basis. One parliamentarian ridiculed an opponent who had warned against

---

[43] *Parliamentary Debates*, fifth series, vol. 8, 1574.   [44] *Ibid.*, vol. 80, 84–6.
[45] Several historians have noted the fluctuation of British fears of decline. See Barry Supple, "Fear of Failing: Economic History and the Decline of Britain," in Peter Clarke and Clive Trebilcock (eds.), *Understanding Decline: Perceptions and Realities of British Economic Performance* (Cambridge, 1997), 9–29, esp. 13–17; David Edgerton, *Science, Technology and British Industrial "Decline," 1870–1970* (Cambridge, 1996), 1–5. For a description of robust national self-confidence in interwar Britain, see Raphael Samuel, "Introduction: Exciting to be English," in *idem* (ed.), *Patriotism*, vol. I, xviii–lxvii. On British foreign-policy debates, see David Reynolds, *Britannia Overruled: British Policy and World Power in the Twentieth Century* (London, 1991), 114–25.
[46] Jon Lawrence, "Forging a Peaceable Kingdom: War, Violence, and Fear of Brutalization in Post-First World War Britain," *Journal of Modern History* 75 (2003), 557–89; Martin Ceadel, *Pacifism in Britain, 1914–1945: The Defining of a Faith* (Oxford, 1980), 62–123.

aerial threats to the nation in the European arena in 1924: "Is the danger to come from Germany? Even the most militant anti-German in this country is compelled to admit that the Treaty of Versailles has effectively disarmed Germany."[47]

The debate gained a new sense of urgency when the scale of the National Socialist armament program began to reveal itself. From 1934 on a growing chorus in Fleet Street and Parliament lobbied the Government to counter the threat emerging from the Third Reich. While secrecy rendered it difficult for British officials to assess the extent of Nazi war preparations, Hitler's announcement that the Luftwaffe had reached "parity" with the British air force in March 1935 implied that Britain now lay within the range of German bombers.[48] In the next exchange on defense in the Commons on 22 May 1935 – by which time Hitler had also proclaimed universal conscription – MPs from all parties devoted particular attention to the danger from the air. Speaking for the Government, Stanley Baldwin stressed the altogether defensive nature of British policy as he outlined plans that included a major pilot-training scheme and the extension of the armaments industry alongside morale-boosting initiatives. Recalling the horrors of the Great War, he could not imagine any British statesman "who, in the light of the experiences of 20 years ago, would dare to let loose machines of destruction by land, on the sea and in the air for selfish reasons of their own." Baldwin insisted that the nation aspired only to military parity with Germany rather than superiority, and that the Government's rearmament policies would go hand in hand with multilateral foreign policy initiatives to preserve peace.[49] Baldwin's proposals were accepted by other MPs outside the National Government, with the exception of Clement Attlee, who characterized British defense plans as "a jumble of inconsistencies in the air" because they did not integrate the British military into an international force under the command of the League of Nations.[50] While criticizing Baldwin for prioritizing national over internationalist military arrangements, however, the Labour leader never accused the Government of pursuing an aggressive policy. Moreover, Attlee's speech did not rule out support for increasing Britain's military force in principle.

Despite dissent, both Government and Opposition respected each other's commitment to the preservation of the Constitution "against all difficulties

---

[47] *Parliamentary Debates*, fifth series, vol. 169, 1677.
[48] Edward L. Homze, *Arming the Luftwaffe: The Reich Air Ministry and the German Aircraft Industry, 1919–1939* (Lincoln, NE, 1976), 98.
[49] *Parliamentary Debates*, fifth series, vol. 302, 359–73. The quote is from col. 365.
[50] *Ibid.*, 376. For an endorsement by leading Liberal Sir Archibald Sinclair, see *ibid.*, 394.

and against all dangers."⁵¹ Disagreement between the Government and Labour thus revolved around the specific policy of ensuring national defense, and did not generate potentially more polarizing accusations of anti-patriotic political behavior, which, given the weighty question at hand, could easily have come to the fore. If such challenges arose, they originated on the political Right, which charged the authorities with irresponsibly leaving the country exposed to threats prior to 1935. The *Daily Mail* and the weekly *The Aeroplane* compared Britain unfavorably with the Italian and German dictatorships on the grounds that the Government allegedly lacked assertiveness in the promotion of military aviation. By the late 1930s, however, the anti-British intentions of German foreign policy could no longer be ignored, and the admiration on the British far Right that underlay calls for emulating the Third Reich's military policy quickly collapsed.⁵²

On the whole, participants in the House of Commons debate in May 1935 shared the conviction that they were preparing for a nightmare without historical precedent: aerial warfare. Memories of the destruction wreaked during the First World War alone do not account for the apocalyptic scenarios that the prospect of renewed conflict conjured up in the British public during the 1930s.⁵³ Virtually everyone expected the next war to involve large-scale air raids on the civilian population, exceeding the horrors of the Great War, whose violence had been largely restricted to the front lines. As part of his strategy to push the defensive motivation of British rearmament to the fore, Baldwin concluded his address in May 1935 with a rhetorical flourish in which he claimed that contemplating "how we can get the mangled bodies of children to the hospitals and how we can keep poison gas from going down the throats of the people" left him "almost physically sick."⁵⁴ Churchill, whose passionate warnings about German rearmament had previously fallen on deaf ears, identified the reason for the manifest sense of confusion: "We are the incredulous, indifferent children of centuries of security . . . not being able to wake up yet to the wonderfully

---

⁵¹ *Ibid.*, 371, 373.
⁵² Charles E. Grey, the editor of the aviation weekly *The Aeroplane*, criticized the Government for not rearming before 1935. In the second half of the 1930s, he repeatedly deemed the measures taken by the Baldwin administration insufficient. See *The Aeroplane*, 1 January 1936, 11–12; 5 February 1936, c1–c2; 6 January 1937, 7–9. On the Right and aviation in more detail, see Edgerton, *England and the Aeroplane*, 47–9; Richard Griffith, *Fellow Travellers of the Right: British Enthusiasts for Nazi Germany, 1933–39* (Oxford, 1983), 137–41, 291–367.
⁵³ For an overview, see Uri Bialer, *The Shadow of the Bomber: The Fear of Air Attack and British Politics, 1932–1939* (London, 1980).
⁵⁴ *Parliamentary Debates*, fifth series, vol. 302, 372–3. On Baldwin's political strategy of further rearmament while seeking to avoid political confrontation with pacifistic opinion, see Philip Williamson, *Stanley Baldwin: Conservative Leadership and National Values* (Cambridge, 1999), 47–9.

transformed condition of the modern world."⁵⁵ Because it fundamentally dislodged the premise that an appropriate defense policy would guarantee territorial invulnerability, the technological feasibility of attacks from the skies dragged into view a topic that left even seasoned politicians struggling for words.

The prospect of air raids conjured up fearsome visions of destruction that strained the imagination of both the British political elite and the wider public. Since the Great War, growing concern about the military uses of aviation had accompanied the success of designers in increasing the range, reliability, and payload of individual machines.⁵⁶ Fear of aerial warfare intensified during the 1930s as systematic bombing campaigns against civilians became an integral element of the Italian "intervention" in Abyssinia and the German involvement in the Spanish Civil War.⁵⁷ After the British government decided to distribute free gas masks to the entire population at the height of the Czechoslovak crisis in September 1938, no one could avoid confronting the prospect of air attacks.⁵⁸ The admission that neither anti-aircraft fire nor fighter planes offered effective protection underlay the public panic about this new form of warfare.⁵⁹ In contrast to previous naval invasion scares, there existed no means of preventing air strikes against British cities. Yet it was not only the conviction that, as Baldwin had famously stated as early as 1932, "the bomber will always get through" that caused severe concern; the speed of the aeroplane, it was assumed, also allowed for surprise raids on London, which, if conducted to deliver a combination of high explosive bombs, incendiary bombs, and chemical weapons, would amount to a military "knock-out blow." London was widely considered "the Achilles' heel of [British] defence," given its population density, its function as a center of traffic and communication, its concentration of industry and services, and its relative proximity to the Continent.⁶⁰ Apocalyptic scenarios were not the preserve of sensational novels and films, but circulated among military planners, who estimated

---

⁵⁵ *Parliamentary Debates*, fifth series, vol. 302, 422.
⁵⁶ For early expressions of the fears of air raids at a time when there was no credible enemy in sight, see Lord Thomson, *Air Facts and Air Problems* (London, 1927), 79–80. Fiction also betrayed unease before 1930. See John Galsworthy, *A Modern Comedy: The Forsyte Chronicles*, vol. 2 (Harmondsworth, 1980 [1926]), 363.
⁵⁷ The attack on Guernica caused particularly strong concerns. See *The Times*, 28 April 1937, 17; 29 April 1937, 12; Paul Preston, *A Concise History of the Spanish Civil War* (London, 1996), 192–4.
⁵⁸ For an account of the reception of these gas masks as an incisive event, see "Gas Masks," in Jan Struther, *Mrs Miniver* (London, 2001 [1939]), 64–6.
⁵⁹ For a detailed examination of this issue, see *The Air Defence of Britain* (Harmondsworth, 1938), 47 59.
⁶⁰ Ibid., 33–6.

in October 1936 that the capital faced a death toll of 150,000 within a week of fully fledged air war.[61]

Once the initial state of shock had passed, experts strove to qualify their assessments of the threat. A group of left-wing Cambridge scientists, who had studied the effects of air raids in Spain and conducted experiments on the diffusion of gases, came to the conclusion that pacifistic pamphlets engaged in irresponsible scare-mongering when they stated that "a single bomb . . . may lay waste to an area of a quarter mile, and burn, poison, or dismember every living thing within it." After a review of available ordnance, the scientists summed up that "the greatest danger is from high explosive bombs, incendiary bombs are of less importance, and the effect of most gas bombs on cities is as yet unknown."[62] Widespread fears about the lethal risks of gas attacks also struck several commentators as exaggerated, since chemical agents would dissipate too rapidly to cause large numbers of casualties.[63] A high-ranking officer of the Royal Air Force, for his part, drew attention to the low accuracy of bombing campaigns.[64] While some argued that the inadequacies of the aiming devices made it unlikely that attacks would obliterate the infrastructure of the nation's capital, other commentators responded that inaccurate raids on strategic targets only made civilian casualties inevitable.[65]

In this context, the establishment of a civil-defense authority in November 1938, under government minister Sir John Anderson, managed to alleviate some fears. In a widely distributed booklet providing an overview of the government-sponsored Air Raid Protection program, the *Daily Telegraph* journalist William Deedes praised the administration's measures because they prevented "every attempt by the enemy to disorganize" the country. Besides discussing the distribution of steel air-raid shelters for low-income families, Deedes covered the training of air-raid wardens, emergency plans for hospitals, and preparations for evacuation.[66] When it uncovered severe gaps between working-class and middle-class areas in expenditure on public shelters, the Left found many social disparities to criticize in government provision.[67] In contrast to 1936, however, in late 1938 few expected the

---

[61] This is the figure given by the Chiefs of Staff Joint Planning Committee in October 1936. See Reynolds, *Britannia Overruled*, 123.
[62] *Fact, No. 13: Air Raid Protection by Ten Cambridge Scientists* (London, 1938), 20, 30.
[63] *The Air Defence of Britain*, 125–7. This had already been argued earlier. See J. M. Spaight, *Air Power and the Cities* (London, 1930), 5.
[64] *The Air Defence of Britain*, 45.
[65] Marcus Samuel, *Aims of Aerial Bombardment: Reprinted from the "Wandsworth Borough News"*, *1 July, 1938* (no place, no date).
[66] William F. Deedes, *A. R. P.: A Complete Guide to Civil Defence Measures* (London, 1939).
[67] *Fact, No. 13: Air Raid Protection*, 58.

country to be "knocked out" by a surprise attack. As Deedes noted, "it is becoming increasingly certain that we shall be physically invulnerable against the worst air raids imaginable; not secure individually, but invulnerable nationally." In addition to focusing on the physical aspects of air warfare, debate now also turned to "the psychological effects of air raids," once the notion of a drawn-out war of attrition, in which Britain herself would seek to undermine public morale and economy in Germany through bombing, began to dominate public expectations.[68] Between the identification of the threat posed by Germany in 1935 and the outbreak of hostilities, then, a formula emerged that uneasily encapsulated Britain's reshaped ethos of military defense while, to put it in Richard Overy's words, an "element of the unknown" ominously continued to linger over the question of aerial warfare.[69] The island nation, it was felt, would withstand outside aggression as a result of government planning and psychological strength, although unprecedented numbers of civilians would inevitably perish during air attacks. Admitting that technological advances had transformed a long-standing fear into a substantial reality by incorporating British territory into the battlefield marked a significant departure from previous notions of national defense and revealed a deeply disturbing aspect of the much-lauded "conquest of space" through the aeroplane.

British assessments of technology's significance for the nation, then, possessed several distinctive features. By emphasizing that innovations consolidated the empire in a cost-effective manner, promoters of shipping and aviation defined Britain as a colonial power with a vital interest in technological change. Furthermore, they stressed that various challenges from European countries compelled the nation to adopt suitable measures to preserve its international leadership position. According to numerous contributors to public debates, Britain reacted to technological developments abroad that called its international standing into question. British efforts to reach, and stay at, the forefront of technological innovation could be publicly interpreted as attempts to preserve and stabilize an international *status quo* that appeared to be jeopardized by initiatives in other countries. From the mid-1930s, the danger of an air war launched by Nazi Germany provided the most dramatic threat, leading to widespread public anxiety and to concrete war preparations. By highlighting the defensive motives behind national drives for technological innovation, its proponents argued

---

[68] Deedes, *A. R. P.*, 26–7. For considerations along these lines, see Spaight, *Air Power*, 14; *The Air Defence of Britain*, 114.
[69] Richard Overy, *The Air War, 1939–1945* (London, 1987), 26.

that they were responding to external pressures for the sake of the nation rather than instigating international competition. Despite charges of irresponsible inactivity from the Right, this stance largely silenced potential Liberal and Labour critics, helping to maintain a fragile public consensus on the political need to innovate. On the whole, British debate considered the preservation rather than the transformation of the international *status quo* as the prime objective of technological innovation. If Britain defined itself as a modern nation by drawing attention to its advanced technologies, this modernity bore a deep imprint of historical continuity which, paradoxically, presupposed a persistent political will to embrace technological change. At the same time, the motifs framing public debate about technology's national importance shifted from an emphasis on counteracting potential technological decline before World War I to combating an existential threat to the leading global power that arose as a result of the dynamic development of aviation. How do these features compare with German assessments of the significance of technology for the nation?

## AGGRESSION AND CONFLICT IN GERMAN DEBATE ABOUT THE NATIONAL IMPORTANCE OF TECHNOLOGY

Given the later acquisition, smaller size, and subsequent loss in 1919 of its overseas possessions, imperial themes played a less important role in technological debates in Germany than in Britain.[70] Far more commonly, speeches and writing about passenger shipping and aviation during the Wilhelminian period emphasized that technological achievements bore out Germany's claims to status on the international stage. When, for example, the Hamburg-American Line celebrated the beginning of the new century with the launch of its liner the *Deutschland* on 10 January 1900, German Chancellor Count Bülow's speech at the launch argued that progress in marine engineering underlay both the nation's economic rise and its new bids in colonial and world politics.[71] Asserting that Germany had become a "world power" (*Weltmacht*), a Liberal Reichstag member similarly reasoned that marine engineering demonstrated the "unimaginable change" (*ungeahnter Wandel*) of the country's international position. In the previous ten or fifteen years, he went on, German shipyards had risen to "rank among the most powerful companies in the world," and German shipping lines consequently ceased to order their vessels from Great Britain.[72] Passenger

---

[70] On the airship as a technology of German colonial rule, see Fritzsche, *A Nation of Fliers*, 37.
[71] *Berliner Tageblatt*, 10 January 1900 (evening edition), 4.
[72] *Verhandlungen des Reichstags*, vol. 288, 4169.

liners made in Germany objectified the country's status as a world power for Conservatives and Liberals alike during the Wilhelmine era. Airships also came to be seen as heralds of German international leadership. In contrast to Britain, where technological innovations in the maritime and aerial sectors triggered anxieties about decline, in the Reich such innovations served as harbingers of national self-confidence.

The outcome of the First World War as well as the stipulations of the Versailles Treaty severely shattered Germany's dreams of global power, which had thrived before 1918 but lacked a long-standing tradition. As Wolfgang Schivelbusch has recently written, in 1918 the defeated nation foresaw falling back into "the abyss of a century-old German misery that victory in 1870/71, the foundation of the Reich, and forty years of power politics appeared to have overcome."[73] The Versailles settlement reinforced widespread public alarm as German lines had to transfer virtually their entire fleet to French, British, and American competitors to compensate for Allied losses from German submarine attacks during the Great War. In addition, the treaty imposed severe restrictions on German aviation. While lengthy negotiations led to the repeal of a ban on airship construction as well as to many regulations concerning civilian aeroplane design in 1926, Germany failed to gain permission to maintain a military air force throughout the Weimar years. The German public repeatedly attacked the technological consequences of the peace for creating an international environment hostile to German aspirations, and reacted by intensifying the narratives of national tenacity that had already shaped a range of tales about technological success during the imperial era.

Promotional literature issued by public-relations departments interacted with Conservative political "propaganda" to generate support for political initiatives that would restore Germany's ability to engage in unrestricted technological development. In 1929, right-wing intellectual Ernst Jünger edited an expensively produced volume for the pressure group *Deutscher Luftfahrerverband* (German Aerial Association) in which former government minister Alexander Dominicus predictably flagellated "the exceptional disadvantages" for German aviation that arose from the ban on military flying.[74] Three years later, Lufthansa sponsored a book featuring contributions by academics, retired war pilots, government officials, and lobbyists, who argued that Germany was forced to travel an exceptionally

---

[73] Wolfgang Schivelbusch, *Die Kultur der Niederlage: Der amerikanische Süden 1865, Frankreich 1871, Deutschland 1918* (Berlin, 2001), 235.
[74] Alexander Dominicus, "Geleitwort," in Ernst Jünger (ed.), *Luftfahrt ist not!* (Leipzig, 1929), 5.

"thorny road" (*Leidensweg*) towards economic and technological recovery.[75] Public-relations material of German shipping lines argued along similar lines.[76] The Conservative press employed metaphors of physical mutilation to convey a myth of victimization when covering technological achievements. One Conservative daily considered the country a "slave of the world whose sinews [had been] severed to beat him into submission," while the organ of the Nazi party wailed that the German aviation industry had been "gagged and raped by the shameful dictates of Versailles (*Versailler Schanddiktat*)."[77] Although the Social Democratic and Liberal press abstained from comparably dramatic language that cast Germany as a tortured martyr, they agreed in principle about the unjust character of the peace settlement. The focus on the obstacles – real and imagined – that Germany faced was forceful enough to sideline the indisputable successes enjoyed by German enterprises during the 1920s. After all, by 1930, Lufthansa had established itself as Europe's leading airline by a wide margin, a German airship crossed the skies around the globe, and the North German Lloyd Line had regained the Blue Riband.

Given the widespread myth of victimization that claimed that German technological innovations had to be achieved in an unfavorable international setting, the press employed the motif of "German" resistance and perseverance in the face of adversity when covering aerial and naval events. The most eloquent rhetorics of national resistance were devoted to airships. The peace treaty originally contained clauses that completely banned airship construction in Germany and threatened the existence of the Zeppelin works in Friedrichshafen. Hugo Eckener, the director of the company, saved the enterprise by negotiating with the US government to accept a new vessel as a reparation. The transfer to the United States of the first German airship built after the Great War, the *ZR 3*, sparked off a host of tales about the national significance of this event in October 1924. When news reached Germany that the craft had arrived at its destination, the moderate Liberal press rejoiced at this "peaceful conquest:" "The German name, disregarded abroad during the long war and postwar years, has a new ring due to this feat of genius."[78] Another article considered the delivery of the vessel to be

---

[75] Wolfgang von Gronau, "Vorwort," in J. B. Malina (ed.), *Luftfahrt voran! Das deutsche Fliegerbuch* (Berlin, 1932), 7.
[76] Hamburg-Amerika Linie, *Wiederaufbau der deutschen Überseeschiffahrt* (Berlin, 1923); DSB library, Norddeutscher Lloyd Bremen, *Columbus* (Hanover, no date), 5.
[77] *Neue Preußische Zeitung*, 16 October 1924 (morning edition), 1; *Völkischer Beobachter*, 15–16 April 1928, 1.
[78] *Berliner Tageblatt*, 15 October 1924 (evening edition), 1.

a "proof of German ability and German will (*Beweis deutschen Könnens und deutschen Wollens*)."⁷⁹

Liberals demanded that initiatives to reform the Versailles settlement should take diplomatically acceptable, peaceful forms. The right-wing language of national resistance, however, resonated with an aggressive revanchism that cast doubts over the Conservatives' commitment to avoid armed conflict. After the *Graf Zeppelin* had circled the globe for the first time in 1929, the Conservative weekly *Die Woche* used the occasion to congratulate Germany on a successful revanche for military defeat. Although the "unprecedented struggle against . . . the entire world" had led to German surrender after "heroic battles" in 1918, the nation's "will and ability [were] unbroken." The airship's world tour finally brought redemption from defeat because it established Germany as a worthy "victor."⁸⁰ According to *Die Woche*, national resistance to international sanctions had proved successful. Further on the Right, *Neue Preußische Zeitung* invoked the obstacles against which German engineers "pondered and worked, fought and gained victory . . . Drenched by hatred and defamation, surrounded by inspectors and spies, exploited and maltreated, we have created a technological wonder under the most difficult conditions."⁸¹ In 1928, the organ of the Nazi party labeled German airships as "a simile of the German fighting spirit."⁸² On the right of the political spectrum, myths of victimization thus gave rise to a variant of technological nationalism that hailed engineering successes in aggressive rhetoric as an effective form of national resistance against unfair international sanctions.⁸³

The martial overtones on the Right provoked criticism from the Left. Although Social Democrats readily agreed on the fundamental injustices of the peace settlement, they maintained that uncompromising and confrontational "clamor" (*Gebrüll*) harmed Germany's international position. Radical nationalism, *Vorwärts* argued, smacked of militarism and played into the hands of Germany's opponents, providing them with additional reasons to insist on severe restrictions on the national economy. Furthermore, Social Democratic anti-militarism ruled out any alignment with the Right: "We decline any coalition with these intellectually and mentally

---

⁷⁹ *Frankfurter Zeitung*, 16 October 1924 (morning edition), 1.   ⁸⁰ *Zeppelin fährt um die Welt*, 5.

⁸¹ *Neue Preußische Zeitung*, 16 October 1924 (morning edition), 1.

⁸² *Völkischer Beobachter*, 17 October 1928, 1. For a similar statement in a trade journal, see *Luftfahrt*, 28 (1924), 276.

⁸³ For a comparative examination of the link between German myths of victimization and the language of aggression in a different context, see Rudy Koshar, *From Monuments to Traces: Artifacts of German Memory, 1870–1990* (Berkeley, 2000), 38–41.

impaired people (*diese geistig Enterbten*)."[84] When erstwhile fighter pilot Hermann Köhl received a triumphant welcome in Berlin after the first successful east–west transatlantic crossing by aeroplane in April 1928, the Social Democrats were quick to acknowledge the national significance of the feat. At the same time, they remained eminently critical of Köhl's membership of the staunchly anti-Republican *Stahlhelm* veterans' association. Writing about a proposed *Stahlhelm* reception in his honor, the SPD daily considered Köhl and his crew to be "ill advised to place themselves at the center of a [celebration] which is organized by a reactionary and subversive association."[85] Such tensions were only indicative of the deeper rifts between the Left and aviation circles. During the 1920s, most flying clubs, for instance, not only functioned as rallying points for former war pilots who typically denounced the Republic; they also organized covert military training for civilian airline pilots as part of secret rearmament schemes. Given these intimate links between anti-Republican circles and nationally minded promoters of aviation, it is hardly surprising that the Social Democrats maintained their own flying clubs with a membership of 20,000 in the late 1920s.[86] If the SPD found it difficult to take pride in aerial achievement, the Communist press sometimes denounced aerial technologies altogether as "symbol[s] of the reawakening of German imperialism," charges that triggered complaints from the Right about "circles totally infected by party politics."[87] German assessments of the national roles of technology thus reflected and enhanced wider ideological divisions. Unlike in Britain, where political tensions over aviation often remained moderate, in Germany open conflicts arose on a national level as the Left challenged Liberal and Conservative outlooks.

Despite considerable political disagreement over national and international issues, Liberals and adherents of the anti-democratic Right shared a common cultural ideal that infused their versions of nationally minded enthusiasm for technology with a staunch spirit of national resistance and resilience. Their admiration for feats of engineering reveals that they subscribed to similar codes of heroic masculinity. Liberals and Conservatives, and later, National Socialists, favored models of manliness that displayed

---

[84] *Vorwärts*, 14 October 1924 (morning edition), 1.   [85] *Vorwärts*, 19 June 1928, 5.
[86] Homze, *Arming the Luftwaffe*, 12, 32; Nils Ferberg, "'Sturmvogel-Flugverband der Werktätigen E. V.' Zur Geschichte einer Arbeitersportorganisation in der Weimarer Republik," *Internationale wissenschaftliche Korrespondenz der deutschen Arbeiterbewegung* 30 (1994), 173–219.
[87] *Rote Fahne*, 12 October 1928, 7. See also *ibid.*, 14 October 1924, 3; *Verhandlungen des Reichstags*, vol. 425, 2452, 2484. The Conservative counter-charge is from *Neue Preußische Zeitung*, 14 April 1928 (evening edition), 1.

an unflagging commitment to aims that had to be achieved despite seemingly insurmountable obstacles. Public assessments of Count Ferdinand von Zeppelin, the most prominent inventor of German airships, bear this out. From the inception of airship construction before the Great War, the figure of Count Zeppelin publicly personified the conviction that persistence and patience would eventually overcome the most improbable odds – in his case, numerous crashes and explosions. Although one Bavarian general condemned Zeppelin's plans as "rubbish" (*Schmarrn*), the Count disregarded such contempt and subsequently rose above ridicule and insult.[88] Descriptions of the inventor changed after a two-day long-distance flight over south-west Germany that was brought to an abrupt end when the vessel exploded outside the village of Echterdingen in 1908. In his steadfast refusal to give up, Zeppelin became a national icon. According to one Conservative newspaper, here was a man "who completely lives up to ideals of German heroism, a man who devotes great energy to a great enterprise and who will not be diverted despite obstacles, resistance, and misfortune."[89] Liberal publications echoed this theme: "Those familiar with Count Zeppelin know that no buffet of fate can discourage him . . . We have to contradict most vehemently those arguing that this last blow . . . has ended the Count's quest. Proponents of this view know neither Count Zeppelin, nor the significance of his enterprise, nor the German people."[90] As noted above, a national collection restored active production of the airships.

During the First World War, the example of Zeppelin, who died in 1916, was invoked to demonstrate how male heroism ensured national survival in times of struggle through its resilience and unshakeable strength of character: "Being like the Count means being German."[91] Count Zeppelin continued to command prominence after the war because, the argument ran, emulating his determination would effect Germany's national regeneration. With an eye to the regulations imposed by the Versailles Treaty, one Liberal paper pointed out that his "nearly legendary figure . . . can teach the German people that one achieves one's aims if one believes in them firmly."[92] In 1929, *Die Woche* thanked the crew who had steered the vessel *Graf Zeppelin* around the world for their display of virility: "The airship

---

[88] *Frankfurter Zeitung*, 5 August 1908 (evening edition), 1.
[89] *Neue Preußische Zeitung*, 6 August 1908 (evening edition), 1.
[90] *B. Z. am Mittag*, 6 August 1908, 1. For similar rhetoric, see *Graf Zeppelin und sein Luftschiff: Luxusausgabe* (Nuremberg, 1908), 7–9; Peter Hoogh, *Zeppelin und die Eroberung des Luftmeeres* (Berlin, 1908), 202.
[91] Adolf Saager (ed.), *Zeppelin: Der Mensch, der Kämpfer, der Sieger* (Stuttgart, 1915), 11. See also *ibid.*, 96–7, 105–6, 120; Arnold Jünke, *Zeppelin im Weltkriege* (Leipzig, 1916), 7.
[92] *Frankfurter Zeitung*, 21 August 1925 (morning edition), 1.

men have shown us that only those who give up are doomed."⁹³ After 1933, National Socialist propaganda took this Liberal and Conservative rhetoric to even higher levels, emphasizing Count Zeppelin's "superhuman stamina."⁹⁴ The heroic figure of Count Zeppelin as a symbol of national regeneration commanded admiration across profound party political and ideological divisions, thereby establishing cultural reference points between Liberals, Conservatives, and National Socialists.⁹⁵

After 1933, National Socialist publications radicalized the emphasis on national resistance in reports about engineering achievements and turned this motif into a core element of their propaganda about technology.⁹⁶ Casting themselves as champions of a "thoroughly modern lifestyle based on the most recent technological advances," the Nazi party took office promising to pursue a "gigantic" modernizing project designed to turn Germany into a highly efficient industrial society at the forefront of the international world.⁹⁷ Motifs of national modernization permeated the public portrayal of a variety of technological initiatives, including the construction of factories, highways, cars, ships, and aviation.⁹⁸ In the context of flight, industrial public-relations departments and the state propaganda apparatus cooperated closely to trumpet new developments as evidence for the country's dynamic transformation, and heralded Hitler and Hermann Göring, the Minister for Aviation, as the "saviors of the aerial cause."⁹⁹ In the party press, Lufthansa attracted specific attention for its increasing passenger figures and an expanding international service that operated a weekly airmail flight to South America from 1934.¹⁰⁰ With respect to airships, the propagandistic

---

⁹³ *Zeppelin fährt um die Welt*, 5.
⁹⁴ Fritz Dettman, *Zeppelin: Gestern und Morgen: Geschichte der deutschen Luftfahrt von Friedrichshafen bis Frankfurt am Main* (Berlin, 1936), 12. See also Ernst Lehmann, *Auf Luftpatrouille und Weltfahrt: Erlebnisse eines Zeppelinführers* (Leipzig, 1936), 146.
⁹⁵ George Mosse noted similar overlaps between conventional and National Socialist ideas about masculinity. See George Mosse, *The Image of Man: The Creation of Modern Masculinity* (New York, 1996), 180.
⁹⁶ For an argument emphasizing the importance of radical nationalism in National Socialism, see Hans-Ulrich Wehler, "Radikalnationalismus – erklärt er das 'Dritte Reich' besser als der Nationalsozialismus?" in *Umbruch und Kontinuität: Essays zum 20. Jahrhundert* (Munich, 2000), 47–64.
⁹⁷ The phrase is taken from Otto Dietrich, *Mit Hitler an die Macht: Persönliche Erlebnisse mit meinem Führer* (Munich, 1934), 65, 72.
⁹⁸ On economic modernization, see Mary Nolan, *Visions of Modernity: American Business and the Modernization of Germany* (New York, 1994). On cars and highways, see Erhard Schütz and Eckhard Gruber, *Mythos Reichsautobahn: Bau und Inszenierung der "Straßen des Führers", 1933–1941* (Berlin, 1996).
⁹⁹ See the volume issued to celebrate Ernst Heinkel's fiftieth birthday, *Kameradschaft der Luft* (Berlin, 1938), 123. Göring received similar praise in *Die Luftreise*, February 1934, 28–32.
¹⁰⁰ For examples, see "10 Jahre Deutsche Lufthansa," in *Völkischer Beobachter*, 5 January 1936, 9; 13 August 1938, 1.

coalition between party and industry focused on the new vessel *Hindenburg*, which secured a coveted "triumph for German technology" in 1936 when it completed its first North Atlantic crossing.[101] Apart from serving as tangible proof of Germany's rapid modernization, aviation technologies became prominent during domestic political campaigns. In the spring of 1936, the airships *Graf Zeppelin* and *Hindenburg* cruised all over Germany to mobilize public support for the referendum on National Socialist rule, an initiative repeated during the opening ceremonies of the Olympic Games later that year.[102] Notwithstanding their penchant for racially charged, ruralist, *Blut-und-Boden* imagery, the Nazis publicly claimed to be forging a symbiotic relationship between the nation and technology.

In order to extend their appeal beyond the immediate party clientele, many National Socialist propagandists employed similar codes and motifs that had shaped public visions of technology before 1933. At the same time, Nazi writers also modified the familiar theme of national resistance in accordance with the racial ideology that lay at the heart of their worldview. They underlined the importance of determination or, in contemporary terms, a strong "will" (*Wille*) as the prerequisite to ensure national survival in an international environment characterized as a Darwinist struggle between individual races. Technology, as an article in *Nationalsozialistische Monatshefte* pointed out in 1935, expressed a race's "will to rule" as well as its "dynamism." If the introduction of racial terminology signaled an ideological departure from established forms of national ardor for things technological, National Socialist commentators further reminded the German public that determination alone did not suffice to secure Germany's existence. The "will to live" (*Lebenswille*) had to manifest itself in the form of specific technologies that would enable the country to assert itself forcefully in conflicts.[103] It was coverage of the rearmament effort that showed most clearly what kind of a new era had dawned.

In a protracted attempt to legitimate their aggressive armament drive as a quest for "national emancipation" (*Gleichberechtigung*), the regime publicly orchestrated the military build-up as an act of self-determination forced upon Germany by European states that were reluctant to dismantle their own arsenals.[104] Insisting on the "liberty to be armed" (*Wehrfreiheit*) – a

---

[101] Ibid., 5 March 1936, 1; 15 May 1936, 2.
[102] For a detailed account of this tour, see Lehmann, *Auf Luftpatrouille und Weltfahrt*, 359–67.
[103] Fritz Nonnenbruch, "Technik, Wirtschaft und Staat," in *Nationalsozialistische Monatshefte*, 6 (1935), 639–43, esp. 640. See also Heinz Orlovius (ed.), *Schwert am Himmel: Fünf Jahre deutsche Luftfahrt* (Berlin, 1940), 36.
[104] For the term *Gleichberechtigung* in this context, see Orlovius, *Schwert am Himmel*, 18; *Kameradschaft der Luft*, 121.

phrase invoking ancient "Germanic" freedoms sanctioned by "history" – military expansion purportedly played an integral part in the Führer's "peace policies," which were designed to mislead international opinion and simultaneously to reassure the German population that the regime was not leading the country into another war.[105] In this context, works of propaganda argued that, given the dominant doctrine of "unlimited aerial warfare," Germany had no alternative to developing its own defensive capabilities.[106] Yet, even as the National Socialists justified their arms build-up publicly as a legitimate measure of defense, the hyperbole and aggression in their heated rhetoric were bound to cause as much alarm in other countries as the concrete weapons of the Luftwaffe themselves. Towards the end of the 1930s, the Nazis no longer sought to hide the aggressive revanchism that stood behind the rearmament program. In 1939, Göring, for instance, took pride in the fact that "National Socialist Germany has fought to attain a place in the sun despite its poverty and the material riches of its opponents. It will continue this fight."[107] The aviation minister also pointed out that the annexation of Austria and of the Sudetenland had been accomplished only because, "during those fateful days . . . the most modern air force a nation can possess" led the international world to react with "utmost caution."[108] As the armament effort proceeded, the Nazi leadership replaced its temporary rhetoric of defense with open gestures of aggression.

Although National Socialist propaganda underlined the crucial role of the air force in providing protection for the Reich in a future war, it left the public in no doubt that the Luftwaffe was incapable of preventing aerial attacks against German territory. Fear of air raids had already loomed large during the Weimar Republic.[109] Less than two months after coming to power, the new regime established the Civil Defense League (*Reichsluftschutzbund*), an organization that grew into one of the largest mass

---

[105] For the term *Wehrfreiheit*, see *Kameradschaft der Luft*, 123; "Aufbau der deutschen Luftfahrt," in *Motor und Sport*, 26 April 1936, 17; Alfred Rosenberg, "Im Zeichen von 1936," in *Völkischer Beobachter*, 1 January 1936, 1. The phraseology of rearmament as part of a wider "peace policy" informed Hitler's speech announcing the reintroduction of universal conscription, which was noted in the debate in the House of Commons referred to above. The speech can be found in Paul Meier-Beneckenstein (ed.), *Dokumente der deutschen Politik*, vol. III (Berlin, 1937), 68–99, esp. 80–4, 89.
[106] Hermann Göring, "Geleitwort," in Dr. Kürbs (ed.), *Die deutsche Luftwaffe* (Berlin, 1936), 5; *Völkischer Beobachter*, 1 January 1936, 2; Orlovius, *Schwert am Himmel*, 36.
[107] Hermann Göring, "Vorwort," in Heinz Bongartz, *Luftmacht Deutschland: Luftwaffe – Industrie – Luftfahrt* (Essen, 1939), vii–xiv, here xiv.
[108] Orlovius, *Schwert am Himmel*, 36.
[109] See Hans Rumpf, *Gasschutz: Ein Handbuch* (Berlin, 1928); Richard Roskoten, *Ziviler Luftschutz: Ein Buch für das deutsche Volk* (Düsseldorf, 1932).

organizations, with 13 million members and 3 million trained air-raid wardens by 1938.[110] With its broad social base, the Civil Defense League became an instrument of social control as its wardens oversaw drills, ordered people to clean out their attics, watched over blackout procedures, and directed building works. Whoever sought to resist these intrusive measures was liable to be charged with endangering the nation through negligence.[111] Most importantly, civil-defense preparations formed a central element in the psychological preparation for war. Virtually every publication on the subject in Nazi Germany pointed out that air raids sought to destroy not only a nation's military and economic base but also its "morale."[112] Public confidence during armed conflict proved a highly sensitive topic, since, like many on the Right, the Nazis fostered the conviction that Germany owed its defeat in the First World War not to military blunders and economic exhaustion but to the fact that the population, "unnerved" by four years of material shortages and enemy propaganda, had abandoned its heroic army. According to this interpretation, military disaster in 1918 resulted from a dissolution of morale – or the loss of the "will to resist" – at the home front, rather than from overwhelming opposition on the battlefield.[113] As one text admonished, the civil-defense measures initiated after 1933 were designed to prevent the recurrence of defeat for similar reasons: "We Germans in particular have learnt in the last war that good morale at the home front is decisive for the outcome of a modern war."[114]

While participating in civil-defense programs became a national "duty" for every German man and woman, party publicists did not resolve the tensions springing from the need, on the one hand, to inform the people about prospective horrors of air raids and, on the other, to display confidence in the effectiveness of precautions.[115] One brochure from 1940 featured an introduction by a general who sought to reassure readers that "civil-defense is neither a problem nor an unusual novelty. Rather it is the most normal topic imaginable." Yet in the same publication Göring limited

---

[110] The Reichsluftschutzbund also published the bi-weekly journal *Die Sirene*, with a circulation of 400,000. See *5 Jahre Reichsluftschutzbund* (Berlin, 1938). See also *ABC des Luftschutzes: herausgegeben vom Präsidium des Reichsluftschutzbundes* (Berlin, 1940), 72–4.

[111] For an overview of activities, see *5 Jahre Reichsluftschutzbund, passim*. For a description of drills, see Fritzsche, *A Nation of Fliers*, 208–12.

[112] See *ABC des Luftschutzes*, 9; E. Meyer and E. Sellien, *Schule und Luftschutz* (Munich, 1934), 40; Heinrich Hunke, *Luftgefahr und Luftschutz* (Berlin, 1935), 44–5.

[113] For a detailed exposition of this myth, see Schivelbusch, *Kultur der Niederlage*, 242–7.

[114] *ABC des Luftschutzes*, 61.

[115] *5 Jahre Reichsluftschutzbund* (pages unnumbered). On women in civil defense, see Hans Riegler, *Deutsches Luftschutz-Jahrbuch* (Berlin, 1940), esp. 33–6, 72–3; Liselotte Klinger, "Deutsche Frau – erfüllst Du Deine Luftschutzpflicht?," *NS-Frauen-Warte* 8 (1939), 186–7.

the purpose of civil-defense measures to merely sparing the nation "from at least the worst consequences of air war" – a phrasing that left much to the imagination.[116] Even in the tightly controlled public sphere of Nazi Germany, it proved impossible to prepare the nation for armed conflict without touching on the issue that "modern air war" obliterated the distinction between combatants and civilians. To escape from this dilemma, propagandists stressed that the prospect of aerial warfare forced all Germans, including women, to transform themselves into "soldiers," whether it was on the battlefield or on the home front that they would see action. In keeping with this demand, the didactic literature on civil defense was replete with images of women handling heavy equipment in heavy boots, army trousers, and steel helmets, a get-up that lent them the appearance of a cross between soldiers and fire fighters.[117] Such masculinized images of women took Nazi enthusiasm for technology to its logical endpoint.[118] The novel conditions of aerial warfare – themselves a product of processes of innovation – demanded a universal militarization of society in the very spirit of the defiance that had formed a long-standing tradition fueling fervor for technology in Germany in the first place.

German assessments of the national importance of technology, in short, bore the following characteristics. After the imperial period, when innovative engineering had publicly underpinned the country's position as a successful international newcomer, technological achievements revealed Germany's unfulfilled political and economic potential in the eyes of many contemporaries. As a result, prominent artifacts expressed a spirit of resilience and resistance that extended from National Liberal circles to the far Right. The National Socialists adopted the rhetoric of defiance through technological excellence, radicalized it, and claimed to have initiated a modernization project that laid the foundation for Germany's renewed ascension to world-power status. Whereas British debate about technology and the nation prioritized notions of defense in the quest to preserve the *status quo*, German nationally minded enthusiasm for technology – most notably on the Right – displayed aggressive traits that sought to effect a fundamental transformation of the international scene. In consequence, German evaluations of technology operated with a temporal concept of modernity that differed from the notions of continuity so prominent in its British equivalent. German technology promised to effect a temporal rupture, thereby ushering

---

[116] *ABC des Luftschutzes*, 3, 61.
[117] See Klinger, "Deutsche Frau," 186; Riegler, *Deutscher Luftschutz*, 72.
[118] Claudia Koonz also notes the imagery of the militarized female. See Claudia Koonz, *Mothers in the Fatherland: Women, the Family, and Nazi Politics* (London, 1988), 395.

in an alternative, unprecedented form of modern times. The hero worship of men who personified unwavering determination in the face of adversity, which found its most prominent expression in the adulation of Count Zeppelin, signaled to Liberals, Conservatives, and National Socialists alike the qualities required to transform the nation's fortunes. Thus, while the aggressiveness of German debate sparked political conflicts between Left and Right, models of manliness and notions of modernity predicated on ideas of a temporal rupture established cultural reference points that extended the nationally minded appeal of innovative artifacts across the deep political and ideological divisions between Liberals and the Right.

### AERIAL WAR HEROES IN GERMANY AND BRITAIN

If public debate linked national identity to technology in peacetime, these connections became most intimate during periods of war, when state authorities exerted greater control over the media. Designed to boost the morale of both civilians and soldiers, British and German propaganda alerted the public to the importance of technology for the fortunes of the nation. Official initiatives drew on nationally specific cultural traditions to praise certain acts of warfare as exemplary displays of heroism. Just as the British and the German legitimizations of violent conduct in war varied markedly, so they accentuated the different national meanings attributed to technology. In both countries, war propaganda devoted considerably more attention to aviation than to civilian shipping. To be sure, pamphlets and brochures highlighted the roles played by passenger liners in military operations as troop transports and as hospital ships during wartime, but their largely non-combative services did not easily lend themselves to the dramatic tales that corresponded to the general situation of the two nations, both engaged in an existential struggle.[119] Air combat correlated much more closely with this theme, and military pilots came to personify national power fantasies that, albeit in highly contrasting ways, legitimated the use of aerial technology during times of armed conflict in both countries.

Despite some experiments with strategic bombing in World War I, technical constraints, including low engine power and light bomb payloads,

---

[119] See Norddeutscher Lloyd Bremen, *Jahrbuch 1914/15: Der Krieg und die Seeschiffahrt unter besonderer Berücksichtigung des Norddeutschen Lloyd* (Bremen, 1915); Cunard Archive, University of Liverpool, University Archives (hereafter Cunard archives), D42/PR3/8/8, *R. M. S. Carmania* (no place, 1919); Cunard archives, D42/PR3/8/4, *To the American Legion* (no place, 1921); Cunard archives, D42/PR3/9/12b, *Cunard Line Peace Day: July 19th, 1919* (no place, 1919); *On War Service* (London, 1919).

as well as the development of effective aerial defenses for major cities such as London and Paris, restricted most war flying to the front lines, where reconnaissance aeroplanes gathered intelligence, taking photographs of enemy positions and troop movements. Both sides systematically developed fighter planes that could be deployed against "scout" aircraft as well as against the tactical bombers that were attacking the infrastructure and ground troops along the fronts.[120] Notwithstanding its initially marginal impact on the overall outcome, aerial warfare arrested the imagination of contemporaries from 1914 on, and "aces" who piloted fast, single-seater fighter planes armed with machine guns became national heroes. "Aces" of the Great War owed their prominence to the fact that, by contrast with ground troops who were locked in mechanized and impersonal "mass slaughter," propaganda cast war pilots as autonomous fighters who determined their own rules of engagement, sometimes even claiming to observe a code of chivalry.[121]

On some occasions, German depictions of fighter pilots conformed to this pattern. For instance, German aces mentioned that they allowed defenseless opponents to escape to safety. It was also widely reported that German authorities honored written requests by French war pilots to receive Catholic funerals, and Richthofen claimed in his memoir to have erected a "nice" tombstone as a tribute to the first opponent he had shot down.[122] These episodes illustrate the fact that German propaganda sometimes nurtured the myth of the knightly aviator, although, during the First World War, aces generally secured their "kills" by attacking technically inferior opponents once squadrons had achieved numerical superiority. Despite a consensus among historians that images of aerial chivalry did not correspond with the prevailing battle tactics, the fact that many German books about the war in the air celebrated pilots as ruthless killers has received little systematic scholarly scrutiny.

In fact, German propaganda emphatically affirmed the relentless use of violence in open defiance of chivalric codes. In biographical and autobiographical writings devoted to Germany's foremost aces, Oswald Bölcke,

---

[120] For an overview of the air war between 1914 and 1918, see John Morrow, *The Great War in the Air: Military Aviation from 1909 to 1921* (London, 1993).

[121] Mosse, *Fallen Soldiers*, 119–25; Morrow, *The Great War in the Air*, xiv; Joanna Bourke, *An Intimate History of Killing: Face-to-Face Killing in Twentieth-Century Warfare* (New York, 1999), 46–7. Robert Wohl and Michael Paris point out that few aces adhered to the code of honor. See Wohl, *A Passion*, 240–1; Michael Paris, *Winged Warfare: The Literature and Theory of Aerial Warfare in Britain, 1859–1917* (Manchester, 1992), 8–9. On the ground war, see Modris Eksteins, *The Rites of Spring: The Great War and the Birth of the Modern Age* (London, 1989), 139–91.

[122] Fritzsche, *A Nation of Fliers*, 87–90; Manfred Freiherr von Richthofen, *Der rote Kampfflieger* (Berlin, 1917), 93.

Max Immelmann, and Manfred von Richthofen appeared neither to show mercy to defeated opponents nor to suffer pricks of conscience after combat. On the contrary, tales by German pilots reveled in the havoc they wreaked. Oswald Bölcke described in a posthumously published letter to his parents in 1917 that he considered the sight of an enemy aircraft going down in flames an aesthetic delight:

> I caught up with the enemy apparatus quickly. The very moment I pulled up my machine over the enemy, I saw him explode. The cloud of smoke hit my face ... Seeing the enemy apparatus break apart just in front of me, bursting into flames and then falling down like a torch was a beautiful spectacle.[123]

Some days earlier, he had proudly described shooting down a defenseless aircraft whose steering had been damaged. A crew member climbed on to a wing to re-establish the balance of the disintegrating machine, and "stared at me in a terrified manner waving his hand. The whole sight looked quite pathetic and I hesitated for a moment to shoot at him." Yet Bölcke quickly squashed these feelings and "aimed some more shots at the pilot."[124] He was by no means the only German war pilot to revel in bringing down opponents. Richthofen recalled "how [his] heart leapt with joy" when hitting his enemies.[125] Another pilot did not suppress his anticipation for the next "nerve-thrilling, wild, and wonderful aerial fight." Looking at the night sky, he issued a martial pledge: "Tomorrow it must rain English pilots' blood."[126]

Gleeful depictions of war violence were not restricted to pilots' descriptions of combat among themselves; their accounts of aerial campaigns against civilian targets showed similar enthusiasm. Rejoicing in the destruction and chaos generated by attacks from the air, pilots prided themselves on selecting targets of little strategic relevance in London, such as Liverpool Street railway station, the Bank of England, Tower Bridge, Russell Square, and residential neighborhoods: "No window pane can be intact in the entire area, it rains glass upon the people who hurry around in wild confusion, trying to salvage belongings where nothing can be saved any more."[127] Triumphantly referring to themselves as "the modern riders of the apocalypse," crew members found bombing raids particularly fulfilling.[128] Richthofen remembered that dropping bombs and seeing their results always left him with the "feeling [of having] accomplished something."[129]

---

[123] *Hauptmann Bölckes Feldberichte, mit einer Einleitung von der Hand des Vaters* (Gotha, 1917), 68.
[124] *Ibid.*, 64.   [125] Richthofen, *Der rote Kampfflieger*, 170. See also *ibid.*, 92.
[126] *Wir Luftkämpfer: Bilder und Berichte deutscher Flieger und Luftschiffer* (Berlin, 1918), 8, 15.
[127] *Zeppeline gegen England* (Berlin, 1916), 9). See also Junke, *Zeppelin*, 120–1.
[128] *Wir Luftkämpfer*, 60.   [129] Richthofen, *Der rote Kampfflieger*, 83.

These displays of military violence, including attacks against defenseless enemies and civilians, clearly did not jeopardize or morally compromise the status of air crews as public heroes, since German propaganda cast these acts in a thoroughly favorable light. German aerial heroes of the First World War embraced a model of military masculinity that, far from seeing military violence as a problem, unconditionally affirmed it. German accounts legitimated bombing campaigns against civilians by arguing that the military were merely reacting to earlier provocations on the part of the Allies. According to this line of thought, raids on British and French cities represented morally justified "punitive vengeance" (*Strafgericht*) that meted out direct reprisals for similar air attacks on towns such as Karlsruhe and Freiburg.[130] Moreover, German propaganda characterized its bombardments as the equivalent of Allied blockade tactics, which, according to one commentator, consciously "blurred the distinction between measures against the military and measures against the civilian population . . . We consider the air war against England as a just retribution that enables us to achieve military aims while disrupting the English economy."[131] Thus German attacks against non-military targets countered and punished Allied initiatives, which had allegedly been the first to break established rules for the conduct of war. In Germany, moral wrath and self-defense legitimized the unconditional use of military force against non-military installations.

In recounting their violent exploits, aces and outside observers often described their acts as analogous to hunting or dueling. Bölcke wrote to his parents that he had "recently taken up the habit of flying to the front . . . to hunt Frenchmen."[132] Hunting metaphors provided a means of highlighting the instinctual aspects of the war in the air. One journalist who visited the German aces Bölcke and Immelmann reported that they "could literally smell when there was some prey to hunt . . . Both feel an appetite for hunting, take off, and immediately find an opponent."[133] Hunting language dehumanized enemies and transformed them into "prey" to be gunned down. Richthofen's memoir, which directly compared the experiences of a fighter pilot with shooting big game, contains the most explicit statement to this effect. As he was waiting for a bison to come within shooting range, Richthofen felt "the very hunting fever that takes possession of me when I am sitting in an aeroplane, spot an Englishman and still have

[130] *Zeppeline gegen England*, 114.   [131] Jünke, *Zeppelin*, 105.
[132] *Hauptmann Bölckes Feldberichte*, 46.
[133] Friedrich Frei (ed.), *Unser Fliegerheld Immelmann: Sein Leben und seine Heldentaten* (Leipzig, 1916), 22.

about five minutes to fly to reach him."[134] Moreover, since hunting was a pastime predominantly pursued by upper-class men, the language of the chase simultaneously glorified fighter pilots' activities and underlined their elite status. Thus the hunting rhetoric affirmed aerial violence in multiple respects, stressing the instinctual aspects of aerial combat, dehumanizing opponents, and ennobling the act of killing. German depictions of aerial combats as "duels" further glorified military violence in class-specific ways. Unlike in Britain, where dueling had been banned in the first half of the nineteenth century, German upper-class men continued to defend their "honor" in duels until the outbreak of the Great War. Many of these confrontations ended with serious injuries and death. Since accounts of war aviation conceived of fights between enemy machines as duels, propaganda narratives capitalized on the positive connotations between masculinity and violence established by this particular form of conflict resolution.[135]

Because German war propaganda remained under their exclusive control, the military authorities were capable of silencing voices that might have challenged these positive accounts of the German war effort.[136] In July 1918, the Social Democratic politician Philipp Scheidemann delivered a speech in the Reichstag that criticized British and German bombing campaigns against civilians as unjustifiable "atrocities," since they amounted to the "cruel murder" of "innocent victims."[137] Similar statements could not be heard outside parliament because of the strict control that the propaganda offices exerted over the national and local press. As a result, interventions that saw the use of military violence as a problem failed to make an active contribution to official German war propaganda. The German military authorities, therefore, were at liberty to construct public images of fighter pilots that glorified their exploits as heroic enactments of an exemplary military masculinity.

Peter Fritzsche is undoubtedly correct in stating that during World War I "the aviator remained a hero, but of a kind that had not been seen before." Still, his claim that the airman's novelty rested primarily in the "new image of the machine-man" is debatable, since both journalists and pilots themselves

---

[134] Richthofen, *Der rote Kampfflieger*, 178.
[135] On the end of dueling in Great Britain, see Donna T. Andrew, "The Code of Honour and its Critics: The Opposition to Duelling in England, 1700–1850," *Social History* 5 (1980), 409–34. On the German duel, see Ute Frevert, *Ehrenmänner: Das Duell in der bürgerlichen Gesellschaft* (Munich, 1995); Kevin McAleer, *Dueling: The Cult of Honor in Fin-de-Siècle Germany* (Princeton, 1994); Peter Gay, *The Cultivation of Hatred: The Bourgeois Experience: Victoria to Freud* (London, 1995), 9–33.
[136] On the swift institution of the German propaganda regime, see David Welch, *Germany, Propaganda and Total War, 1914–1918* (London, 2000), 20–36. Richthofen's memoir was censored by the press office of the Air Service. See Wohl, *A Passion for Wings*, 225.
[137] *Verhandlungen des Reichstags*, vol. 313, 5706–7.

only rarely resorted to the sober language of technological efficiency in their descriptions of flying during the war.[138] Between 1914 and 1918, the figure of the war pilot usually derived its exceptional character from other iconic elements. While his relative autonomy in battle elevated him above other military personnel, it was the uncompromising celebration of brutality that distinguished the airman from other German military heroes. Walter Flex's bestseller *Wanderer zwischen den Welten* (*The Wanderer Between the Worlds*, 1916) depicted its protagonist Ernst Wurche as an infantry soldier who sacrificed himself for the nation out of a romantic impetus that had been fostered by the prewar youth movement, and contrasted sharply with the grim, impersonal conduct of ground war.[139] If trench warfare could give rise to nostalgic yearnings, comparable sentiments did not find their way into accounts of the air war, which celebrated aviators as ruthless, independent military agents.

In contrast to the violent traits that marked the image of the "ace" during the war, however, the figure of the noble aerial knight flourished in Weimar Germany. Its medieval overtones not only provided one of the many historicizing motifs through which Germans, like other Europeans, memorialized the war; it also underpinned a domestic myth about the moral integrity of the German military during the hostilities, which the Right was particularly keen to propagate to the outside world. Encasing the aerial warrior in the iconic armor of chivalry emphasized the German armed forces' adherence to an internationally accepted military ethos, thereby challenging propositions that the preponderance of a particularly aggressive version of militarism in German culture justified the harsh conditions in the peace treaty.[140] A book entitled *Letters to a Young Girl by a German Fighter Pilot*, which appeared posthumously in 1930 and contained the amorous correspondence between a nurse and a military aviator, is indicative of this trend. In it, the ace emphasized the "knightly or, to put it in more modern terms, sportive" character of the "duels" he fought.[141] Similarly, when *Wings* – an American movie about two American fighter pilots stationed

---

[138] Fritzsche, *A Nation of Fliers*, 96–7.
[139] Flex's book had sold 250,000 copies by the end of the war. For a brief appreciation, see Thomas Medicus, "Jugend in Uniform: Walter Flex und die deutsche Generation von 1914," in Ursula Breymayer, Bernd Ulrich, and Karin Wieland (eds.), *Willensmenschen: Über deutsche Offiziere* (Frankfurt, 1999), 94–108.
[140] On the importance of traditional language in the formulation of memory after the war, see Winter, *Sites of Memory*; Stefan Goebel, *Medievalism in the Commemoration of the Great War in Britain and Germany, 1914–1939* (Cambridge, PhD dissertation, 2001). On German atrocities in Belgium in the First World War and their denial in the interwar years, see John Horne and Alan Kramer, *German Atrocities 1914: A History of Denial* (New Haven, 2001).
[141] Johannes Werner (ed.), *Briefe eines deutschen Kampffliegers an ein junges Mädchen* (Leipzig, 1930), 67. See also *ibid.*, 7, 59, 90, 171. For another example along these lines, see *Luftfahrt Voran*, 36–8.

in France in 1918 – hit the screens of Berlin in 1929, the Conservative press applauded its depiction of German aces as "chivalrous," while the critic for the Social Democratic evening paper praised the film's technical accomplishment and its realism (*Naturwahrheit*).[142] Thus, images of aerial chivalry not only played up the honor of German soldiers but also avoided fueling conflict between the Right and the Left in Weimar Germany, since they lacked the overtly aggressive overtones that caused political concern in Social Democratic circles. Compared with accounts produced during the war itself, the figure of the chivalrous war pilot was more prominent in the Weimar Republic for both domestic and foreign political reasons.

After 1933, military airmen personified the dynamicism and vigor of National Socialist Germany. To create its version of a new and modern Germany, the regime promised to revive and transform the military virtues that, while allegedly having sustained national power before 1918, had been suppressed during the Weimar years. Göring, himself the last commander of the fighter squadron that had counted Bölcke and Richthofen among its leaders, declared in a speech in 1939 that he expected the German air force to continue "the heroic tradition" whose origins had been laid by the fighter pilots of the First World War. To underline this "indestructible connection with the glorious past," he named three of the initial fighter squadrons of the newly established Luftwaffe after Germany's most famous aces, Immelmann, Bölcke, and Richthofen.[143]

Although the "aces" of the Great War figured among National Socialism's military ancestry, significant shifts sidelined the knightly ideal in the iconography of the military airman because its historicizing overtones clashed with the rhetoric of radical innovation that accompanied the expansion of the German air force. While all the aerial heroes of the First World War had belonged to the exclusive and individualistic order of fighter pilots, National Socialist propaganda, taking up the idea of the supposedly classless *Volksgemeinschaft*, placed these men within the "community of fighters" (*Kampfgemeinschaft*) that also included air-crew members such as radio operators and navigators as well as ground personnel.[144] The cult of the military aviator in the Third Reich did not produce individual stars comparable to Richthofen or Bölcke during the Second World War. Apart from

---

[142] *Neue Preußische Zeitung*, 13 January 1929, 10; *Der Abend*, 10 January 1929, 3.
[143] Orlovius, *Schwert am Himmel*, 24, 38; Karl Georg von Stackelberg, *Legion Condor: Deutsche Freiwillige in Spanien* (Berlin, 1939), 1.
[144] The term *Kampfgemeinschaft* is from Dr. Eichelbaum, *Das Buch von der Luftwaffe* (Berlin, 1941), 43. The hierarchy is described in an article in the party paper detailing the training of air crews. See *Völkischer Beobachter*, 17 January 1937, 11.

potentially fostering envy among the "community" of pilots, this practice would have entailed having to inform the public of a hero's death, which would have subverted their morale if it happened too often. As a result, the military pilot of the Third Reich remained a largely anonymous, albeit heroic, figure that exemplified the new masculinity deemed necessary to enact the demands of Nazi ideology.[145]

"Toughness" and "hardness" (*Härte*) provided the core of the aerial fighter's personality, as outlined both in the run-up to and during the Second World War.[146] According to official depictions, an unshakeable physical and mental frame allowed aerial "soldiers" to release their "steely will of annihilation" (*stählerner Vernichtungswille*) on to their enemies, who, in the words of Adolf Hitler, would be forced to "realize that they are now facing a different Germany from the one in 1914."[147] Toughness not only turned the German pilots into the "terror for all opponents" but also, in keeping with National Socialist cults celebrating the dead, prepared them to sacrifice themselves willingly for *Volk und Führer*.[148] In its depictions of military aviators in action, the Nazi pornography of violence focused on two traits that constituted this exemplary soldier. First, these accounts told of pilots openly rejoicing in the destruction they sowed. For instance, after dropping his payload over a Polish railway station, the pilot of a *JU 87* dive bomber wanted to "shout with delight" as he observed the chaos he had created on the tracks.[149] Similarly, a fighter pilot who saw action in the Spanish Civil War claimed to have attacked enemy positions "for the fun of it," even once they had been rendered inactive.[150] Such sadistic texts about the military aviators of the Third Reich fit seamlessly into the tradition of celebrating gratuitously violent actions that had originated during the First World War.

Second, reports alternatively emphasized how calm and detached the heroes of the air remained during their missions. Even with "every nerve" "tense," pilots gained "trust from their own strength" as well as from the "steely hum of the engines, the loaded drums of the machine guns, [and] the hatch full of bombs." Depictions of commanders as "clear and calm" also

---

[145] On National Socialist notions of masculinity, see Mosse, *The Image of Man*, 154–80.
[146] For descriptions of soldiers as hard, see "Die Antwort des deutschen Soldaten," *NS-Frauen-Warte* 8 (1939), 192; von Stackelberg, *Legion Condor*, 47; Orlovius, *Schwert am Himmel*, 91.
[147] Orlovius, *Schwert am Himmel*, 59.
[148] On the cult of the dead, see Sabine Behrenbeck, *Der Kult um die toten Helden: Nationalsozialistische Mythen, Rituale und Symbole* (Vierow, 1996), 178–84, 214–41.
[149] Hermann Adler, *Unsere Luftwaffe in Polen: Erlebnisberichte* (Berlin, 1940), 49.
[150] Von Stackelberg, *Legion Condor*, 56. A similar passage can be found in Peter Supf, *Luftwaffe von Sieg zu Sieg: Von Norwegen bis Kreta* (Berlin, 1942), 15.

conveyed an atmosphere of self-confidence and discipline.[151] A sparse documentary style further underlined an image of air crews as single-mindedly dedicated to their military duties. A pilot's account from Poland includes the following typical passage: "I see gigantic fountains of smoke and sand beneath me. Houses explode. Rails are bent like thin wires as if by an invisible hand as a result of the air pressure. A train stands on a parallel track – direct hit. The direction of the carriages has changed by ninety degrees."[152] In addition to reflecting the physical and emotional detachment of the military aviator from his targets on the ground, the pervasive imagery of calm discipline indicated that a spirit of sobriety or "*Sachlichkeit*" dominated the Luftwaffe.

From the 1920s, observers on the Right had considered the embrace of an ethos of sobriety to be a precondition for Germany's return to world-power status.[153] In this context, praise for uncompromising, calm determination consistently went hand in hand with a disowning of all emotional responses that could have been construed as symptoms of weakness by reason of their threatening overtones of defeat. Ernst Jünger was not alone in casting military pilots as epitomes of sobriety, which became a popular alternative to depicting them as aerial knights.[154] As Ernst von Brandenburg, the architect of the secret aerial rearmament program during the Weimar years, wrote in 1925, only a new kind of man, who "weighed the facts coldly, soberly, intelligently and without flitting nervousness," would be in a position to help the nation assert itself in a hostile international environment. According to Brandenburg, there was "no better school for the education of the mind and soul" to attain this outlook than flying.[155] After 1933, as military recruitment schemes incorporated psychological tests to select candidates who promised to make sober officers with strong willpower, a host of propaganda texts correspondingly impressed upon readers that the new pilots epitomized the figure of the ruthless, death-defying fighter whose "steel-like" physical and mental frame allowed him to maintain his composure

---

[151] Von Stackelberg, *Legion Condor*, 28–9.   [152] Adler, *Unsere Luftwaffe in Polen*, 61.
[153] On *Sachlichkeit*, see Ulrich Herbert, *Best: Biographische Studien über Radikalismus, Weltanschauung und Vernunft, 1903–1989* (Bonn, 1996), 42–51; Helmut Lethen, *Cool Conduct: The Culture of Distance in Weimar Germany* (Berkeley, 2001).
[154] For such a celebration of aerial fighters by Jünger, see Ernst Jünger, "Einleitung," in *idem* (ed.), *Luftfahrt ist not!*, 9–13, esp. 12. On Jünger and his relationship with the far Right in general and the Nazis in particular, see Thomas Nevin, *Ernst Jünger and Germany: Into the Abyss, 1914–1945* (Durham, NC, 1996); Nikolaus Wachsmann, "Marching Under the Swastika? Ernst Jünger and National Socialism, 1918–1933," *Journal of Contemporary History* 33 (1998), 573–89.
[155] Ernst von Brandenburg, "Wer soll fliegen lernen?" in Johannes Poeschel (ed.), *Einführung in die Luftfahrt* (Leizpig, 1925), 159–61.

in dangerous situations.¹⁵⁶ Just as proclaiming a novel symbiosis between men and machines was a prominent motif in artistic modernism, Nazi imagery linked pilots' control over their emotions with their power over a technological instrument of destruction.

These characteristics come together to create a warrior figure whose ruthlessness the Nazis had idolized since their modest beginnings.¹⁵⁷ Above all, this fighter was supposed to possess one central trait: he was entirely devoid of any qualms or ethical considerations when dealing with the enemy, following the maxim that "the German sword is sharp and where it strikes, all life must end."¹⁵⁸ In keeping with this amoral outlook, war correspondents noted with pride that German dive bombers operating in Poland left behind "a picture of complete dissolution, complete destruction from which those escaping alive and pale-faced emerged paralyzed with fear, entering into captivity voluntarily."¹⁵⁹ From the very outset, German propaganda openly celebrated the Second World War as a campaign of relentless annihilation in which aerial warriors displayed the sober qualities of steel-like toughness that promised German success on the battlefield. Thus emphatic German affirmations of aerial technology's violent potential culminated in descriptions that cast the military pilot as an operator using the aeroplane as an efficient tool of destruction. Even if one refrains from speculating to what extent military propaganda contributed to a cultural atmosphere conducive to genocide, National Socialist writing about the air war makes chilling reading for its gruesome, contemptuous, and cynical tone.¹⁶⁰

How do these characteristics compare with the British imagery surrounding aerial warriors? In Britain, the figure of the aerial knight provided a more important motif for the public celebration of airmen during the First World War than was the case in Germany. A 1917 pamphlet entitled *The Romance of Air-Fighting* hailed the air war as "the revival of ancient chivalry" and likened dogfights to "sublime" and "picturesque"

---

¹⁵⁶ Thomas Flemming, "'Willenspotentiale': Offizierstugenden als Gegenstand der Wehrmachtspsychologie," in Ursula Breymayer *et al.* (eds.), *Willensmenschen* 111–22. For descriptions of war pilots as calm in situations of danger, see Supf, *Von Sieg zu Sieg*, 83; von Stackelberg, *Legion Condor*, 22, 28, 30–1, 46.

¹⁵⁷ Eve Rosenhaft, "Gewalt in der Politik: Zum Problem des 'Sozialen Militarismus,'" in Klaus-Jürgen Müller (ed.), *Militär und Militarismus in der Weimarer Republik* (Düsseldorf, 1978), 237–58; Eve Rosenhaft, "Links gleich rechts? Militante Straßengewalt um 1930," in Thomas Lindenberger and Alf Lüdtke (eds.), *Physische Gewalt: Studien zur Geschichte der Neuzeit* (Frankfurt, 1995), 238–75; Bernd Weisbrod, "Gewalt in der Politik: zur politischen Kultur in Deutschland zwischen den Weltkriegen," *Geschichte in Wissenschaft und Unterricht* 16 (1990), 391–404.

¹⁵⁸ Adler, *Unsere Luftwaffe in Polen*, 116.    ¹⁵⁹ Orlovius, *Schwert am Himmel*, 98.

¹⁶⁰ Omer Bartov argues that increasingly radical army propaganda on the eastern front created a climate among troops that predisposed them to participate in genocide. See Omer Bartov, *Hitler's Army: Soldiers, Nazis and War in the Third Reich* (New York, 1992).

medieval "tournaments."¹⁶¹ Reports also paid homage to pilots' fairness and mutual respect, stating, for instance, that aces dropped wreaths behind enemy lines to honor the deaths of prominent opponents. Moreover, pilots allegedly waved to one another to acknowledge their enemies' skill and courage when "duels" ended in stalemate.¹⁶² By emphasizing the theme of medieval chivalry, British war propaganda cast fighter pilots as commendable military heroes who upheld in the new conditions of the First World War the same code of honor that had been instrumental in the celebration of Victorian army leaders such as General Havelock, who was killed in the Indian "mutiny" in 1857, and General Gordon, who died during an uprising in the Sudan in 1885.¹⁶³

At the same time, British depictions of aerial warfare displayed a deep sense of ambivalence about the ethical implications of combat. Many accounts of British fighter pilots in the First World War rendered military heroism problematic because battle strained the moral foundations of existing speech about elite masculinity. The officially authorized biography of Albert Ball, an ace who shot down forty-three enemies before his death at the age of only twenty, provides the best illustration of the moral pitfalls of combat. Chief of the Air Staff Hugh Trenchard, Commander-in-Chief Douglas Haig, and Prime Minister David Lloyd George contributed prefaces to this account, which was based on letters the pilot had written to his family about life at the front. Haig's preface underlined Ball's chivalry and gallantry, while Trenchard praised his modesty and sense of responsibility.¹⁶⁴ Lloyd George singled Ball out as a "stirring example of heroic simplicity" because of his dutiful commitment to "freedom, home and country." The Prime Minister was particularly impressed by Ball's moral integrity:

> In all his fighting record there is no trace of resentment, revenge or cruelty. What he says in one of his letters, 'I hate this game, but it is the only thing one must do just now,' represents . . . the conviction of those vast armies who, realising what is at stake, have risked all that liberty may be saved.¹⁶⁵

Ball stood out as a role model because he fought in a just war and escaped moral destruction during the violence of battle. The biography told how he resembled

---

¹⁶¹ B. Wherry Anderson, *The Romance of Air-Fighting* (London, 1917), 12–13.
¹⁶² *The Times*, 4 November 1916, 5; Walter A. Briscoe and H. Russel Stannard, *Captain Ball V.C.* (London, 1918), 213.
¹⁶³ See Mark Girouard, *The Return to Camelot: Chivalry and the English Gentleman* (New Haven, 1981); Graham Dawson, *Soldier Heroes: British Adventure, Empire and the Imagining of Masculinities* (London, 1994).
¹⁶⁴ Briscoe and Stannard, *Captain Ball*, ix.   ¹⁶⁵ *Ibid.*, vii.

a young knight of gentle manners who learnt to fly and kill at a time when all the world was killing, and who . . . remained a good-natured happy boy, a little saddened by the great tragedy that had come into the world and made him a terrible instrument of Death.[166]

Under the circumstances of the Great War, Ball appeared as a tragic hero because the violence of war strained the chivalrous and honest moral qualities ascribed to him. The biography quoted a letter Ball had written to his parents about the qualms he suffered as a result of battle engagements: "Oh, it was a good fight and the Huns were fine sports . . . but oh! I do get tired of always living to kill, and I am really beginning to feel like a murderer."[167] Other passages further underscored Ball's moral qualities. References to his young age were phrased as illustrations of childlike innocence. He thanked his family for food supplies, and once took off while eating a piece of cake sent by his sister, a detail which, according to his biographers, revealed Ball's pure character: "In everything he did, Ball was half schoolboy, half soldier, but wholly loveable."[168] The authors also admired Ball's embodiment of respectable middle-class domesticity, as reflected in his letters about his garden: "Well, my garden is going fine, peas are now four inches, and I had the first crop of mustard and cress in my mess today. It did make the chaps laugh, but they liked it."[169] In short, Ball was conveyed as struggling for everyday normality and integrity despite his duty to kill for the nation.

The biography of Ball was not the only public text to address the problematic heroism of military pilots. Ball's colleague James McCudden also described how his battle activities haunted him. Although he considered shooting down Huns to be "fun," he was at the same time "conscience-stricken" after having set an enemy alight: "As soon as I saw it I thought 'poor devil,' and felt really sick. It was at that time very revolting to see any machine go down, especially when it was done by my own hands."[170] Similarly, the autobiography of bomber pilot Paul Bewsher contains passages in which its author recalls shivering at the thought of killing. Inspecting his equipment before take-off, Bewsher thought of his potential victims in a "phase of pity: I shall kill tonight . . . Is it true that those plump yellow bombs . . . are destined to rip flesh and blood?"[171] Narratives about the moral burden of aerial killing even slipped into the account by the Canadian William Bishop, who claimed to have found his life's vocation as a fighter pilot. Bishop's memoir from 1918 contained no references to

---

[166] *Ibid.*, 23.   [167] *Ibid.*, 21.   [168] *Ibid.*, 61.   [169] *Ibid.*, 157.
[170] James McCudden, *Five Years in the Royal Flying Corps* (London, 1919), 264, 220.
[171] Paul Bewsher, *Green Balls: The Adventures of a Night-Bomber* (Edinburgh, 1919), 51–2.

gentlemanly chivalry in the air, and he enjoyed watching enemies go down in flames with "great satisfaction."[172] He asserted that bringing "down a machine did not seem like killing a man," because his primary aim consisted in "destroying a mechanical target." However, when he reminded himself "that a live man had been piloting the machine," he "worried a bit" and lost sleep for the night because, he maintained, he did not "relish the idea of killing Germans."[173] Despite his dedication to aerial fighting, Bishop's mental and moral equilibrium was by no means unshakeable.

British military aviators claimed to struggle with the moral implications of serving one's country in war. Accounts of fighter pilots constructed an image of British military personnel that was in sharp contrast with British official representations of German soldiers as inhuman "evil Huns" who habitually committed atrocities.[174] British fighter pilots, therefore, underscored the moral superiority of British forces over their enemies. Moreover, emphases on the moral impact of war service on the individual reflected a wider public justification for Britain's participation in the conflict. British politicians often presented the country's involvement in the Great War as a moral obligation to restore a European order based on international law, which German aggression had disrupted. Thus flying aces personified the moral dilemmas that arose from the tension between violent means and ethical ends, which the British nation confronted as a whole as it sought to re-establish justice through military force.

In addition, British depictions of fighter pilots' masculinity drew heavily on the idea of the gentleman, as the biography of Ball illustrates. While this model of British masculinity underwent significant transformations during the late Victorian and Edwardian periods and increasingly foregrounded demands for toughness and physical strength, older conceptions of the gentleman as "brave, loyal, true to his word, courteous, generous, and merciful" – as Mark Girouard has summarized them – continued to remain influential.[175] The public image of British fighter pilots conformed to a form of masculinity whose norms tended to regard the use of violence as a problem rather than to affirm it. Finally, British war propaganda

---

[172] W. A. Bishop, *Winged Warfare: Hunting the Huns in the Sky* (London, 1918), 102. See also *ibid.*, 45–6, 96.

[173] *Ibid.*, 175.

[174] On British stereotyping of the German army during the Great War, see Cate Haste, *Keep the Home Fires Burning: Propaganda in the First World War* (London, 1977), 79–107.

[175] Girouard, *The Return to Camelot*, 16. On British concepts of masculinity during the Great War, see also Ilana R. Bet-El, "Men and Soldiers: British Conscripts, Concepts of Masculinity and the Great War," in Billie Melman (ed.), *Borderlines: Genders and Identities in War and Peace, 1870–1939* (London, 1998), 73–94. On the renegotiation of elite masculinities since the late nineteenth century, see James A. Mangan, *Athleticism in the Victorian and Edwardian Public School: The Emergence and Consolidation of an Educational Ideology* (Cambridge, 1981).

was never conceived under the sole direction of the military establishment, which might have encouraged a more emphatic rhetoric of combat violence.[176] As Lloyd George's contribution to Ball's biography illustrates, British war propaganda represented a collaborative effort between the military and civilians. In fact, the political and military leadership recruited leading press figures and newspaper proprietors to render its attempts to influence mass opinion more convincing. Since many of the civilians involved in public-relations initiatives, like the Prime Minister, harbored an ambivalent attitude to violence and military conflict, British propaganda highlighted the moral dilemmas faced by aerial soldiers. In contrast to German accounts that emphatically celebrated the ruthless use of military violence, British publications, then, cast war pilots as ambivalent heroes who exemplified the moral problems that arose from the need to exert military force in pursuit of a legitimate political aim. In this respect, the figure of the air ace supported other official legitimizations of the British war effort that played up the country's moral commitment to the restoration of international law.

The First World War thus produced an iconography of the military pilot that stretched gentlemanly heroism in Britain to its limits. Although this did little to diminish public interest in flying aces and their feats during the interwar years, the reception of feature films about the war in the air demonstrates that unqualified eulogies of military pilots remained rare. When *Wings* opened in London in early 1928, the British press unanimously praised the production for its camera work and sound effects. The journalist who attended the first screening for *The Times* found it difficult to describe "in so many words the marvels of photography the picture achieves; the formation of the squadrons, the manoeuvres in the air, the preparations for battle, the fighting itself, the soaring and wheeling and planning and somersaulting of the machines" exerted an overwhelming and disconcerting effect that struck the dazzled reporter as both "magnificent and terrible."[177] Declaring the film a "cinema sensation," the *Daily Herald*'s critic also struck a note that betrayed his unease at the topic despite his praise for the movie's dramatic qualities: "If you can find pleasure in seeing men, made into giants by the Magnascope, dying in agony at the wheel, then you will enjoy this picture. If you thrill at the sight of bombed villages, then 'go to it.'"[178] Both the Conservative and the left-wing newspaper found it necessary to point out that, as a movie about the air war, the feature dealt with a subject that

---

[176] On the organization of British war propaganda, see M. L. Sanders and Philip M. Taylor, *British Propaganda During the First World War, 1914–1918* (London, 1982), 15–100; Gary S. Messinger, *British Propaganda and the State in the First World War* (Manchester, 1992).
[177] *The Times*, 27 March 1928, 14.   [178] *The Daily Herald*, 27 March 1928, 5.

resisted glorification. Films that appeared to celebrate the military aviator uncritically encountered mixed receptions. In early 1939, a reviewer of *Dawn Patrol*, a feature about a Royal Flying Corps unit that suffered heavy losses as a result of inept aerial strategy in France, considered it a "good" film since neither its "humour" nor its "heroism" appeared "forced." Writing of the equanimity with which some of the fictional British pilots met their fate, the journalist argued that "it is thus, we instinctively feel, that Englishmen would and do behave under the stress of mortal danger." Another critic, however, considered the treatment of this theme as disingenuous because it merely "flatter[ed] our national pride by emphasizing what are known as the masculine virtues." In fact, this reviewer denounced the film, calling it as anachronistic as Tennyson's *The Light Brigade*, the classic Victorian poem on soldierly sacrifice.[179]

Military aviators did not command as much public reverence in Britain as did the figure of the honorable aerial knight in the Weimar Republic. Of course, significant sections of the British public continued to cherish displays of military heroism such as war movies and the annual shows staged by the RAF in Hendon. Furthermore, literary characters with a penchant for violent behavior, such as Bulldog Drummond, enjoyed a wide readership in interwar Britain. Yet the decline of British organized militarism, which had been vibrant in the Edwardian period, after the First World War left British promoters of the military without the framework of associations that supported right-wing public-relations efforts in Germany.[180] In part, the dissolution of organized militarism reflected a wider sense of disillusion about the outcome of the First World War. In the wake of the protracted battles of attrition that had marked the Great War, eulogies of individual bravery sounded unconvincing to the ears of many contemporaries. Moreover, peacetime itself, with its increased social conflicts, unemployment, and intensified international tensions in the 1930s, eroded any hope that the conflagration of the Great War had justified its human cost by improved political and social conditions in its aftermath. Promoting military heroes remained a difficult enterprise as long as "the 'Big Words' – duty, honour, country – had a hollow ring for many people."[181] It was not until the outbreak of the Second World War

---

[179] The reviews are from *The Movies*, 12 February 1939; *The Cinema*, 3 March 1939. They can be found in the micro-jacket collection at the British Film Institute in London.
[180] On the decline of militaristic organizations, see Summers, "Edwardian Militarism," 254–6; Paris, *Warrior Nation*, 155–66.
[181] Winter, *Sites of Memory*, 8. See also Samuel Hynes, *A War Imagined: The First World War and English Culture* (London, 1991).

that the image of the military aviator established itself less problematically again.

Military aviators gained undisputed status as heroic defenders of the nation in the "Battle of Britain," when the Hurricane and Spitfire pilots repelled German aerial attacks during the summer and autumn of 1940. In Winston Churchill's turn of phrase, made famous by the opening passage of an official account that sold over a million copies, "the gratitude of every home in our island, in our Empire and indeed throughout the world . . . goes out to the British airmen . . . Never in the field human conflict was so much owed by so many to so few."[182] Other publications took up this theme, hailing the pilots as the "young champions of civilization."[183] Despite its ready acknowledgment of their pivotal role in the war effort, state propaganda did not single out individual pilots, "because the RAF wanted to avoid the pitfalls of glamorizing a few heroes at the expense of the rest of the force."[184] Nonetheless, their selfless "teamwork" became a prominent symbol of their self-image in wartime.[185]

With Germany the open aggressor, the United Kingdom found itself in exactly the defensive position that had underpinned British reflections about the national significance of technology since the turn of the century. Assertions of German numerical superiority, exaggerated though they may have been, added to the Luftwaffe's aggressive image and placed Britain in the role of David confronting Goliath.[186] As they faced an opponent with seemingly inexhaustible supplies, some British pilots admitted to feeling despondent, especially at the beginning of the Battle of Britain.[187] Others described the fear that combat instilled in them.[188] As they confronted the enemy, pilots not only worried about death; the prospect of injury, particularly from burns, was even more gruesome. Air crews left the public in no doubt about the terrible price some men paid while serving their country. It was only in an elliptic style whose incomplete syntax underscored the terror of the air war that bomber pilot Leonard Cheshire could recall the

---

[182] *The Battle of Britain, August–October 1940: An Air Ministry Account of the Great Days from 8$^{th}$ August –31$^{st}$ October 1940* (London, 1941), 3. On the circulation of the pamphlet, see Richard Overy, *The Battle* (Harmondsworth, 2000), 118–19.

[183] J. M. Spaight, *The Battle of Britain, 1940* (London, 1940), 55.

[184] Overy, *The Battle*, 117.    [185] *The Battle of Britain*, 8.

[186] On German material superiority, see *The Battle of Britain: Abridged by Permission from the Air Ministry Account of the Great Days, August to October 1940* (Harmondsworth, 1941), 29; *Forever England: A Young People's Story of the Battle of Britain and the Men Who Fought It* (London, 1941), 24. Overy emphasizes that forces were actually less mismatched than was thought at the time. Overy, *The Battle*, 50–4; 72–4.

[187] For an account of dejection, see D. M. Crook, *Spitfire Pilot* (London, 1942), 31–2.

[188] Richard Hillary, *The Last Enemy* (London, 1978 [1942]), 2; Crook, *Spitfire Pilot*, 23.

wounds his radio operator had suffered during a mission: "I pray to God I may never have to see such a sight again. Instead of a face, a black, crusted mask streaked with blood, and instead of two eyes, two vivid scarlet pools. 'I'm going blind, sir, I'm going blind.' I didn't say anything. I could not have if I had wanted to."[189] Such accounts of inexpressible suffering were central to British conceptions of the air war. Characteristically, Richard Hillary, who chronicled his recovery from burns through protracted plastic surgery in a bestselling memoir, and visibly bore the traces of his injury, became the most prominent Spitfire pilot during the war.[190] The "agony" suffered by him and other pilots provided an illustration of the human cost arising from the German violation of British sovereignty that was just as powerful as accounts of the bombing raids of the Luftwaffe against the civilian population.

If narratives of suffering emphasized Britain's position as a victim of outside aggression, the courage and dedication of the pilots also provided assurance that the nation was no helpless victim. Reports praised those who, despite having been shot down earlier in the day, survived and flew "again in the afternoon."[191] Moreover, even severe injuries would not induce some aviators to abandon their posts.[192] "Grim determination" characterized their mood in critical situations, an outlook that propagandists also ascribed to the Londoners who continued to attend work despite the Blitz.[193] Given the dedication their duty required and the nature of the threat posed by Germany, pilots made no secret of the considerable satisfaction they derived from killing enemies. After shooting down his first enemy, Richard Hillary, for instance, recalled a "feeling of the essential rightness of it all."[194] Others took this theme further and recounted how war flying granted "gloriously exciting" experiences. A bomber pilot, for his part, admitted that he had "never felt such a thrill" as when he first released his charge over a German target.[195] While accounts of the "fascination" of war flying signaled the pleasures of revenge, the heroes of the air nonetheless remained reluctant to celebrate their violent feats in an unqualified manner. Writing in 1943, Leonard Cheshire admitted that "an inescapable note of sadness" tainted his memories because the outcome of encounters with the enemy depended on unpredictable luck.[196] Unlike in the First World War, however, moral

[189] Leonard Cheshire, *Bomber Pilot* (London, 1943), 82.
[190] For the account of his injuries, see Hillary, *The Last Enemy*, 3–8.
[191] *Battle of Britain: Abridged*, 19.   [192] *Forever England*, 30–1.
[193] Crook, *Spitfire Pilot*, 22. On Londoners as "not too much afraid to go to work," see *Battle of Britain: Abridged*, 23.
[194] Hillary, *The Last Enemy*, 121.   [195] Crook, *Spitfire Pilot*, 26–9; Cheshire, *Bomber Pilot*, 57.
[196] Cheshire, *Bomber Pilot*, 101.

qualms did not plague British pilots, since the war had been caused by blatant German aggression.

Thus the public image of British aerial warriors differed markedly from the depictions of military aviators in Germany during World War II. To begin with, far less coherence characterized the public appearance of British fighter pilots, who described the experience of war as confusing and contradictory. Fear, exhilaration, unspeakable suffering, despair, cruelty – all these diverse phenomena provided the public face of British military aviators because the British state did not impose a standardized propaganda image. While the German authorities retained a firm grasp over the public sphere during the Second World War, the British pursued a more circumscribed propaganda policy. Military authorities may have controlled the circulation of sensitive information, and the Ministry of Information may have aimed to influence public debate through pamphlets and press releases, but pluralism remained a defining feature of British public life to such an extent that George Orwell marveled at the fact that in December 1940 "newspapers abusing the Government... are being sold in the streets, almost without interference." In fact, the British press successfully resisted several overt censorship initiatives during the war.[197]

This comparatively liberal media environment provided the framework that allowed British pilots to retain an aura of individualism rather than conform to a state-sanctioned mold of military heroism. While war pilots remained nameless in German propaganda, British military aviators such as Richard Hillary published their accounts under civilian names, thereby marking their narratives as singular experiences. Hillary characteristically recalled his horror of organizations, and claimed to have joined war flying because it promised an "individual" and "disinterested" form of warfare.[198] Similarly, one Spitfire pilot retrospectively chided himself and his comrades for their "conceited selves" because they had boycotted drill routines during basic training as part of a reflex-like disdain of organized activity.[199] Of course, all these elements established a powerful contrast between the ethos that guided British military aviators and the prevalent image of "German mass psychology."[200] Moreover, when British military pilots admitted to feelings of fear and despair, they cast themselves as the polar opposites of

---

[197] George Orwell, "The Lion and the Unicorn," in *Essays* (Harmondsworth, 1994), 138–88, 149. This section of the essay was originally published in December 1940. Of course, Orwell claimed that wartime pluralism stemmed from a general disregard of public opinion on the part of the political elite. On propaganda in Britain and Germany, see Michael Balfour, *Propaganda in War, 1939–1945: Organisations, Policies and Publics in Britain and Germany* (London, 1979), esp. 53–102, 195–200.
[198] Hillary, *The Last Enemy*, 16.   [199] Crook, *Spitfire Pilot*, 9–10.
[200] Hillary, *The Last Enemy*, 71.

the hard and passionate Nazi warriors, who remained devoid of any trace of weakness.

Even more importantly in this context, their comparatively unideological guise distinguished British pilots favorably from those of the Luftwaffe. According to British observers, the latter had been "poisoned by the dope of Nazi doctrines," with the result that their actions displayed "an almost romantic diabolism" that intended "to make the worst fantasies of the sensational novelists come true."[201] Setting himself apart from such fanaticism, Richard Hillary echoed a widespread sentiment when he declared that he maintained "basically a suspicion of anything radical."[202] A skeptical attitude towards rigid political pronouncements not only established a firm political contrast between British and German military aviators; it also demonstrated that Britain itself avoided regressing back into the "orgy of licensed lunacy" of verbal Hun-bashing that, as contemporaries declared during the interwar period, had disgraced British propaganda in the Great War.[203] Even during the intensive bombing campaigns during the second half of the war, the language of British propaganda remained remarkably staid when compared with German pronouncements.[204] Of course, British pilots acknowledged that they fought to preserve liberty and freedom but, on the whole, they refrained from overtly ideological statements.[205] Hillary's memoir *The Last Enemy* provides the best case in point, charting its protagonist's conversion from self-confessed "selfish swine," who initially enlisted in the Royal Air Force for the prospect of adventure, to committed defender of civilization. Yet, even after identifying his cause, Hillary shied away from political terminology and stated that rather than fighting the Nazis or "the German mentality," he considered his role to combat "Evil itself" – a highly abstract and deeply moral justification for going to war.[206] Thus, during World War II, moral considerations again left a deep imprint on the public

---

[201] Spaight, *Battle of Britain*, 79; Wing Commander, *Bombers' Battle: Bomber Command's Three Years of War* (London, 1943), 15.

[202] Hillary, *The Last Enemy*, 12. Arthur Koestler identified Hillary's politically skeptical outlook and distrust of "big words and slogans" as a trait explaining his popular appeal. See Arthur Koestler, "In Memory of Richard Hillary," in *The Yogi and the Commissar and Other Essays* (London, 1964), 46–67, 51. This essay originally appeared in April 1943.

[203] The quote is from Mrs. Miniver, the fictional embodiment of female common sense writing in *The Times* in the late 1930s. Struther, *Mrs Miniver*, 65. On Mrs. Miniver, see Alison Light, *Forever England: Femininity, Literature and Conservatism between the Wars* (London, 1991), 113–55.

[204] The coverage of the large air raids on Hamburg in 1943 and Dresden in 1945 in *The Times* contained virtually no triumphalism. On the contrary, an editorial conceded that the bombing of Dresden was "a new and terrifying prodigy of air power." *The Times*, 16 February 1945, 5. See also the restrained descriptions of violence in Flying Officer X (H. E. Bates), *There's Something in the Air* (New York, 1943) which was first published in Britain.

[205] *Battle of Britain*, 8; *Battle of Britain: Abridged*, 3; Hillary, *The Last Enemy*, 102–12.

[206] Hillary, *The Last Enemy*, 56, 107, 215–16.

iconography of the British aerial soldier, whose ostentatious distrust of political radicalism corresponded to an ethos of individualism and went hand in hand with an uncompromising commitment to national defense.

German and British images of aerial fighters thus legitimized the military use of aerial technology in markedly contrasting ways. During the First World War, most German propaganda praised pilots as violent and brutal fighters, a tradition the Nazis radicalized in celebrations of military aviators as cruel, amoral warriors whose effectiveness rested on a passion for killing and a sober frame of mind. In fact, National Socialist propaganda cast pilots as disciplined operators whose efficiency replicated traits widely associated with technology. In contrast, the multiple transformations undergone by the British image of the military aviator between 1914 and 1945 supports work that contradicts the recent suggestion that a relatively stable "warrior myth" existed in twentieth-century Britain.[207] While British pilots of the Great War often displayed chivalrous and tragic traits, those of the Second World War appeared as politically skeptical individualists who, despite temporary despair, never compromised their commitment to national defense. While both countries publicly revered military aviators, the iconography of the aerial soldier led to the use of the aeroplane as an instrument of military force being regarded as a problem in Britain, but emphatically affirmed in Germany.

In sum, then, British and German assessments of the national importance of technology reveal fundamentally different ways of publicly promoting innovation despite certain similarities. Technological artifacts became potent national symbols because they bestowed prestige in both countries and lent material expression to the creative potential that was considered indispensable for national self-assertion in competitive environments that were crucially shaped and driven by Anglo-German rivalry. Numerous celebrations that rallied the population around ships, airships, and aeroplanes provided points where support was crystalized, and further enhanced technology's role as an indicator of each country's economic, political, and military power before and after the First World War. Although the Great War demonstrated technology's destructive potential, and enhanced people's fears about innovation, fundamental public opposition to technological change remained limited during the interwar years because most

---

[207] Bourke, *An Intimate History of Killing*, 33–56, argues for a consistent warrior myth. Although not directly addressing Bourke's argument, a recent article by David French casts doubt on her thesis. See David French, "'You Cannot Hate the Bastard Who is Trying to Kill You . . .' Combat and Ideology in the British Army in the War Against Germany, 1939–1945," *Twentieth Century British History* 11 (2000), 1–22.

commentators acknowledged its national importance. To be sure, the Left launched a sustained critique of technology's military uses, but it also affirmed the civilian benefits of innovation for the nation. The further commentators were situated along the right of the political spectrum, the more likely they were to embrace technology's civilian *and* military dimensions on national grounds. Rather than undermining national enthusiasm for technology, then, the Great War deepened political conflicts about the appropriate uses of innovation.

Despite their properties as national symbols, technological artifacts conveyed significantly different public meanings in Britain and Germany, and these became particularly palpable after the Great War. Debating issues of empire and national defense, the British public conceived of innovations as instruments for the preservation of a favorable international *status quo* in which it played a leading role. Britain's embrace of technology for national reasons thus reveals a conservative notion of modernity predicated on a desire for continuity. Fired by a spirit of resilience and resistance, German public discussion saw new technologies as tools to challenge a world order that frustrated Germany's claims to international status, especially after 1918. This sentiment found its most virulent articulation under the Nazi regime, which advanced the most radical demands for a break with the present to establish an alternative, unparalleled modernity.

Dissimilar notions of modernity thus underpinned British and German forms of nationally minded enthusiasm for technology. In addition to underscoring Britain's vital dependency on innovation as an imperial power, public debate emphasized that challenges from abroad obliged the country to maintain technological leadership to secure her international standing. War revealed technology's national importance in the starkest manner when both Britain's international position and her liberal political constitution underwent existential challenges. Combining praise for the aeroplane as an instrument of war with moral ambivalence and skepticism towards radical ideologies, military aviators publicly personified an ethos of national defense that struggled to maintain a balance between civilian norms and the requirements of military service. The British military pilot was thus an icon of a "liberal militarism" that linked the readiness to take up arms with a commitment to the dominant norms of the domestic political order.[208] In Germany, national ardor for technology displayed more aggressive traits, which expressed frustration at the

---

[208] For the term, see David Edgerton, "Liberal Militarism and the State"; Summers, "Edwardian Militarism," 237.

country's diminished international role after the Great War. Men such as Count Zeppelin, who was widely credited with unflagging determination in the face of obstacles, embodied the spirit of resilience and resistance that promised to transform the nation's fortunes. Often generating vehement political conflicts, the aggressive streak in German assessments of technology's national significance found its most poignant manifestation in a hero worship of aerial warriors that emphatically affirmed battle violence, thereby illustrating profoundly illiberal dimensions in German enthusiasm for aviation. The iconography of the military aviator reached its radical climax in National Socialist propaganda that celebrated pilots as ruthless killers. Contingent upon their vastly different political cultures, British and German views about technological innovation were thus inflected with dreams of the modern that took on nationally specific, highly divergent dimensions and point to the different roles that destructive impulses played in the promotion of technological innovation – an issue to which we shall return in the final chapter.

CHAPTER 9

# *Conclusion*

Between the late nineteenth century and World War II, innovative technologies figured as prominent symbols of modernity in Germany and in Britain. Aviation, passenger shipping, and film generated passionate public debates that, even at the height of political and cultural acrimony, created a cultural climate conducive to innovation. While the technologies in question acquired multiple and, at times, fervently contested public meanings, disputes about them only rarely led to a rejection of novel artifacts. Nonetheless, public support for innovations oscillated between fascination and fear, because most Britons and Germans confronted new mechanisms from a position of ignorance: they lacked a detailed understanding of how novel contraptions worked and how they were produced. Although they admired inventions and ardently welcomed many innovations, contemporaries also felt intellectually and emotionally overwhelmed. The formula that characterized technology as a "modern wonder" dramatized this ambivalence, which neither repeated references to technology's scientific foundations nor a persistent belief in progress could fully eradicate. A fundamental knowledge problem thus underpinned debate about technology.

As the proliferation of new mechanisms gave rise to vague anxieties, public reflections on the risks of the "modern wonders" had the potential to grow into a forthright technophobia that would undermine support for innovation. Debates about aviation and passenger liners concentrated on the physical dangers that revealed themselves in shipwrecks, aeroplane crashes, and airship explosions, while, according to a broad range of middle-class moralists, the cinema posed a major cultural danger to society. Although they reserved the right to abandon those aerial and maritime technologies that did not improve their safety records, the British and the German public was more concerned about the transformation of the media landscape as a consequence of the rise of film than about novel transport technologies. Unlike aviation and passenger shipping, whose physical dangers primarily affected discrete social groups such as travelers and soldiers, film seemed to constitute a much larger risk imposition on society as the cinema quickly developed a mass following. Furthermore, as discussion about flight and shipping accidents embraced a dialectical notion of progress, according to which mishaps laid foundations for further safety improvements that would eliminate existing physical dangers, it was difficult to maintain a similarly optimistic attitude with respect to film. As long as it proved impossible to explain how the movies achieved their make-belief, no one could authoritatively estimate the effects of film consumption on audiences. Although defenders of film praised the medium's artistic, propagandistic, and entertaining qualities in the interwar years, others regarded the movies as an elusive source of deception throughout our period. Public debate about film, however, did not give rise to anti-technological tirades, because film was hardly ever viewed as a technology in the first place. If film discourse remained particularly fractured, it was also contained in a wider debate about commercial mass culture whose dominant terms directed attention away from the medium's mysterious technological properties and thus blocked explicit charges that technology posed a grave cultural danger.

Given people's deep-seated ambivalence about the "modern wonders" of aviation, passenger ships, and film, as well as concern about their risks, active public support for technological innovations could never be taken for granted. Social fantasies, or dreams revealing widely shared desires, played a crucial role in generating enthusiasm for new devices in both countries. For a start, the iconography of the solo pilot promoted aviation not so much despite, as because of, its risks. The British and the German public held its breath as long-distance aviators appeared to challenge and conquer death in daring transatlantic and transcontinental flights. At a time when both countries still mourned the numerous fallen soldiers of the First World War, the successful pilots of the 1920s and early 1930s willingly entered into, re-emerged from, and told tales about death zones similar to the ones from which millions had failed to return between 1914 and 1918. While life triumphed over perdition in the transcendent image of the pilot, a shrewd and protracted public-relations initiative of the shipping industry transformed the public understanding of the ocean liner at the end of the nineteenth century. The notion of the "floating palace" cast ships as orderly worlds of luxury consumption and leisure under paternalistic regimes, while remaining silent about the gruesome labor conditions on board. This sanitized and glamorous vision tapped into pervasive dreams of affluence, and replaced the previously dominant view of ships as spaces of danger and disease with one that ultimately aestheticized them as lavish works of art. Finally, film technology came to be seen in a thoroughly positive light as amateur

cameras and projectors were praised as perfect tools for self-realization.
Rather than being a commercialized, dubious source of deception, amateur film offered a wholesome spare-time activity that fortified the individual and strengthened the family against the manifold pressures associated with modern life. Fantasies of self-fulfillment and individual autonomy, then, buttressed the public approval of amateur film equipment from the mid-1920s.

Ambivalence, fears of cultural as well as physical dangers, and prominent social fantasies shaped public evaluations of aviation, passenger shipping, and film in Germany and Britain alike between the 1890s and 1933, bringing to light transnational cultural similarities that rarely receive detailed attention in comparative Anglo-German historical scholarship. While ambivalence and assessments of physical dangers reveal strong continuities in British and German attitudes towards technology at the time, similarities also persisted well into the interwar period, when discussions took new turns as debates about the cultural dangers of film and celebrations of daring pilots illustrate. If these aspects highlight conceptions of technology that were shared in most of Western Europe at the time, British and German assessments of the significance of technology for the nation contrasted sharply. These differences had already taken shape in the decades before World War I, but became particularly pronounced after 1918. Although horror at new forms of warfare did not turn into technophobia, the First World War marked a political turning point that accentuated a significant divergence between British and German debates about technology. Not even in the context of aviation did dynamic developments in the 1930s trigger demands for restrictions on flight technologies, after it had transpired that the next war would subject civilians to previously unimaginable terrors.[1] At the same time, it needs to be stressed that the intensity of domestic political conflicts about technological innovation increased in the interwar years especially in Germany. While leaving intact the public's overwhelming support for technological change, the course and outcome of World War I enhanced strongly contrasting political evaluations of innovation in Britain and in Germany.

The British public emphasized the presentation of the empire and of the country's leading international status as the main reason why the world's leading global player had to remain at the technological forefront. In contrast, German speakers cast their nation as an international newcomer stifled

---

[1] In 1932 Stanley Baldwin proposed the international abolition of air forces to the Cabinet but quickly withdrew his proposal. See Philip Williamson, *Stanley Baldwin: Conservative Leadership and National Values* (Cambridge, 1999), 305.

by the results of the First World War. In Weimar Germany, technology symbolized the spirit of national resistance and resilience that, it was claimed, sustained drives towards innovation with the ultimate aim of transforming a seemingly hostile international environment. German appraisals of technology's national importance thus possessed more aggressive traits than did comparable statements in the United Kingdom, which conceived of their nation's pursuit of technological leadership in defensive terms. This difference became particularly palpable in the propaganda campaigns during the Great War. While depictions of British aerial warriors tended to regard the military duty to kill as a source of moral dilemmas, German portrayals glorified aerial violence in emphatic terms. This contrast sprang, to a significant extent, from the different notions of masculinity that underpinned British and German assessments of technology's national importance. Unwavering and uncompromising toughness provided the notional core of the aerial warrior in Germany, while British heroes of the air often dwelt on tensions between the requirements of military service and the wish to maintain a sense of moral integrity and individuality. The fact that German aviatrixes found it much more difficult to rise to national prominence than their British counterparts, who confronted less rigid notions of masculine toughness, further underlined the varying gender concepts that informed debates about aviation in the two countries. Different gender models thus overarched the pronounced political dissimilarities between British and German nationally motivated support for technological change.

While dissimilarities between technology's public meanings in Britain and in Germany emerged primarily in political and military contexts until the early 1930s, this changed after 1933. National Socialist understandings of technology diverged fundamentally from a variety of aspects that had dominated previous German debate, thereby, of course, also deepening the semantic contrasts with prevailing British reflections. The closest links between debates of the Weimar and the Nazi eras surfaced in statements about the national importance of technology. In this context, Nazi propagandists radicalized and militarized the rhetoric of national resilience and resistance through technological exertion. Indicative of Nazi ideology's general tendency to view human existence as a relentless existential struggle, this interpretation culminated in celebrations of aerial warriors either as bloodthirsty killers or as sober technicians of death who coolly finished off opponents. Few dimensions of public reflection on aviation, passenger shipping, and film escaped major modification by the National Socialists. When praising technology as a "modern wonder," the Nazis downplayed the importance of science and, in attempts to reconcile technology and anti-rationalism, cast engineering drives as the assertion of an existential "will" to survive. The risks associated with the technologies under consideration also underwent characteristic reinterpretation after 1933. The Nazis exhibited a schizophrenic attitude towards the physical dangers of flight when they strove to hide most aerial accidents from the public eye and simultaneously eulogized crash victims as heroes who had voluntarily sacrificed themselves on the altar of technological progress. While couching mistrust of the alleged deceptive properties of film in anti-Semitic rhetoric, Goebbels also dedicated himself to building a film industry that, he hoped, would entertain and employ the cinematic medium's power to influence people to further the party's propagandistic aims. Turning their attention to maritime technology, the National Socialists recast the "floating palace" as a palace for the racially pure and supposedly classless *Volksgemeinschaft*, divesting ocean travel of many socially exclusive overtones. Finally, although they mistrusted the ethos of individualism and personal independence that was so prominent in the hobby film movement, the National Socialists nonetheless endeavored to claim the hand-held camera for their own purposes by integrating amateur societies into the regime's propaganda apparatus. The long-distance solo pilot was the only eminent icon that the Nazis did not transform in keeping with their ideological preoccupations: as in Britain and other countries, these stars were gradually pushed to the public sidelines as the image of the responsible airline pilot rose to dominance during the 1930s.

Thus, with one exception, the National Socialists ideologically appropriated the major themes in public discourse about technology, going far beyond the espousal of technology for military and productivist purposes that have previously caught the attention of historians. The National Socialist embrace of modernity can thus hardly be dismissed as a mere façade because, as the example of technology demonstrates, National Socialism advanced, not a piecemeal, but a comprehensive alternative vision of Germany as a modern society.[2] Fervent support for engineering as the expression of a will to fight, the celebration of sacrificial risk-taking for the nation, a productivist economy that also generated new forms of consumption and entertainment open to wide sections of the *Volksgemeinschaft*, and a militaristic state that pursued aggressive policies, were central elements

---

[2] Hans Mommsen argues most vocally against suggestions to this effect. See Hans Mommsen, "Nationalsozialismus als vorgetäuschte Modernisierung," in Walter Pehle (ed.), *Der Nationalsozialismus und die deutsche Gesellschaft* (Reinbeck, 1991), 31–46. See also Norbert Frei, "Wie modern war der Nationalsozialismus?" *Geschichte und Gesellschaft* 19 (1993), 367–87.

in the vision of a National Socialist modernity that, moving beyond technology, also incorporated positive appraisals of racial science and comprehensive, tightly controlled, political mass mobilization. Although Jeffrey Herf suggests the contrary, it was *not* "paradoxical to reject the Enlightenment and embrace technology at the same time," and, rather than labeling the Nazis as "reactionary modernists," we should place them squarely and unambiguously among the myriad proponents of modernity during the first half of the twentieth century.[3] Of course, there existed a considerable discrepancy between propagandistic representations and actual developments, especially where National Socialist economic policies were concerned, but profound gulfs between image and reality also existed in other countries – the Stakhanovite movement in the Soviet Union springs to mind – that trumpeted their own modernity.[4] As a matter of fact, the very obsession to extol Germany's modernity reveals more than anything else that the National Socialists wished to be included among those who were creating a new world, even when their policies did not measure up to public pronouncements.

At the same time, it needs to be stressed that, while they initiated significant semantic shifts, the Nazis largely adhered to the dominant themes that had structured public debate about aviation, passenger shipping, and film in Germany and Britain before 1933. By adapting existing debate to their ideological tenets, the National Socialists did not effect a fundamental break with established ways of making sense of innovation. Seen in this light, Nazi writings and speeches fit into the wider context of public promotions of technological modernity in Europe between the late nineteenth century and World War II. German eulogies of engineering before 1933, National Socialist party propaganda, and British considerations throughout our entire period all shared, not just a passionate fascination for aviation, passenger shipping, and film, but also, despite their ideological dissimilarities, a view of technology as a "modern wonder" that posed a stubborn knowledge problem requiring constant examination and interpretation.

---

[3] Jeffrey Herf, *Reactionary Modernism: Technology, Culture, and Politics in Weimar and the Third Reich* (Cambridge, 1984), 3. As several historians have shown, the supposed "paradox" arises only if one normatively stipulates that a pro-democratic stance needs to underpin an embrace of technology and science. See Mario Biagoli, "Science, Modernity, and the 'Final Solution'," in Saul Friedlander (ed.), *Probing the Limits of Representation: Nazism and the Final Solution* (Cambridge, MA, 1992), 185–205; Michael T. Allen, "Modernity, the Holocaust, and Machines Without History," in Michael T. Allen and Gabrielle Hecht (eds.), *Technologies of Power: Essays in Honor of Thomas Parke Hughes and Agatha Chipley Hughes* (Cambridge, MA, 2001), 175–214.
[4] On the Stakhanovites, see Jeffrey Brooks, *Thank You Comrade Stalin: Soviet Public Culture from the Revolution to the Cold War* (Princeton, 2000), 83–105; Lewis H. Siegelbaum, "The Making of Stakhanovites, 1935–36," *Russian History/Histoire Russe* 13 (1986), 259–92.

Placing Hitler and his followers among other German and British promoters of modernity, however, raises the question: what logical connection existed between enthusiasm for technology and the unprecedented brutality unleashed by the National Socialists?

A venerable and profoundly pessimistic line of reasoning has related the growth of specialized technological and scientific knowledge to the devastating violence of European history in the first half of the twentieth century. Max Horkheimer and Theodor W. Adorno are well-known intellectual patrons of interpretations that stress the destructiveness of the enlightened "reason that rules over disenchanted nature." In their quest for knowledge, Horkheimer and Adorno famously argued, in *Dialektik der Aufklärung* (*The Dialectic of Enlightenment*) in 1944, that "what men want to learn from nature is how to use it in order wholly to dominate it and other men." Noting that "the essence of such knowledge is technology," Horkheimer and Adorno viewed this kind of instrumental knowledge as inherently destructive because its power drive annihilated the "critical elements" that had initially provided the emancipatory impulses of the Enlightenment project. According to *Dialektik der Aufklärung*, the enlightened world "glowed in triumphant catastrophe" by mid-century, having turned on to itself its ability to produce ever more powerful instruments.[5]

The founders of the Frankfurt School maintain that dynamic technological and scientific progress resulted in a disastrous epistemic problem as practical knowledge overwhelmed its creators. Other scholars have pursued related lines of reasoning, often casting Nazi Germany as prototypical of the pathology of modernity, which manifested itself in wanton destruction. Reflecting on the "major transformations in demographic patterns and social organization, in politics and industry, and in science and technology," Omer Bartov has recently argued that their "immense impact" in times of war did not solely stem from the material enhancement of "war's destructive potential"; the prospect of unprecedented martial violence, Bartov goes on to claim, also triggered a barrage of the "wildest fantasies and most nightmarish visions" that demonstrate that "ours is a century in which man's imagination has been conducting a desperate race with the practice of humanity." According to Bartov, the inability of "the mind" to "catch up with man-made reality" led to "visions of the future that surpassed all known ... destruction and thereby created the preconditions for even greater suffering, pain, and depravity."[6] Bartov contributes

---

[5] Max Horkheimer and Theodor W. Adorno, *Dialektik der Aufklärung: Philosophische Fragmente* (Frankfurt, 1971), 2 ff.

[6] Omer Bartov, *Mirrors of Destruction: War, Genocide, and Modern Identity* (New York, 2000), 10.

to the influential interpretational tradition that Horkheimer and Adorno launched, but modifies it in a crucial respect: he posits a catastrophic link between acts of the imagination and unparalleled means of implementing them, rather than ascribing inherently fatal characteristics to technological knowledge *per se*. Horkheimer/Adorno and Bartov are at the opposite yet related ends of a spectrum of inquiries that argue that technological innovation either produced self-destructive expert knowledge or unleashed irrational, murderous fantasies that could be turned into reality with the help of new machinery. Scholars as diverse as Zygmunt Bauman, Götz Aly, Susanne Heim, and Detlev Peukert have recently contended that scientific and technological advances eroded the ethical foundations of societies that embraced modernity by either producing intrinsically destructive forms of rational knowledge or delivering the tools to implement previously inconceivable crimes.[7]

The findings presented here cast doubt on the charges of those who, claiming to follow in Horkheimer and Adorno's immediate intellectual footsteps, condemn "Enlightenment rationalism through guilt by association" with National Socialism, as Michael Allen has recently put it.[8] First of all, the Nazis repeatedly praised the supposedly irrational foundations of technology when they considered novel artifacts as material expressions of the will to survive rather than as products of reason. Furthermore, there is little evidence to suggest that an inherently destructive logic in technological developments drove the Third Reich to commit its crimes. Rather, as we saw in the sections on aerial warfare, the National Socialists claimed authority over technology, using it consciously for destructive purposes and openly celebrating their atrocities. For the German case, Omer Bartov's charge that, among other developments, technological change stoked irrational desires that, in turn, exploded in acts of destruction committed with the help of technology is incisive.

Nonetheless, we also ought to note that British public debate proved more resistant to the temptation to extol brutal military undertakings that relied on recent inventions. To be sure, the British public did not hesitate to praise technology's potential for expanding and consolidating imperial

---

[7] Zygmunt Bauman, *Modernity and the Holocaust* (Ithaca, 1989), 68. See also *idem*, *Modernity and Ambivalence* (Oxford, 1991), esp. 18–52. Other inquiries along related lines include Detlev J. K. Peukert, "The Genesis of the 'Final Solution' from the Spirit of Science," in Thomas Childers and Jane Caplan (eds.), *Reevaluating the Third Reich* (New York, 1993), 234–52; Götz Aly and Susanne Heim, *Vordenker der Vernichtung* (Frankfurt, 1993).

[8] Allen, "Modernity, the Holocaust, and Machines Without History," 183. The history of cancer research has given rise to similar considerations. See Robert N. Proctor, *The Nazi War on Cancer* (Princeton, 1999), 6, 248–52.

rule. Moreover, in World War II many British pilots shed the restraint that had governed earlier propaganda on war flying and began to take pride in the destruction they wreaked among their German enemies. Yet British positive acknowledgments of aerial violence gained public currency only against a background of German aggression. On the whole, the liberalism that suffused Britain's public culture more thoroughly than Germany's both assigned to technology a central military significance and simultaneously restrained fantasies of national power that involved the use of technology with an open disregard for individual life. Put differently, a comparative examination of British and German public debate thus shows that the causal link between an embrace of technological change and fantasies of destruction, which laid the mental foundations for actual atrocities, is more strongly contingent upon a nation's political culture than the works cited above allow.

Finally, a focus on technology's undeniable destructive potential captures only part of the public fascination for innovation at the time. Even during the highly disruptive, violent period between the end of the nineteenth century and World War II, a host of social fantasies and attitudes towards risk, as well as a specific form of ambivalence, were necessary to create the passionate support for technology whose fervent enthusiasm stands in contrast with the sober, at times somber, present-day public attitudes in Britain and Germany. The new tone of public debate about engineering is indicative of the major transformation in European public evaluations of technology since World War II. While inventions such as television, supersonic flight, space exploration, and the computer have continued to elicit delighted wonder, a rising number of critical voices have made themselves heard and marked new levels of public resistance to innovation, especially since the 1960s. Partly, this growing skepticism has resulted from the demise of narratives about the national significance of technology, which had generated public enthusiasm in Britain and Germany before the Second World War. As both countries settled into secondary global roles in the postwar era, backers of costly innovations found it increasingly difficult to garner public support by arguing that their countries required certain technologies to stabilize or regain a position of world leadership. Furthermore, the nuclear threat created fears that dynamic developments in engineering would lead to Armageddon.[9] Apprehensions about nuclear annihilation, in turn, intensified anxiety about the physical dangers of technology in everyday life. To

---

[9] Meredith Veldman, *Fantasy, the Bomb and the Greening of Britain: Romantic Protest, 1945–1980* (Cambridge, 1994). For an example from US history, see Paul Boyer, *By the Bomb's Early Light: American Thought and Culture at the Dawn of the Atomic Age* (Chapel Hill, 1994).

name but two examples, denunciations of civilian nuclear technology and environmental fears have revolved around the conviction that technological risks no longer affect specific groups, but pervade society in its entirety.[10] If worries about pervasive risk impositions sometimes take the form of veritable paranoia, they have also significantly lowered risk tolerance, since mishaps and accidents are no longer viewed primarily as stepping stones towards future improvements but as confirmations of the omnipresence of danger.[11] In short, the postwar years have witnessed a significant erosion of the public belief in technological modernity in Europe despite continuing innovation, and it is for this reason that the Second World War represents this book's chronological endpoint.

[10] Martin Bauer (ed.), *Resistance to New Technology: Nuclear Power, Information Technology and Biotechnology* (Cambridge, 1995). On nuclear technology in Germany, see Joachim Radkau, *Technik in Deutschland: Vom 18. Jahrhundert bis zur Gegenwart* (Frankfurt, 1989), 351–7.
[11] Ulrich Beck, *Risikogesellschaft: Auf dem Weg in eine andere Moderne* (Frankfurt, 1986).

# Bibliography

## PRIMARY SOURCES

### ARCHIVAL SOURCES

British Film Institute Library
    Microjacket collection
Brooklands Museum, Weybridge, Technical Archives
Deutsches Museum, Munich, archive
    brochure collection
    Junkers archive
    Nachlaß Etzdorf
    Nachlaß Reitsch
    Luft- und Raumfahrtdokumentation (LRD)
Deutsches Schiffahrtsmuseum Bremerhaven
    library
    archive
Ellis Island Museum, New York City, Ellis Island Oral History Project
Hochschule für Fernsehen und Film München Library
    Privatarchiv Eberhard von Berswordt
Royal Air Force Archive, Hendon
    file AC 77 (Amy Johnson)
Stiftung Deutsche Kinemathek Berlin,
    Schriftgutarchiv
University of Cambridge, Centre of South Asian Studies, film archive
University of Liverpool, University Archives
    Cunard archive, file D42
Yorkshire Film Archive, York
    Cine film collection

### PUBLISHED MATERIAL

#### Newspapers and periodicals

*The Aeroplane*
*Amateur Ciné World*

*The Amateur Photographer and Cinematographer*
*Arbeiter-Illustrierte Zeitung*
*Arbeiterbühne und Film*
*Arbeitertum*
*Berliner Tageblatt*
*The Bioscope*
*British Gaumont News*
*B. Z. am Mittag*
*Ciné-Kodak-Magazin*
*Daily Express*
*Daily Herald*
*Daily Mail*
*Film für Alle*
*Film und Volk*
*Film Welt*
*Flight*
*Frankfurter Zeitung*
*Imperial Airways Gazette*
*Kino-Amateur*
*The Kodak Magazine*
*Labour Leader*
*Luftfahrt*
*Die Luftreise*
*Manchester Guardian*
*Motor und Sport*
*Nationalsozialistische Monatshefte*
*Neue Preußische Zeitung*
*New York Times*
*NS-Frauen-Warte*
*Pathé Cine Journal*
*Punch*
*Rote Fahne*
*The Times*
*Ufa-Magazin*
*Völkischer Beobachter*
*Vorwärts*

Books and articles

*5 Jahre Reichsluftschutzbund* (Berlin, 1938).
*1909: General Catalogue of Classified Subjects: "Urban," "Eclipse," "Radios" Film Subjects and "Urbanora" Educational Series* (London, 1909).
*ABC des Luftschutzes: herausgegeben vom Präsidium des Reichsluftschutzbundes* (Berlin, 1940).
Adler, Hermann, *Unsere Luftwaffe in Polen: Erlebnisberichte* (Berlin, 1940).

*The Air Defence of Britain* (Harmondsworth, 1938).
Altenloh, Emilie, *Zur Soziologie des Kinos* (Jena, 1914).
Anderson, B. Wherry, *The Romance of Air-Fighting* (London, 1917).
Arnheim, Rudolf, *Film als Kunst* (Berlin, 1932).
*Die Auswandererhallen der Hamburg-Amerika-Linie* (Hamburg, 1907).
*Aviaticus: Jahrbuch der deutschen Luftfahrt, 1931* (Berlin, 1931).
Balazs, Bela, *Der Geist des Films* (Halle, 1930).
Barry, Iris, *Let's Go to the Pictures* (London, 1926).
*The Battle of Britain, August–October 1940: An Air Ministry Account of the Great Days from 8th August–31st October 1940* (London, 1941).
*The Battle of Britain: Abridged by Permission from the Air Ministry Account of the Great Days, August to October 1940* (Harmondsworth, 1941).
Beaumont, André, *My Three Big Flights* (London, 1912).
Beinhorn, Elly, *180 Stunden über Afrika* (Berlin, 1933).
  *Ein Mädchen fliegt um die Welt* (Berlin, 1932).
Bewsher, Paul, *Green Balls: The Adventures of a Night-Bomber* (Edinburgh, 1919).
Bishop, W. A., *Winged Warfare: Hunting the Huns in the Sky* (London, 1918).
Bongartz, Heinz, *Luftmacht Deutschland: Luftwaffe – Industrie – Luftfahrt* (Essen, 1939).
Breuhaus de Groot, Fritz, *Bauten und Räume* (Berlin, 1935).
Briscoe, Walter A., and Stannard, H. Russel, *Captain Ball VC* (London, 1918).
Broadley, A. M., *The Ship Beautiful: Art and the Aquitania* (Liverpool, 1914).
Brunner, Karl, *Das neue Lichtspielgesetz im Dienst der Volks- und Jugendwohlfahrt* (Berlin, 1920).
Buchanan, Andrew, *Films: The Way of the Cinema* (London, 1932).
Buckle, Gerard Ford, *The Mind and the Film: A Treatise on the Psychological Factors in the Film* (London, 1926).
Burgess, Marjorie A. Lovell, *A Popular Account of the Development of the Amateur Ciné Movement in Great Britain* (London, 1932).
Burnett, R. G., and Martell, E. D., *The Devil's Camera: The Menace of a Film-Ridden World* (London, 1932).
Busch, Karl (ed.), *Nach den glücklichen Inseln: Mit KdF-Flaggschiff "Robert Ley" nach der farbenprächtigen Welt von Madeira und Teneriffa* (Berlin, 1940).
Chatterton, E. Keble, *Aquitania: The Making of a Mammoth Liner* (London, 1914).
Cheshire, Leonard, *Bomber Pilot* (London, 1943).
Cobham, Alan, *Australia and Back* (London, 1926).
  *Skyways* (London, 1925).
Courtenay, William, *Airman Friday* (London, 1937).
Crook, D. M., *Spitfire Pilot* (London, 1942).
*The Cunard Passenger Logbook* (Glasgow, 1897).
*Cunard Passenger Logbook: A Short History of the Cunard Steamship Company and a Description of the Royal Mail Steamers Campania and Lucania* (Glasgow, 1904).
Davy, Charles (ed.), *Footnotes to the Film* (London, 1937).

Deedes, William F., *A.R.P.: A Complete Guide to Civil Defence Measures* (London, 1939).
Dettman, Fritz, *Zeppelin: Gestern und Morgen: Geschichte der deutschen Luftfahrt von Friedrichshafen bis Frankfurt am Main* (Berlin, 1936).
*Die deutsche Luftfahrzeugindustrie auf der Deutschen Verkehrsausstellung München 1925* (Munich, 1925).
Dietrich, Otto, *Mit Hitler an die Macht: Persönliche Erlebnisse mit meinem Führer* (Munich, 1934).
*Deutsche Lufthansa Summer Time Table 1938* (no place, no date).
Dixon, Charles, *Amy Johnson – Lone Girl Flyer* (London, 1930).
Dykes, Robert, *The Amateur Cinematographer's Handbook on Movie-Making* (London, 1931).
Eckener, Hugo, *Die Amerikafahrt des "Graf Zeppelin,"* ed. Rolf Brandt (Berlin, 1928).
Eichelbaum, Dr., *Das Buch von der Luftwaffe* (Berlin, 1941).
*Die Entwickelung des Norddeutschen Lloyd Bremen* (Bremen, 1912).
Etzdorf, Marga v., *Kiek in die Welt: Als deutsche Fliegerin über drei Erdteilen* (Berlin, 1931).
*Fact, No. 13: Air Raid Protection by Ten Cambridge Scientists* (London, 1938).
*Film in National Life, The, Being the Report of an Enquiry Conducted by the Commission on Educational and Cultural Films into the Service which the Cinematograph May Render to Education and Social Progress* (London, 1932).
Flying Officer X (H. E. Bates), *There's Something in the Air* (New York, 1943).
*Forever England: A Young People's Story of the Battle of Britain and the Men Who Fought It* (London, 1941).
Frei, Friedrich (ed.), *Unser Fliegerheld Immelmann: Sein Leben und seine Heldentaten* (Leipzig, 1916).
Frerk, Friedrich Willy, *Der Kino-Amateur: Ein Lehr- und Nachschlagebuch* (Berlin, 1926).
(ed.), *Jahrbuch des Kino-Amateurs, 1931–32* (Berlin, 1931).
Fröhlich, Elke (ed.), *Die Tagebücher von Joseph Goebbels: Sämtliche Fragmente*, Teil I: *Aufzeichnungen 1924–1941*, vol. III (Munich, 1987).
Fülöp-Miller, Rene, *Die Fantasiemaschine: Eine Saga der Gewinnsucht* (Berlin, 1931).
Furniss, Harry, *Our Lady Cinema* (Bristol, 1914).
Gad, Urban, *Der Film: Seine Mittel – Seine Ziele* (Berlin, 1921).
Galsworthy, John, *A Modern Comedy: The Forsyte Chronicles*, vol. II (Harmondsworth, 1980 [1926]).
Gaupp, Robert, "Über moralisches Irresein und jugendliches Verbrechertum," *Juristisch-psychiatrische Grenzfragen: Zwanglose Abhandlungen* 2 (1904), 51–68.
Geisenheyner, Max, *Mit "Graf Zeppelin" um die Welt: Ein Bild-Buch* (Frankfurt, 1929).
*Graf Zeppelin und sein Luftschiff: Luxusausgabe* (Nuremberg, 1908).
Grahame-White, Claude, *Aviation* (London, 1912).
Grahame-White, Claude, and Harper, Harry, *The Aeroplane: Past, Present and Future* (London, 1911).

Graves, Robert, *Goodbye to All That* (Harmondsworth, 1960 [1926]).
Gross-Talmon, Paul, *Filmrezepte für den Hausgebrauch: Kleine Drehbücher für jedermann* (Halle, 1939).
Hamburg-Amerika Linie, *Wiederaufbau der deutschen Überseeschiffahrt* (Berlin, 1923).
*Hauptmann Bölckes Feldberichte, mit einer Einleitung von der Hand des Vaters* (Gotha, 1917).
Heath, Lady, and Wolfe Murray, Stella, *Woman and Flying* (London, 1929).
Heiber, Helmut (ed.), *Goebbels Reden*, II: *1939–1945* (Düsseldorf, 1972).
Herrnkind, Paul, *Die Schmalfilm-Kinematographie* (Vienna and Leipzig, 1929).
Hillary, Richard, *The Last Enemy* (London, 1978 [1942]).
Hirth, Hellmuth, *20,000 Kilometer im Luftmeer* (Berlin, 1913).
Hitler, Adolf, *Mein Kampf*, 218th edition (Munich, 1936).
Hobbs, Edward, *Cinematography for Amateurs: A Simple Guide to Motion Picture Taking, Making and Showing* (London, 1930).
Hoeppner, Ernst von, *Deutschlands Krieg in der Luft: Ein Rückblick auf die Entwicklung und die Leistungen unserer Heeres-Luftstreitkräfte im Weltkriege* (Leipzig, 1921).
Hoogh, Peter, *Zeppelin und die Eroberung des Luftmeeres* (Berlin, 1908).
Hubert, Ali, *Hollywood: Legende und Wirklichkeit* (Leipzig, 1930).
Hunke, Heinrich, *Luftgefahr und Luftschutz* (Berlin, 1935).
*Illustrated Souvenir of the Cunard Steamship Co., Limited* (Liverpool, 1902).
*Imperator auf See: Gedenkblätter an die erste Ausfahrt des Dampfers Imperator am 11. Juni 1913* (Hamburg, 1913).
Internationale Studiengesellschaft zur Erforschung der Arktis mit dem Luftschiff (ed.), *Das Luftschiff als Forschungsmittel in der Arktis: Eine Denkschrift* (no place, 1924).
Italiaander, Rolf, *Drei deutsche Fliegerinnen: Elly Beinhorn, Thea Rasche, Hanna Reitsch* (Berlin, 1940).
Johnson, Amy, *Sky Roads of the Air* (London, 1939).
Jünger, Ernst (ed.), *Luftfahrt ist not!* (Leipzig, 1929).
Jünke, Arnold, *Zeppelin im Weltkriege* (Leipzig, 1916).
*Kameradschaft der Luft* (Berlin, 1938).
Kipling, Rudyard, *Gunga Din and Other Favorite Poems* (New York, 1990).
Klebinder, Paul, *Der deutsche Kaiser im Film* (Berlin, 1912).
Klemperer, Victor, *Leben sammeln, nicht fragen wozu und warum: Tagebücher 1925–1932* (Berlin, 1996).
Koestler, Arthur, "In Memory of Richard Hillary," in *The Yogi and the Commissar and Other Essays* (London, 1964), 46–67.
Köhl, Hermann, Fitzmaurice, James C., and Hünefeld, Günther v., *The Three Musketeers of the Air: Their Conquest of the Atlantic from East to West* (New York, 1928).
Kracauer, Siegfried, *Kino: Essays, Studien, Glossen zum Film* (Frankfurt, 1974).
Kraszna Krausz, Andor (ed.), *Kurble! Ein Lehrbuch des Filmsports* (Halle, 1929).
Kurbs, Dr. (ed.), *Die deutsche Luftwaffe* (Berlin, 1936).

Lange, Helmut, *Filmmanuskripte und Film-Ideen: 121 Ideen für den Kino-Amateur* (Berlin: Photokino Verlag, 1931).
*Wie entsteht ein Amateurfilm? Ein Amateur plaudert von seinen Filmarbeiten* (Berlin, 1930).
*Der neue Schmalfilmer* (Berlin, 1941).
Langlands, Thomas, *Popular Cinematography: A Book for the Camera* (London, 1926).
Lehmann, Ernst, *Auf Luftpatrouille und Weltfahrt: Erlebnisse eines Zeppelinführers* (Leipzig, 1936).
Lejeune, C. A., *Cinema* (London, 1931).
Lemke, Hermann, *Die Kinematographie der Gegenwart, Vergangenheit und Zukunft: Eine kulturgeschichtliche und industrielle Studie* (Leipzig, 1911).
Liesegang, Paul, *Das lebende Lichtbild: Entwicklung, Wesen und Bedeutung des Kinematographen* (Düsseldorf, 1910).
Lindbergh, Charles A., *We: The Famous Flier's Own Story of His Life and His Transatlantic Flight* (New York, 1927).
Long, S. H., *In the Blue* (London, 1920).
Lore, Colden, *The Modern Photoplay and its Construction* (London, 1923).
Ludendorff, Erich, *Kriegführung und Politik* (Berlin, 1922).
*Meine Kriegserinnerungen, 1914–1918* (Berlin, 1919).
Luftschiffbau Zeppelin G.m.b.H., *LZ 126: 20 Originalphotographien vom Amerikaluftschiff* (Friedrichshafen, 1924).
Lummerzheim, Heinz, *Das AGFA-Schmalfilm-Handbuch* (Harzburg, 1935).
Malina, J. B. (ed.), *Luftfahrt voran! Das deutsche Fliegerbuch* (Berlin, 1932).
Mann, Thomas, "Unterwegs," in *Bemühungen* (Berlin, 1925), 257–67.
McCudden, James T. B., *Five Years in the Royal Flying Corps* (London, 1919).
Meier-Beneckenstein, Paul (ed.), *Dokumente der deutschen Politik*, vol. III (Berlin, 1937).
Menzies, H. Stuart, *All Ways by Airways* (London, 1932).
Messel, Rudolph, *This Film Business* (London, 1928).
Meyer, E., and Sellien, E., *Schule und Luftschutz* (Munich, 1934).
Milch, Erhard, *Die Sicherheit im Luftverkehr auf Grund der Betriebsergebnisse der Deutschen Luft Hansa, 1926–1928* (Berlin, 1929).
Mollison, James, *Death Cometh Soon or Late* (London, 1932).
*Playboy of the Air* (London, 1937).
Moreck, Curt, *Sittengeschichte des Kinos* (Dresden, 1926).
Münzenberg, Willi, *Erobert den Film: Winke aus der Praxis für die Praxis proletarischer Filmpropaganda* (Berlin, 1925).
National Council of Public Morals, *The Cinema: Its Present Position and Future Possibilities* (London, 1917).
Neumann, Carl, Belling, Curt, and Betz, Hans-Walther, *Film-"Kunst", Film-Kohn, Film-Korruption: ein Streifzug durch vier Filmjahrzehnte* (Berlin, 1937).
Norddeutscher Lloyd Bremen, *Die Geschichte des Norddeutschen Lloyd* (Bremen, 1901).
*Volkstümliche Reisen zur See* (Bremen, 1925).

North German Lloyd, *Round the World in Ninety Days in the Bremen* (Boston, 1937).
N.-S. Gemeinschaft Kraft durch Freude, *Hochsee- und Landreisen im Urlaubsjahr 1938* (Bayreuth, 1937).
*Jahresprogramm 1935 Gau Franken* (Nuremberg, 1934).
*NS-Presseanweisungen der Vorkriegszeit: Edition und Dokumentation*, vol. v: 1937, ed. Karen Peter (Munich, 1998).
Orlovius, Heinz (ed.), *Schwert am Himmel: Fünf Jahre deutsche Luftfahrt* (Berlin, 1940).
Orwell, George, "The Lion and the Unicorn," in *Essays* (Harmondsworth, 1994 [1940]).
*Parliamentary Debates: House of Commons*, fourth and fifth series (London).
Plüschow, Günther, *Der Flieger von Tsingtau* (Berlin, 1927).
Poeschel, Johannes (ed.), *Einführung in die Luftfahrt* (Leipzig, 1925).
Porten, Henny, *Vom "Kintopp" zum Tonfilm: Ein Stück miterlebter Filmgeschichte* (Dresden, 1932).
Priestley, J. B., *English Journey* (London, 1994 [1933]).
*R 101: The Airship Disaster, 1930* (London, 1999 [1931]).
Rawnsley, H. D., *The Child and the Cinematograph Show and the Picture Post-Card Evil* (London, 1913).
*Das Reichslichtspielgesetz vom 12. Mai 1920: Für die Praxis erläutert von Dr. jur. Ernst Seeger* (Berlin, 1932).
*Report of the Colonial Films Committee*, Cmd. 3630 (London, 1930).
*Report on the Loss of the "Titanic,"* Cmd. 8194 (London, 1912).
Richthofen, Manfred Freiherr von, *Der rote Kampfflieger* (Berlin, 1917).
Riegler, Hans, *Deutscher Luftschutz: Jahrbuch* (Berlin, 1940).
Robson, E. W. and M. M., *The Film Answers Back: An Historical Appreciation of the Cinema* (London, 1939).
Roskoten, Richard, *Ziviler Luftschutz: Ein Buch für das deutsche Volk* (Düsseldorf, 1932).
Rotha, Paul, *Documentary Film*, 2nd edition (London, 1939).
  *The Film Till Now: A Survey of the Cinema* (London, 1930).
*Royal Mail Triple-Screw Steamers "Olympic" and "Titanic"* (Liverpool, 1911).
Rumpf, Hans, *Gasschutz: Ein Handbuch* (Berlin, 1928).
Rüst, Ernst, *Der praktische Kinoamateur* (Stuttgart, 1925).
Saager, Adolf (ed.), *Zeppelin: Der Mensch, der Kämpfer, der Sieger* (Stuttgart, 1915).
Saint-Exupéry, Antoine de, *Night Flight* (New York, 1938).
Schaffner, Jakob, *Volk zu Schiff: Zwei Seefahrten mit der 'KdF'-Hochseeflotte* (Hamburg, 1936).
Schultze, Ernst, *Der Kinematograph als Bildungsmittel: Eine kulturpolitische Untersuchung* (Halle, 1911).
Shand, P. Morton, *Modern Theatres and Cinemas: The Architecture of Pleasure* (London, 1930).
Spaight, J. M., *Air Power and the Cities* (London, 1930).
  *The Battle of Britain, 1940* (London, 1940).

Spengler, Oswald, *Der Mensch und die Technik: Beitrag zu einer Philosophie des Lebens* (Munich, 1931).
Stackelberg, Karl Georg von, *Legion Condor: Deutsche Freiwillige in Spanien* (Berlin, 1939).
Steer, Valentia, *The Secrets of the Cinema* (London, 1920).
Stepun, Fedor, *Theater und Kino* (Berlin, 1932).
Strasser, Alex, *Kind und Kegel vor der Kamera* (Halle, 1932).
Streicher, Julius (ed.), *Reichstagung in Nürnberg 1934* (Berlin, 1934).
Struther, Jan, *Mrs Miniver* (London, 2001 [1939]).
Stüler, A., *Filmen leicht gemacht*, 2nd edition (Stuttgart, 1931).
Supf, Peter, *Luftwaffe von Sieg zu Sieg: Von Norwegen bis Kreta* (Berlin, 1942).
Talbot, Frederick A., *Moving Pictures: How They Are Made and Worked* (London, 1912).
Thielmann, Ewald, *Amateur-Kinematographie* (Leipzig, 1925).
Thomson, Lord, *Air Facts and Air Problems* (London, 1927).
Tilgenkamp, E. (ed.), *Do X* (Zürich, 1931).
*Verhandlungen des Reichstags: Stenographische Berichte* (Berlin).
Werner, Johannes (ed.), *Briefe eines deutschen Kampfpiloten an ein junges Mädchen* (Leipzig, 1930).
Wing Commander, *Bombers' Battle: Bomber Command's Three Years of War* (London, 1943).
*Wir Luftkämpfer: Bilder und Berichte deutscher Flieger und Luftschiffer* (Berlin, 1918).
*Zeppelin fährt um die Welt: Ein Gedenkbuch der "Woche"* (Berlin, 1929).
*Zeppeline gegen England* (Berlin, 1916).

## SECONDARY SOURCES

Abel, Richard, *The Ciné Goes to Town: French Cinema, 1896–1914* (Berkeley, 1994).
  (ed.), *Silent Film* (London, 1996).
Adas, Michael, *Machines as the Measure of Man: Science, Technology, and Ideologies of Western Dominance* (Ithaca, 1990).
Albrecht, Gerd, *Nationalsozialistische Filmpolitik: Eine soziologische Untersuchung über die Spielfilme des Dritten Reiches* (Stuttgart, 1969).
Aldrich, Mark, *Safety First: Technology, Labor, and Business in the Building of American Work Safety* (Baltimore, 1997).
Allen, Michael T., "Modernity, the Holocaust, and Machines Without History," in Michael T. Allen and Gabrielle Hecht (eds.), *Technologies of Power: Essays in Honor of Thomas Parke Hughes and Agatha Chipley Hughes* (Cambridge, MA, 2001), 175–214.
  "The Puzzle of Nazi Modernism: Modern Technology and Ideological Consensus in an SS-Factory at Auschwitz," *Technology and Culture* 37 (1996), 527–71.
Allen, Robert C., "*Traffic in Souls*," *Sight and Sound* 44 (1975), 50–2.
Altick, Richard, *The Shows of London* (Cambridge, MA, 1978).
Aly, Götz, and Heim, Susanne, *Vordenker der Vernichtung* (Frankfurt, 1993).

Amstädter, Rainer, *Der Alpinismus* (Vienna, 1996).
Andersen, Arne, and Ott, René, "Risikoperzeption im Industriezeitalter am Beispiel des Hüttenwesens," *Archiv für Sozialgeschichte* 28 (1988), 75–109.
Anderson, Joseph and Barbara, "Motion Perception in Motion Pictures," in Teresa de Lauretis (ed.), *The Cinematic Apparatus* (London, 1980), 76–95.
Andrew, Donna T., "The Code of Honour and its Critics: The Opposition to Duelling in England, 1700–1850," *Social History* 5 (1980), 409–34.
Ankum, Katharina von (ed.), *Women and the Metropolis: Gender and Modernity in Weimar Culture* (Berkeley, 1997).
Armes, Roy, *A Critical History of British Cinema* (London, 1978).
Ascheid, Antje, "Nazi Stardom and the 'Modern Girl': The Case of Lilian Harvey," *New German Critique* 74 (1998), 57–89.
Auerbach, Jeffrey, *The Great Exhibition of 1851: A Nation on Display* (New Haven, 1999).
Augustine, Dolores L., *Patricians and Parvenus: Wealth and High Society in Wilhelmine Germany* (Oxford, 1994).
Bailey, Peter, "Conspiracies of Meaning: Music-Hall and the Knowingness of Popular Culture," *Past and Present* 144 (1994), 139–70.
Balfour, Michael, *Propaganda in War, 1939–1945: Organisations, Policies and Publics in Britain and Germany* (London, 1979).
Baranowski, Shelley, "Strength through Joy: Tourism and National Integration in the Third Reich," in Shelley Baranowski and Ellen Furlough (eds.), *Being Elsewhere: Tourism, Consumer Cultures, and Identity in Modern Europe and North America* (Ann Arbor, 2001), 213–36.
Barbian, Jan-Pieter, "Politik und Film in der Weimarer Republik: Ein Beitrag zur Kulturgeschichte der Weimarer Republik," *Archiv für Kulturgeschichte* 80 (1998), 213–45.
Barnett, Correlli, *The Collapse of British Power* (Gloucester, 1984).
Barthes, Roland, *Camera Lucida: Reflections on Photography* (New York, 1981).
Barton Michael, "Journalistic Gore: Disaster Reporting and Emotional Discourse in the *New York Times*, 1852–1956," in Peter Stearns and Jan Lewis (eds.), *An Emotional History of the United States* (New York, 1998), 155–72.
Bartov, Omer, *Hitler's Army: Soldiers, Nazis, and War in the Third Reich* (New York, 1992).
*Mirrors of Destruction: War, Genocide, and Modern Identity* (New York, 2000).
Bauer, Martin (ed.), *Resistance to New Technology: Nuclear Power, Information Technology and Biotechnology* (Cambridge, 1995).
Bauman, Zygmunt, *Modernity and Ambivalence* (Oxford, 1991).
*Modernity and the Holocaust* (Ithaca, 1989).
Bausinger, Hermann, "Bürgerlichkeit und Kultur," in Jürgen Kocka (ed.), *Bürger und Bürgerlichkeit im 19. Jahrhundert* (Göttingen, 1987), 121–42.
Beck, Ulrich, *Risikogesellschaft: Auf dem Weg in eine andere Moderne* (Frankfurt, 1986).
Becker, Gay, *Disrupted Lives: How People Create Meaning in a Chaotic World* (Berkeley, 1997).

Bédarida, François, *A Social History of England, 1815–1990* (London, 1991).
Behe, George, *Titanic: Psychic Forewarnings of a Tragedy* (Wellingborough, 1988).
Behrenbeck, Sabine, *Der Kult um die toten Helden: Nationalsozialistische Mythen, Riten und Symbole, 1923 bis 1945* (Vierow, 1996).
Berger, Stefan, *The British Labour Party and the German Social Democrats, 1900–1931* (Oxford, 1994).
Bering, Dietz, "Jews and the German Language: The Concept of *Kulturnation* and Anti-Semitic Propaganda," in Norbert Finzsch and Dietmar Schirmer (eds.), *Identity and Intolerance: Nationalism, Racism, and Xenophobia in Germany and the United States* (Cambridge, 1998), 251–91.
Berman, Marshall, *All That Is Solid Melts Into Air: The Experience of Modernity* (London, 1983).
Bessel, Richard (ed.), *Fascist Italy and Nazi Germany: Comparisons and Contrasts* (Cambridge, 1996).
Biagini, Eugenio (ed.), *Citizenship and Community: Liberals, Radicals, and Collective Identities in the British Isles, 1865–1931* (Cambridge, 1996).
Biagoli, Mario, "Science, Modernity, and the 'Final Solution,'" in Saul Friedlander (ed.), *Probing the Limits of Representation: Nazism and the Final Solution* (Cambridge, MA, 1992), 185–205.
Bialer, Uri, *The Shadow of the Bomber: The Fear of Air Attack and British Politics, 1932–1939* (London, 1980).
Biel, Steven, *Down With the Old Canoe: A Cultural History of the Titanic Disaster* (New York, 1996).
Bird, Jay W., *To Die for Germany: Heroes in the Nazi Pantheon* (Bloomington, 1990).
Blackbourn, David, "'Taking the Waters': Meeting Places of the Fashionable World," in Martin H. Geyer and Johannes Paulmann (eds.), *The Mechanics of Internationalism: Culture, Society, and Politics from the 1840s to the First World War* (Oxford, 2001), 435–57.
Blackbourn, David, and Eley, Geoff, *The Peculiarities of German History: Bourgeois Society and Politics in Nineteenth-Century Germany* (Oxford, 1984)
Bölkow, Ludwig, *Ein Jahrhundert Flugzeuge: Geschichte und Technik des Fliegens* (Düsseldorf, 1990).
Bottomore, Stephen, "The Panicking Audience? Early Cinema and the Train Effect," *Historical Journal of Film, Radio and Television* 19 (1999), 177–216.
Bourdieu, Pierre, *Outline of a Theory of Practice* (Cambridge, 1984).
Bourke, Joanna, *Dismembering the Male: Men's Bodies, Britain, and the Great War* (London, 1996).
  *An Intimate History of Killing: Face-to-Face Killing in Twentieth-Century Warfare* (New York, 1999).
Bouvet, Vincent, "Méwès: une révolution esthétique," *Monuments Historiques* 130 (1984), 44–8.
Boyer, Paul, *By the Bomb's Early Light: American Thought and Culture at the Dawn of the Atomic Age* (Chapel Hill, 1994).
Brayer, Elizabeth, *George Eastman: A Biography* (Baltimore, 1996).

Bredekamp, Horst, *Antikensehnsucht und Maschinenglauben: Die Geschichte der Kunstkammer und die Zukunft der Kunstgeschichte* (Berlin, 1993).
Brendon, Piers, *Thomas Cook: 150 Years of Popular Tourism* (London, 1991).
Breuer, Stefan, *Anatomie der Konservativen Revolution* (Darmstadt, 1993).
Breuilly, John, "National Peculiarities?" in *Labour and Liberalism in Nineteenth-Century Europe: Essays in Comparative History* (Manchester, 1992), 273–95.
Breymayer, Ursula, Ulrich, Bernd, and Wieland, Karin (eds.), *Willensmenschen: Über deutsche Offiziere* (Frankfurt, 1999).
Briggs, Asa, *The History of Broadcasting in the United Kingdom*, 5 vols. (Oxford, 1961–95).
Brock, W. H., *Science for All: Studies in the History of Victorian Science and Education* (Aldershot, 1996).
Brooks, Jeffrey, *Thank You Comrade Stalin: Soviet Public Culture from the Revolution to the Cold War* (Princeton, 2000).
Budde, Gunilla, *Auf dem Weg ins Bürgerleben: Kindheit und Erziehung in englischen und deutschen Bürgerfamilien, 1840–1914* (Göttingen, 1994).
Buettner, Elizabeth, *Empire Families: Britons in Late Imperial India* (Oxford, 2004).
Bullen, J. B. (ed.), *Post-Impressionists in England* (London, 1988).
Burke, Edmund, *A Philosophical Enquiry into the Sublime and Beautiful and Other Pre-Revolutionary Writings*, ed. David Womersley (Harmondsworth, 1998).
Burton, Antoinette, *Burdens of History: British Feminists, Indian Women, and Imperial Culture* (Chapel Hill, 1994).
Cadogan, Mary, *Women with Wings: Female Flyers in Fact and Fiction* (London, 1992).
Caine, Barbara, *English Feminism, 1780–1980* (Oxford, 1997).
Campbell, Joan, *The German Werkbund: The Politics of Reform in the Applied Arts* (Princeton, 1978).
Cannadine David, "War and Death, Grief and Mourning in Modern Britain," in Joachim Whaley (ed.), *Mirrors of Mortality: Studies in the Social History of Death* (London, 1981), 187–242.
Carter, Ian, *Railways and Culture in Britain: The Epitome of Modernity* (Manchester, 2001).
Castle, Terry, "Phantasmagoria: Spectral Technology and the Metaphorics of Modern Reverie," *Critical Inquiry* 15 (1988), 26–61.
Ceadel, Martin, *Pacifism in Britain, 1914–1945: The Defining of a Faith* (Oxford, 1980).
Collini, Stefan, *Public Moralists: Political Thought and Intellectual Life in Britain, 1850–1930* (Oxford, 1993).
Collins, Marcus, *Modern Love: An Intimate History of Men and Women in Twentieth Century Britain* (London, 2002).
Conekin, Becky, Mort, Frank, and Waters, Chris (eds.), *Moments of Modernity: Reconstructing Britain, 1945–1964* (London, 1999).
Confino, Alon, *The Nation as a Local Metaphor: Württemberg, Imperial Germany, and National Memory, 1871–1918* (Chapel Hill, 1997).

Corbin, Alain, *The Lure of the Sea: The Discovery of the Seaside, 1750–1840* (Harmondsworth, 1995).
Corn, Joseph C., *The Winged Gospel: America's Romance with Aviation, 1900–1950* (New York, 1983).
Cowan, Ruth Schwartz, *More Work for Mother: The Ironies of Household Technology from the Open Hearth to the Microwave* (New York, 1983).
Crary, Jonathan, *Techniques of the Observer: On Vision and Modernity in the Nineteenth Century* (Cambridge, MA, 1992).
Curran, James, and Seaton, Jean, *Power Without Responsibility: The Press and Broadcasting in Britain* (London, 1981).
Czarnowski, Gabriele, *Das kontrollierte Paar: Ehe- und Sexualpolitik im Nationalsozialismus* (Weinheim, 1991).
Dahrendorf, Ralf, *Society and Democracy in Germany* (Garden City, 1969).
Daston, Lorraine, and Parks, Katharine, *Wonders and the Order of Nature, 1150–1750* (New York, 1998).
Daston, Lorraine, and Galison, Peter, "The Image of Objectivity," *Representations* 40 (1992), 81–128.
Daum, Andreas W., *Wissenschaftspopularisierung im 19. Jahrhundert: Bürgerliche Kultur, naturwissenschaftliche Bildung und deutsche Öffentlichkeit* (Munich, 1998).
Daunton, Martin J., *Royal Mail: The Post Office Since 1840* (London, 1985).
Daunton, Martin J., and Rieger, Bernhard (eds.), *Meanings of Modernity: Britain from the Late-Victorian Era to World War II* (Oxford, 2001).
Davidoff, Leonore, *The Best Circles: Society, Etiquette and the Season* (London, 1986).
Dawson, Graham, *Soldier Heroes: British Adventure, Empire and the Imagining of Masculinities* (London, 1994).
Dedman, Martin, "Baden-Powell, Militarism, and the 'Invisible Contributors'," *Twentieth-Century British History* 4:3 (1993), 201–23.
Dewerpe, Alain, "Miroirs d'usines: Photographie industrielle et organisation du travail à l'Ansaldo, 1900–1920," *Annales ESC* 42 (1987), 1079–114.
Dickinson, Margaret, and Street, Sarah, *Cinema and the State: The Film Industry and the British Government, 1927–1984* (London, 1985).
Dintenfass, Michael, *The Decline of Industrial Britain, 1870–1980* (London, 1992).
Doty, William G., *Mythography: The Study of Myths and Rituals* (Tuscaloosa, 1986).
Douglas, Mary, *Purity and Danger: An Analysis of the Concepts of Pollution and Taboo* (London, 1996).
Douglas, Mary, and Wildavsky, Aaron, *Risk and Culture: An Essay on the Selection of Technological and Environmental Dangers* (Berkeley, 1983).
Douglas, Susan J., *Radio and the American Imagination: From Amos 'n' Andy and Edward R. Murrow to Wolfman Jack and Howard Stern* (New York, 1999).
Droste, Magdalena, *Bauhaus, 1919–1933* (Cologne, 1998).

Düding, Dieter, "Die Kriegervereine im Wilhelminischen Reich und ihr Beitrag zur Militarisierung der deutschen Gesellschaft," in Jost Dülffer and Karl Holl (eds.), *Bereit zum Krieg: Kriegsmentalität im Wilhelminischen Deutschland, 1890–1914* (Göttingen, 1986), 99–121.
During, Simon, *Modern Enchantments: The Cultural Power of Secular Wonder* (Cambridge, MA, 2002).
Dyer, Richard, *Heavenly Bodies: Film Stars and Society* (London, 1985).
Eaton, John P., and Haas, Charles A., *Falling Star: Misadventures of White Star Line Ships* (Wellingborough, 1989).
*Titanic: Triumph and Tragedy* (Wellingborough, 1986).
Edgerton, David, *England and the Aeroplane: An Essay on a Militant and Technological Nation* (Basingstoke, 1991).
"Liberal Militarism and the British State," *New Left Review* 185 (1991), 138–69.
*Science, Technology and British Industrial "Decline"* (Cambridge, 1996).
Eksteins, Modris, *The Rites of Spring: The Great War and the Birth of the Modern Age* (Boston, 1989).
Eley, Geoff, "Army, State, and Civil Society: Revisiting the Problem of German Militarism," in *From Unification to Nazism: Re-thinking the German Past* (Boston, 1986), 85–109.
"Nations, Publics, and Political Cultures: Placing Habermas in the Nineteenth Century," in Nicholas B. Dirks, Geoff Eley, and Sherry B. Ortner (eds.), *Culture/Power/History: A Reader in Contemporary Social Theory* (Princeton, 1994), 297–335.
(ed.), *Society, Culture and the State in Germany, 1870–1930* (Ann Arbor, 1996).
Elkins, James, *The Object Stares Back: On the Nature of Seeing* (San Diego, 1997).
Ellis, John, "Stars as a Cinematic Phenomenon," in Jeremy G. Butler (ed.), *Star Texts: Image and Performance in Film and Television* (Detroit, 1991), 300–15.
Elmsley, Clive, "'Mother, What Did Policemen Do When There Weren't Any Motors?' The Law, the Police and the Regulation of Motor Traffic in England, 1900–1939," *Historical Journal* 36 (1993), 357–81.
Epkenhans, Michael, *Die wilhelminische Flottenrüstung 1908 bis 1914. Weltmachtstreben, industrieller Fortschritt, soziale Integration* (Munich, 1991).
Eyles, Allen, *Gaumont British Cinemas* (London, 1996).
*Odeon Cinemas*: vol. 1: *Oscar Deutsch Entertains Our Nation* (London, 2002).
Ezra, Elizabeth, *Georges Méliès: The Birth of the Auteur* (Manchester, 2000).
Falkenberg, Regine, "Untergang des Auswandererschiffs 'Austria,'" in *Zeitzeugen: Ausgewählte Objekte aus dem Deutschen Historischen Museum* (Berlin, 1992), 171–4.
Fearon, Peter, "Aircraft Manufacturing," in Neil Buxton and Derek Aldcroft (eds.), *British Industry Between the Wars: Instability and Development* (London, 1979), 216–40.
"The Growth of Aviation in Britain," *Journal of Contemporary History* 20 (1985), 21–40.

Ferberg, Nils, "'Sturmvogel-Flugverband der Werktätigen E. V.' Zur Geschichte einer Arbeitersportorganisation in der Weimarer Republik," *Internationale wissenschaftliche Korrespondenz der deutschen Arbeiterbewegung* 30 (1994), 173–219.
Fisher, Philip, *The Vehement Passions* (Princeton, 2002).
*Wonder, the Rainbow, and the Aesthetics of Rare Experiences* (Cambridge, MA, 1998).
Foucault, Michel, *The Archaeology of Knowledge* (New York, 1972).
Fraser, Nancy, "Politics, Culture, and the Public Sphere: Toward a Postmodern Conception," in Linda Nicholson and Steven Seidman (eds.), *Social Postmodernism: Beyond Identity Politics* (Cambridge, 1995), 287–302.
Freedgood, Elaine, *Victorian Writings about Risk: Imagining a Safe England in a Dangerous World* (Cambridge, 2000).
French, David, "'You Cannot Hate the Bastard Who is Trying to Kill You . . .' Combat and Ideology in the British Army in the War Against Germany, 1939–1945," *Twentieth Century British History* 11 (2000), 1–22.
Frevert, Ute, *Ehrenmänner: Das Duell in der bürgerlichen Gesellschaft* (Munich, 1995).
*Women in German History: From Bourgeois Emancipation to Sexual Liberation* (New York, 1989).
Friedberg, Aaron L., *The Weary Titan: Britain and the Experience of Relative Decline, 1895–1905* (Princeton, 1988).
Friedlander, Saul, *Nazi Germany and the Jews*: vol. I: *The Years of Persecution, 1933–1939* (New York, 1997).
Fritzsche, Peter, *Germans into Nazis* (Cambridge, MA, 1998).
*A Nation of Fliers: German Aviation and the Popular Imagination* (Cambridge, MA, 1992).
"Nazi Modern," *Modernism/Modernity* 3 (1996), 1–22.
*Reading Berlin 1900* (Cambridge, MA, 1996).
Führer, Christian, Hickethier, Knut, and Schildt, Axel, "Öffentlichkeit-Medien-Geschichte: Konzepte der modernen Öffentlichkeit und Zugänge zu ihrer Erforschung," *Archiv für Sozialgeschichte* 41 (2001), 1–38.
Führer, Karl Christian, "A Medium of Modernity? Broadcasting in Weimar Germany, 1923–1932," *Journal of Modern History* 69 (1997), 722–53.
Fussel, Paul, *The Great War and Modern Memory* (Oxford, 1975).
Gardiner, Robert (ed.), *The Advent of Steam: The Merchant Steamship before 1900* (London, 1993).
Gassert, Philipp, *Amerika im Dritten Reich: Ideologie, Propaganda und Volksmeinung 1933–1945* (Stuttgart, 1997).
"Amerikanismus, Antiamerikanismus, Amerikanisierung: Neue Literatur zur Sozial-, Wirtschafts- und Kulturgeschichte des amerikanischen Einflusses in Deutschland und Europa," *Archiv für Sozialgeschichte* 39 (1999), 531–61.
Gay, Peter, *The Cultivation of Hatred: The Bourgeois Experience, Victoria to Freud* (London, 1995).
Gennep, Arnold van, *The Rites of Passage* (Chicago, 1960).

Gernsheim, Helmut, *The History of Photography: From the Earliest Uses of the Camera Obscura in the Eleventh Century to 1914* (London, 1969).
Gerstenberger, Heike, and Welke, Ulrich, *Vom Wind zum Dampf: Sozialgeschichte der deutschen Handelsschiffahrt im Zeitalter der Industrialisierung* (Münster, 1996).
Geyersbach, Michael, *Wie verkauf' ich meine Tante? Corporate Design bei Junkers, 1892 bis 1933* (Dessau, 1996).
Giddens, Anthony, *Modernity and Self-Identity: Self and Society in the Late Modern Age* (Stanford, 1991).
Gillis, John R., "Making Time for Family: The Invention of Family Time(s) and the Reinvention of Family History," *Journal of Family History* 21 (1996), 4–21.
 *A World of Their Own Making: Myth, Ritual, and the Quest for Family Values* (New York, 1996).
Girouard, Mark, *The Return to Camelot: Chivalry and the English Gentleman* (New Haven, 1981).
Gispen, Kees, *Poems in Steel: National Socialism and the Politics of Inventing from Weimar to Bonn* (New York, 2002).
Glaser, Hermann, *Bildungsbürgertum und Nationalismus: Politik und Kultur im Wilhelminischen Deutschland* (Munich, 1993).
Goebel, Stefan, *Medievalism in the Commemoration of the Great War in Britain and Germany, 1914–1939* (Cambridge, PhD dissertation, 2001).
Gollin, Alfred, *No Longer an Island: Britain and the Wright Brothers, 1902–1909* (Stanford, 1984).
Gosser, H. Mark, "The *Bazar-de-la-Charité* Fire: The Reality, the Aftermath, the Telling," *Film History* 10 (1998), 70–89.
Graves, Robert, and Hodge, Alan, *The Long Weekend: A Social History of Great Britain, 1919–1939* (London, 1991 [1940]).
Grazia, Victoria de, "Mass Culture and Sovereignty: The American Challenge to European Cinema, 1920 to 1960," *Journal of Modern History* 61 (1989), 53–87.
Grazia, Victoria de, and Furlough, Ellen (eds.), *The Sex of Things: Gender and Consumption in Historical Perspective* (Berkeley, 1996).
Griffith, Richard, *Fellow Travellers of the Right: British Enthusiasts for Nazi Germany, 1933–39* (Oxford, 1983).
Groh, Dieter, *"Vaterlandslose Gesellen": Sozialdemokratie und Nation, 1860–1990* (Munich, 1992).
Gross, Charles G., *Brain, Vision, Memory: Tales in the History of Neuroscience* (Cambridge, MA, 1999).
Grossmann, Atina, "*Girlkultur* or Thoroughly Rationalized Female: A New Woman in Weimar Germany?" in Judith Friedlander *et al.* (eds.), *Women in Culture and Politics: A Century of Change* (Bloomington, 1986), 62–80.
Gunning, Tom, "The Cinema of Attraction: Early Film, Its Spectators and the Avant-Garde," *Wide Angle* 8:3 (1986), 63–70.
Gwynn-Jones, Terry, *Farther and Faster: Aviation's Adventuring Years, 1909–1939* (Washington, 1991).

Habermas, Jürgen, *The Structural Transformation of the Public Sphere: An Inquiry into a Category of Civil Society* (London, 1989).
Hake, Sabine, *The Cinema's Third Machine: Writing on German Cinema, 1907–1933* (Lincoln, NE, 1993).
*Popular Cinema of the Third Reich* (Austin, 2001).
Hall, Leslie, "Impotent Ghosts from No Man's Land, Flappers' Boyfriends, or Cryptopatriarchs? Men, Sex and Social Change in 1920s Britain," *Social History* 21 (1996), 54–70.
Hansen, Miriam, *Babel and Babylon: Spectatorship in American Silent Film* (Cambridge, MA, 1991).
Hansen, Peter, "Albert Smith, the Alpine Club, and the Invention of Mountaineering in Mid-Victorian Britain," *Journal of British Studies* 34 (1995), 300–24.
Hård, Mikael, "German Regulation: The Integration of Modern Technology into National Culture," in Mikael Hård and Andrew Jamison (eds.), *The Intellectual Appropriation of Technology: Discourses on Modernity, 1900–1939* (Cambridge, MA, 1998), 33–67.
Harding, Colin, "A Transatlantic Emanation: The Kodak Comes to Britain," in David E. Nye and Mick Gidley (eds.), *American Photographs in Europe* (Amsterdam, 1994), 109–29.
Hardtwig, Wolfgang, "Kunst, liberaler Nationalismus und Weltpolitik: Der deutsche Werkbund 1907–1914," in *Nationalismus und Bürgerkultur in Deutschland, 1500–1914* (Göttingen, 1994), 246–73.
Harris, Jose, *Private Lives, Public Spirit: Britain, 1870–1914* (Harmondsworth, 1994).
Harvey, David, *The Condition of Postmodernity: An Enquiry into the Origins of Cultural Change* (Oxford, 1989).
Haste, Cate, *Keep the Home Fires Burning: Propaganda in the First World War* (London, 1977).
Haupt, Heinz-Gerhard, and Kocka, Jürgen, "Historischer Vergleich: Methoden, Aufgaben, Probleme: Eine Einleitung," in Heinz-Gerhard Haupt and Jürgen Kocka (eds.), *Geschichte und Vergleich: Ansätze und Ergebnisse international vergleichender Geschichtsschreibung* (Frankfurt, 1996), 9–45.
Hazelgrove, Jenny, *Spiritualism and British Society Between the Wars* (Manchester, 2000).
Headrick, Daniel R., *The Invisible Weapon: Telecommunications and International Politics, 1851–1945* (New York, 1991).
*The Tools of Empire: Technology and European Imperialism in the Nineteenth Century* (Oxford, 1981).
Hecht, Gabrielle, *The Radiance of France: Nuclear Power and National Identity after World War II* (Cambridge, MA, 1998).
Herbert, Ulrich, *Best: Biographische Studien über Radikalismus, Weltanschauung und Vernunft* (Bonn, 1996).
Herf, Jeffrey, *Reactionary Modernism: Technology, Culture and Politics in Weimar and the Third Reich* (Cambridge, 1984).
Higonnet, Anne, *Pictures of Innocence: The History and Crisis of Ideal Childhood* (London, 1998).

Hiley, Nicholas, "'No Mixed Bathing': The Creation of the British Board of Film Censors in 1913," *Journal of Popular British Cinema* 3 (2000), 5–19.
Hobsbawm, Eric J., *Industry and Empire* (Harmondsworth, 1986).
Hobsbawm, Eric, and Ranger, Terence (eds.), *The Invention of Tradition* (Cambridge, 1983).
Hoffmann, Stefan Ludwig, "Sakraler Monumentalismus um 1900: Das Leipziger Völkerschlachtsdenkmal," in Reinhart Koselleck and Michael Jeismann (eds.), *Der politische Totenkult: Kriegerdenkmäler in der Moderne* (Munich, 1994), 249–80.
Homze, Edward L., *Arming the Luftwaffe: The Reich Air Ministry and the German Aircraft Industry, 1919–1939* (Lincoln, NE, 1976).
Hopwood, Nick, "Producing a Socialist Popular Science in the Weimar Republic," *History Workshop Journal* 41 (1996), 117–53.
Horkheimer, Max, and Adorno, Theodor W., *Dialektik der Aufklärung: Philosophische Fragmente* (Frankfurt, 1971).
Horne, John, and Kramer, Alan, *German Atrocities 1914: A History of Denial* (New Haven, 2001).
Horowitz, Roger (ed.), *Boys and Their Toys? Masculinity, Technology, and Class in America* (New York, 2001).
Horowitz, Roger, and Mohun, Arwen P., *His and Hers: Gender, Consumption, and Technology* (Charlottesville, 1998).
Howe, Stephen, *Anticolonialism in British Politics: The Left and the End of Empire, 1918–1964* (Oxford, 1993).
Howells, Richard, *The Myth of the "Titanic"* (Basingstoke, 1999).
Hübinger, Gangolf, and Mommsen, Wolfgang J. (eds.), *Intellektuelle im Deutschen Kaiserreich* (Frankfurt, 1993).
Hughes, Thomas P., *Networks of Power: Electrification in Western Society* (Baltimore, 1983).
Huyssen, Andreas, "Mass Culture as Woman: Modernism's Other," in *After the Great Divide: Modernism, Mass Culture, Postmodernism* (Houndmills, 1988), 44–62.
"Monumental Seduction," *New German Critique* 69 (1996), 181–200.
Hynes, Samuel, *A War Imagined: The First World War and English Culture* (London, 1991).
Jacobsen, Wolfgang, Kaes, Anton, and Prinzler, Hans Helmut (eds.), *Geschichte des deutschen Films* (Stuttgart, 1993).
James, Harold, "Die Frühgeschichte der Lufthansa: Ein Unternehmen zwischen Banken und Staat," *Zeitschrift für Unternehmensgeschichte* 42 (1997), 4–13.
Jameson, Fredric, *A Singular Modernity: Essay on the Ontology of the Present* (London, 2002).
Jantzen, Eva, and Niehuss, Merith, *Das Klassenbuch: Geschichte einer Frauengeneration* (Reinbeck, 1997).
Jay, Martin, *Downcast Eyes: The Denigration of Vision in Twentieth-Century French Thought* (Berkeley, 1993).

Jones, Caroline A., and Galison, Peter (eds.), *Picturing Art, Producing Science* (New York, 1998).
Jung, Uli (ed.), *Der deutsche Film: Aspekte seiner Geschichte von den Anfängen bis zur Gegenwart* (Trier, 1993).
Kaplan, Wendy (ed.), *Designing Modernity: The Arts of Reform and Persuasion* (New York, 1995).
Kasson, John F., *Civilizing the Machine: Technology and Republican Values in America, 1776–1900* (New York, 1999).
Keitz, Christine, *Reisen als Leitbild: Die Entstehung des modernen Massentourismus in Deutschland* (Munich, 1997).
Kemner, Gerhard, and Eisert, Gelia, *Lebende Bilder: Eine Technikgeschichte des Films* (Berlin, 2000).
Kennedy, Paul, *The Rise of Anglo-German Antagonism, 1860–1914* (London, 1980).
Kern, Stephen, *The Culture of Time and Space* (Cambridge, MA, 1983).
Kershaw, Ian, and Levin, Moshe (eds.), *Stalinism and Nazism: Dictatorships in Comparison* (Cambridge, 1997).
King, Tony, *The History of Salford Cinemas* (Manchester, 1987).
Kingsley Kent, Susan, *Making Peace: The Reconstruction of Gender in Interwar Britain* (Princeton, 1993).
Kiupel, Uwe, "Selbsttötung auf bremischen Dampfschiffen: Die Arbeits- und Lebensbedingungen der Feuerleute, 1880 bis 1914," in Wiltrud Drechsel and Heide Gerstenberger (eds.), *Arbeitsplätze: Schiffahrt, Hafen, Textilindustrie, 1880 bis 1933* (Bremen, 1983), 15–104.
Kline, Ronald, "Construing 'Technology' as 'Applied Science.' Public Rhetoric of Scientists and Engineers in the United States, 1880–1945," *Isis* 86 (1995), 194–221.
Kludas, Arnold, *Geschichte der deutschen Passagierschiffahrt*, vols. I–III (Hamburg, 1986–8).
Knowles, Scott Gabriel, "Lessons from the Rubble: The World Trade Center and the History of Disaster Investigation in the United States," *History and Technology* 19 (2003), 9–28.
Kocka, Jürgen, "The Middle Classes in Europe," *Journal of Modern History* 67 (1995), 783–806.
König, Wolfgang, and Weber, Wolfhard, *Netzwerke, Stahl und Strom: 1840 bis 1914* (Berlin, 1997).
Koonz, Claudia, *Mothers in the Fatherland: Women, the Family, and Nazi Politics* (London, 1988).
Koselleck, Reinhart, *Futures Past: On the Semantics of Historical Time* (Cambridge, MA, 1985).
 *Zeitschichten: Studien zur Historik* (Frankfurt, 2000).
Koshar, Rudy, *Germany's Transient Pasts: Preservation and National Memory in the Twentieth Century* (Chapel Hill, 1998).
 *From Monuments to Traces: German Artifacts of Memory, 1870–1990* (Berkeley, 2000).

*German Travel Cultures* (Oxford, 2000).
"'What Ought to Be Seen': Tourist Guidebooks and National Identities in Modern Germany and Europe," *Journal of Contemporary History* 33 (1998), 323–40.
Kratzsch, Gerhard, *Kunstwart und Dürerbund: Ein Beitrag zur Geschichte der Gebildeten im Zeitalter des Imperialismus* (Göttingen, 1969).
Kreimeier, Klaus, *The Ufa Story: A History of Germany's Greatest Film Company, 1918–1945* (New York, 1996).
Kuball, Michael, *Familienkino: Geschichte des Amateurfilms in Deutschland*, 2 vols. (Reinbeck, 1980).
Kuhn, Annette, "Cinema-going in Britain in the 1930s: Report of a Questionnaire Survey," *Historical Journal of Film, Radio and Television* 19 (1999), 531–41.
Kuzniar, Alice A., *The Queer German Cinema* (Stanford, 2000).
Lang, Christoph, "'Herren im Hause': Die Unternehmer," in Richard van Dülmen (ed.), *Industriekultur an der Saar: Leben und Arbeit einer Industrieregion, 1840 bis 1914* (Munich, 1989), 132–45.
Lawrence, Jon, "Forging a Peaceable Kingdom: War, Violence, and Fear of Brutalization in Post-First World War Britain," *Journal of Modern History* 75 (2003), 557–89.
Leed, Eric J., *No Man's Land: Combat and Identity in World War I* (Cambridge, 1981).
Lees, Andrew, *Cities Perceived: Urban Society in European and American Thought, 1820–1940* (Manchester, 1985).
LeMahieu, Dan, *A Culture for Democracy: Mass Communication and the Cultivated Mind in Britain Between the Wars* (Oxford, 1988).
"*The Gramophone*: Recorded Music and the Cultivated Mind in Britain Between the Wars," *Technology and Culture* 23 (1982), 372–91.
Lerman, Nina E., Oldenziel, Ruth, and Mohun, Arwen P., "The Shoulders We Stand on and the View From Here: Historiography and Directions for Research," *Technology and Culture* 38 (1997), 9–30.
Lerner, Paul F., *Hysterical Men: War, Psychiatry, and the Politics of Trauma in Germany, 1890–1930* (Ithaca, 2003).
Lethen, Helmut, *Cool Conduct: The Culture of Distance in Weimar Germany* (Berkeley, 2001).
Linse, Ulrich (ed.), *Zurück, o Mensch, zur Mutter Erde: Landkommunen in Deutschland, 1890–1933* (Munich, 1983).
Löfgren, Orvar, *On Holiday: A History of Vacationing* (Berkeley, 1999).
Loiperdinger, Martin, "The Kaiser's Cinema: An Archaeology of Attitudes and Audiences," in Thomas Elsaesser (ed.), *A Second Life: German Cinema's First Decades* (Amsterdam, 1996), 41–50.
Lomax, Judy, *Women of the Air* (London, 1986).
Low, Rachel, *The History of the British Film, 1906–1914*, 2nd edition (London, 1973).
Luckin, Bill, *Questions of Power: Electricity and the Environment in Interwar Britain* (Manchester, 1990).

Lüdtke, Alf, "'Ehre der Arbeit:' Industriearbeiter und die Macht der Symbole: Zur Reichweite symbolischer Orientierungen im Nationalsozialismus," in Klaus Tenfelde (ed.), *Arbeiter im 20. Jahrhundert* (Stuttgart, 1991), 292–343.

Maase, Kaspar, *Grenzenloses Vergnügen: Der Aufstieg der Massenkultur, 1850–1970* (Frankfurt, 1997).

Maase, Kaspar, and Kaschuba, Wolfgang (eds.), *Schund und Schönheit: Populäre Kultur um 1900* (Cologne, 2001).

McAleer, Kevin, *Dueling: The Cult of Honor in Fin-de-Siècle Germany* (Princeton, 1994).

MacKenzie, Donald, *Inventing Accuracy: A Historical Sociology of Nuclear Missile Guidance* (Cambridge, MA, 1990).

Maier, Charles, "Between Taylorism and Technocracy: European Ideologies and the Vision of Industrial Productivity in the 1920s," *Journal of Contemporary History* 5 (1970), 27–61.

Mandler, Peter, "Against 'Englishness': English Culture and the Limits of Rural Nostalgia, 1850–1940," *Transactions of the Royal Historical Society*, sixth series, 7 (1997), 155–75.

Mangan, James A., *Athleticism in the Victorian and Edwardian Public School: The Emergence and Consolidation of an Educational Ideology* (Cambridge, 1981).

Marssolek, Inge, "Radio in Deutschland 1923–1960: Zur Sozialgeschichte eines Mediums," *Geschichte und Gesellschaft* 27 (2001), 207–39.

Marvin, Carolyn, *When Old Technologies Were New: Thinking About Electrical Communication in the Late Nineteenth Century* (New York, 1988).

Marx, Karl, and Engels, Friedrich, *Werke, vol.* XXIII: *Das Kapital*, vol. 1 (Berlin, 1974).

Matlock, Jann, *Scenes of Seduction: Prostitution, Hysteria, and Reading Difference in Nineteenth-Century France* (New York, 1994).

Matsuda, Matt K., *The Memory of the Modern* (New York, 1996).

Mazower, Mark, *Dark Continent: Europe's Twentieth Century* (New York, 1998).

Meikle, Jeffrey, *Twentieth-Century Limited: Industrial Design in America. 1925–1939* (Philadelphia, 1979).

Melman, Billie, *Women and the Popular Imagination in the Twenties: Flappers and Nymphs* (Houndmills, 1988).

—— (ed.), *Borderlines: Genders and Identities in War and Peace, 1870–1930* (New York, 1998).

Messinger, Gary S., *British Propaganda and the State in the First World War* (Manchester, 1992).

—— "An Inheritance Worth Remembering: The British Approach to Official Propaganda during the First World War," *Historical Journal of Film, Radio and Television* 13 (1993), 117–27.

Metcalf, Thomas R., *Ideologies of the Raj* (Cambridge, 1995).

Moeller, Felix, *Der Filmminister: Goebbels und der Film im Dritten Reich* (Berlin, 1998).

Mohun, Arwen P., "Designed for Thrills and Safety: Amusement Parks and the Commodification of Risk, 1880–1929," *Journal of Design History* 14 (2001), 291–306.

Mommsen, Hans, "Nationalsozialismus als vorgetäuschte Modernisierung," in Walter H. Pehle (ed.), *Der historische Ort des Nationalsozialismus* (Frankfurt, 1990), 31–46.

Mommsen, Wolfgang J., "Die Kultur der Moderne im Deutschen Kaiserreich," in Wolfgang Hardtwig and Harm-Hinrich Brandt (eds.), *Deutschlands Weg in die Moderne: Politik, Gesellschaft und Kultur im 19. Jahrhundert* (Munich, 1993), 254–74.

Montgomery-Massingberd, Hugh, and Watkin, David, *The London Ritz: A Social and Architectural History* (London, 1989).

Moore-Gilbert, B. J., *Kipling and "Orientalism"* (London, 1986).

Morrow, John H., *The Great War in the Air: Military Aviation from 1909 to 1921* (London, 1993).

"Knights of the Sky: The Rise of Military Aviation," in Frans Coetzee and Marilyn Shevin-Coetzee (eds.), *Authority, Identity and the Social History of the Great War* (Oxford, 1995), 305–24.

Mosse, George L., *Fallen Soldiers: Reshaping the Memory of the Wars* (New York, 1991).

*The Image of Man: The Creation of Modern Masculinity* (New York, 1996).

"The Knights of the Sky and the Myth of the War Experience," in Robert A. Hinde and Helen E. Watson (eds.), *War: A Cruel Necessity? The Bases of Institutional Violence* (London, 1995), 133–42.

Müller, Corinna, *Frühe deutsche Kinematographie: Formale, wirtschaftliche und kulturelle Entwicklungen* (Stuttgart, 1994).

Murray, Bruce, *Film and the German Left in the Weimar Republic: From Caligari to Kuhle Wampe* (Austin, 1990).

Nava, Mica, and O'Shea, Alan (eds.), *Modern Times: Reflections on a Century of English Modernity* (London, 1996).

Neufeld, Michael, "Weimar Culture and Futuristic Technology: The Rocketry and Spaceflight Fad, 1923–1933," *Technology and Culture* 31 (1990), 725–52.

Neumann, Kerstin-Luise, "Idolfrauen oder Idealfrauen? Kristina Söderbaum und Zarah Leander," in Thomas Koebner (ed.), *Idole des deutschen Films: Eine Gallerie von Schlüsselfiguren* (Munich, 1997), 231–43.

Nevin, Thomas, *Ernst Jünger and Germany: Into the Abyss, 1914–1945* (Durham, NC, 1996).

Nicholson, Heather Norris, "Seeing How It Was? Childhood Geographies and Memories in Home Movies," *Area* 33 (2001), 128–40.

Nipperdey, Thomas, *Deutsche Geschichte, 1866–1918*, vol. II: *Machtstaat vor der Demokratie* (Munich, 1992).

*Wie das Bürgertum die Moderne fand* (Berlin, 1988).

Nolan, Mary, *Visions of Modernity: American Business and the Modernization of Germany* (New York, 1994).

Nye, David E., *American Technological Sublime* (Cambridge, MA, 1994).

*Electrifying America: Social Meanings of a New Technology, 1880–1940* (Cambridge, MA, 1990).
O'Connell, Sean, *The Car and British Society: Class, Gender and Motoring, 1896–1939* (Manchester, 1998).
Olsen, Donald J., *The City as a Work of Art: London, Paris, Vienna* (New Haven, 1986).
Omissi, David E., *Air Power and Colonial Control: The Royal Air Force, 1919–1939* (Manchester, 1990).
"The Hendon Air Pageant, 1920–37," in John MacKenzie (ed.), *Popular Imperialism and the Military* (Manchester, 1992), 198–220.
Oppenheim, Janet, *The Other World: Spiritualism and Psychological Research in England, 1850–1914* (Cambridge, 1988).
*Shattered Nerves: Doctors, Patients, and Depression in Victorian England* (New York, 1991).
Osietzki, Maria, "Körpermaschinen und Dampfmaschinen: Vom Wandel der Physiologie und des Körpers unter dem Einfluß von Industrialisierung und Thermodynamik," in Philipp Sarasin and Jakob Tanner (eds.), *Physiologie und industrielle Gesellschaft: Studien zur Verwissenschaftlichung des Körpers im 19. und 20. Jahrhundert* (Frankfurt, 1998), 313–46.
Overy, Richard, *The Air War, 1939–1945* (London, 1987).
*The Battle* (Harmondsworth, 2000).
Paech, Anne, *Kino zwischen Stadt und Land. Geschichte des Kinos in der Provinz: Osnabrück* (Marburg, 1985).
Paret, Peter, *The Berlin Secession: Modernism and its Enemies in Imperial Germany* (Cambridge, MA, 1980).
Paris, Michael, *From the Wright Brothers to Top Gun: Aviation, Nationalism and Popular Cinema* (Manchester, 1995).
*Warrior Nation: Images of War in British Popular Culture, 1850–2000* (London, 2000).
*Winged Warfare: The Literature and Theory of Aerial Warfare in Britain, 1859–1917* (Manchester, 1992).
Parsons, Talcott, *The System of Modern Societies* (Englewood Cliffs, 1971).
Paulmann, Johannes, "Internationaler Vergleich und interkultureller Transfer: Zwei Forschungsansätze zur europäischen Geschichte des 18. bis 20. Jahrhunderts," *Historische Zeitschrift* 267 (1998), 649–85.
Payne, Stanley G., *A History of Fascism* (Madison, 1995).
Perkin, Harold, *The Rise of Professional Society: England since 1880* (London, 1989).
Perrow, Charles, *Normal Accidents: Living with High-Risk Technologies* (Princeton, 1999).
Peukert, Detlev J. K., "The Genesis of the 'Final Solution' from the Spirit of Science," in Thomas Childers and Jane Caplan (eds.), *Reevaluating the Third Reich* (New York, 1993), 234–52.
*Inside Nazi Germany: Conformity, Opposition, and Racism in Everyday Life* (New Haven, 1987).

*Die Weimarer Republic: Krisenjahre der Klassischen Moderne* (Frankfurt, 1987).
Phillips, Richard, *Mapping Men and Empire: A Geography of Adventure* (London, 1996).
Pollard, Sidney, and Robertson, Paul, *The British Shipbuilding Industry, 1870–1914* (Cambridge, MA, 1979).
Preston, Paul, *A Concise History of the Spanish Civil War* (London, 1996).
Procida, Mary, "Good Sports and Right Sorts: Guns, Gender, and Imperialism in British India," *Journal of British Studies* 40 (2001), 454–88.
Proctor, Robert N., *The Nazi War on Cancer* (Princeton, 1999).
Pursell, Carroll, "Domesticating Modernity: The Electrical Association for Women, 1924–1986," *British Journal for the History of Science* 32 (1999), 47–67.
Rabinbach, Anson, *The Human Motor: Energy, Fatigue, and the Origins of Modernity* (Berkeley, 1992).
"Nationalsozialismus und Moderne: Zur Technik-Interpretation im Dritten Reich," in Wolfgang Emmerich and Carl Wege (eds.), *Der Technik-Diskurs in der Hitler-Stalin-Ära* (Stuttgart, 1995), 94–113.
"Social Knowledge, Social Risk, and the Politics of Industrial Accidents in Germany and France," in Dietrich Rueschemeyer and Theda Skocpol (eds.), *States, Social Knowledge, and the Origins of Modern Social Policies* (Princeton, 1996), 48–89.
Radkau, Joachim, *Technik in Deutschland: Vom 18. Jahrhundert bis zur Gegenwart* (Frankfurt, 1989).
Radkau, Wolfgang, "Die Wilhelminische Ära als nervöses Zeitalter, oder: Die Nerven als Netz zwischen Tempo- und Körpergeschichte," *Geschichte und Gesellschaft* 20 (1994), 211–41.
*Das Zeitalter der Nervosität: Deutschland zwischen Bismarck und Hitler* (Berlin, 2000).
Ramusack, Barbara N., "Cultural Missionaries, Maternal Imperialists, Feminist Allies: British Women Activists in India, 1865–1945," in Nupur Chauduri and Margaret Strobel (eds.), *Western Women and Imperialism: Complicity and Resistance* (Bloomington, 1992), 119–36.
Rappaport, Erika Diane, *Shopping for Pleasure: Women in the Making of London's West End* (Princeton, 2000).
Read, Donald, *England, 1868–1914: The Age of Urban Democracy* (London, 1988).
Readman, Paul, "Landscape Preservation, 'Advertising Disfigurement' and English National Identity, c. 1890–1914," *Rural History* 12 (2001), 61–83.
Reeves, Nicholas, "Cinema, Spectatorship and Propaganda: 'Battle of the Somme' (1916) and Its Contemporary Audience," *Historical Journal of Film, Radio and Television* 17 (1997), 5–28.
Reitsch, Hanna, *Fliegen – Mein Leben* (Stuttgart, 1956).
Renneberg, Monika, and Walker, Mark (eds.), *Science, Technology and National Socialism* (Cambridge, 1994).
Rentschler, Eric, *The Ministry of Illusion: Nazi Cinema and its Afterlife* (Cambridge, MA, 1996).

Repp, Kevin, *Reformers, Critics, and the Paths of German Modernity: Anti-Politics and the Search for Alternatives, 1890–1914* (Cambridge, MA, 2000).
Reynolds, David, *Britannia Overruled: British Policy and World Power in the Twentieth Century* (London, 1991).
Richards, Jeffrey, *The Age of the Dream Palace: Cinema and Society in Britain, 1930–1939* (London, 1984).
"The British Board of Film Censors and Content Control in the 1930s: Images of Britain," *Historical Journal of Film, Radio and Television* 1 (1981), 95–116.
Ringer, Fritz K., *The Decline of the German Mandarins: The German Academic Community, 1890–1933* (Cambridge, MA, 1969).
Roberts, Mary Louise, *Civilization Without Sexes: Reconstructing Gender in Postwar France, 1917–1927* (Chicago, 1994).
Robertson, James C., *The Hidden Cinema: British Film Censorship in Action, 1913–1972* (London, 1989).
Rohkrämer, Thomas, "Antimodernism, Reactionary Modernism and National Socialism: Technocratic Tendencies in Germany, 1890–1945," *Contemporary European History* 8 (1999), 29–50.
 *Eine andere Moderne? Zivilisationskritik, Natur und Technik in Deutschland 1880–1933* (Paderborn, 1999).
 *Der Militarismus der "kleinen Leute": Die Kriegervereine im Deutschen Kaiserreich 1871–1914* (Munich, 1990).
Rose, Sonya O., *Limited Livelihoods: Gender and Class in Nineteenth-Century England* (Berkeley, 1992).
Rosenhaft, Eve, "Gewalt in der Politik: Zum Problem des 'Sozialen Militarismus,'" in Klaus-Jürgen Müller (ed.), *Militär und Militarismus in der Weimarer Republik* (Düsseldorf, 1978), 237–58.
 "Links gleich rechts? Militante Straßengewalt um 1930," in Thomas Lindenberger and Alf Lüdtke (eds.), *Physische Gewalt: Studien zur Geschichte der Neuzeit* (Frankfurt, 1995), 238–75.
Rossell, Deac, *Living Pictures: The Origins of the Movies* (Albany, 1998).
Rostow, Walt W., *The Stages of Economic Growth* (Cambridge, 1960).
Rother, Rainer, *Leni Riefenstahl: Die Verführung des Talents* (Berlin, 2000).
Rubinstein, W. D., *Capitalism, Culture and Decline in Britain, 1750–1990* (London, 1994).
Ryan, Deborah, *The Ideal Home through the Twentieth Century* (London, 1997).
Sachs, Wolfgang, *For Love of the Automobile: Looking Back into the History of Our Desires* (Berkeley, 1992).
Saler, Michael, *The Avant-Garde in Interwar England: Medieval Modernism and the London Underground* (New York, 1999).
Samuel, Raphael, *Theatres of Memory*, vol. 1: *The Past and Present in Contemporary Culture* (London, 1994).
 (ed.), *Patriotism: The Making and Unmaking of British National Identity*, vol. 1: *History and Politics* (London, 1989).
Sanders, M. L., and Taylor, Philip M., *British Propaganda During the First World War, 1914–1918* (London, 1982).

Sarasin, Philipp, "Subjekte, Diskurse, Körper: Überlegungen zu einer diskursanalytischen Kulturgeschichte," in Wolfgang Hardtwig and Hans-Ulrich Wehler (eds.), *Kulturgeschichte Heute* (Göttingen, 1996), 131–64.
Saunders, Thomas J., *Hollywood in Berlin: American Cinema and Weimar Germany* (Berkeley, 1994).
Schäfer, Hans-Dieter, *Das gespaltene Bewußtsein: Über deutsche Kultur und Lebensbewußtsein, 1933–1945* (Munich, 1982).
Schildt, Axel, "NS-Regime, Modernisierung und Moderne: Anmerkungen zur Hochkunjunktur einer andauernden Diskussion," *Tel Aviver Jahrbuch für Geschichte* 23 (1994), 3–22.
Schivelbusch, Wolfgang, *Die Kultur der Niederlage: Der amerikanische Süden 1865, Frankreich 1871, Deutschland 1918* (Berlin, 2001).
*The Railway Journey: The Industrialization of Time and Space in the Nineteenth Century* (Berkeley, 1986).
Schlör, Joachim, *Nachts in der großen Stadt: Paris, Berlin, London, 1840 bis 1930* (Munich, 1994).
Schnädelbach, Herbert, *Philosophy in Germany, 1831–1933* (Cambridge, 1984).
Schnall, Uwe (ed.), *Auf Auswandererseglern: Berichte von Zwischendecks- und Kajütpassagieren* (Bremerhaven, 1976).
Schöck, Eva Cornelia, *Arbeitslosigkeit und Rationalisierung: Die Lage der Arbeiter und die kommunistische Gewerkschaftpolitik, 1920–1928* (Frankfurt, 1977).
Schoenbaum, David, *Hitler's Social Revolution: Class and Status in Nazi Germany, 1933–1939* (Garden City, 1966).
Schöllgen, Gregor, *Imperialismus und Gleichgewicht: Deutschland, England und die orientalische Frage* (Munich, 1992).
Schöttler, Peter, "Mentalities, Ideologies, Discourses: On the 'Third Level' as a Theme in Social-Historical Analysis," in Alf Lüdtke (ed.), *The History of Everyday Life: Reconstituting Experiences and Ways of Life* (Princeton, 1995), 72–115.
Schulte-Sasse, Linda, *Entertaining the Third Reich: Illusions of Wholeness in Nazi Cinema* (Durham, NC, 1996).
Schütz, Erhard, and Gruber, Eckhard, *Mythos Reichautobahn: Bau und Inszenierung der "Straßen des Führers", 1933–1941* (Berlin, 1996).
Searle, Geoffrey, *The Quest for National Efficiency* (Oxford, 1971).
Sennett, Richard, *The Corrosion of Character: The Personal Consequences of Work in the New Capitalism* (New York, 1998).
Showalter, Elaine, *The Female Malady: Women, Madness, and English Culture, 1830–1980* (London, 1987).
Sieferle, Rolf-Peter, *Fortschrittsfeinde: Opposition gegen Technik und Industrie von der Romantik bis zur Gegenwart* (Munich, 1984).
*Die Konservative Revolution: Fünf biographische Skizzen* (Frankfurt, 1995).
Siegelbaum, Lewis H., "The Making of Stakhanovites, 1935–36," *Russian History/Histoire Russe* 13 (1986), 259–92.
Sinha, Mrinalini, *Colonial Masculinity: The "Manly Englishman" and the "Effeminate Bengali" in the Late Nineteenth Century* (Manchester, 1995).

Smith, Constance Babington, *Amy Johnson* (London, 1967).
Smith, Malcolm, *British Air Strategy Between the Wars* (Oxford, 1984).
Smith, Merrit Roe, and Marx, Leo (eds.), *Does Technology Drive History? The Dilemma of Technological Determinism* (Cambridge, MA, 1994).
Smith, Sidonie, *Moving Lives: 20th-Century Women's Travel Writings* (Minneapolis, 2001).
Smith, T., *Making the Modern: Industry, Art and Design in America* (Chicago, 1993).
Somers, Margaret R., "What's Political or Cultural about Political Culture and the Public Sphere? Toward an Historical Sociology of Concept Formation," *Sociological Theory* 13 (1995), 113–44.
Southall, Derek J., *Magic in the Dark: The Cinemas of Manchester and Ardwick Green* (Manchester, 1999).
Spackman, Barbara, *Fascist Virilities: Rhetoric, Ideology, and Social Fantasy in Italy* (Minneapolis, 1996).
Spalding, Frances, *Duncan Grant* (London, 1997).
Spode, Hasso, "Arbeiterurlaub im Dritten Reich," in Carola Sachse *et al.* (eds.), *Angst, Belohnung, Zucht und Ordnung: Herrschaftsmechanismen im Nationalsozialismus* (Opladen, 1982), 275–328.
Stafford, Barbara Maria, "Introduction: Visual Pragmatism for a Virtual World," in *Good Looking: Essays on the Virtue of Images* (Cambridge, MA, 1997), 2–18.
Stamp, Shelley, *Movie-Struck Girls: Women and Motion-Picture Culture after the Nickelodeon* (Princeton, 2000).
Stansky, Peter, *On or About December 1910: Early Bloomsbury and its Intimate World* (Cambridge, MA, 1996).
Stargardt, Nicholas, *The German Idea of Militarism: Radical and Socialist Critics, 1866–1914* (Cambridge, 1994).
Stark, Gary D., "Cinema, Society and the State: Policing the Film Industry in Imperial Germany," in Gary D. Stark and Bede Karl Lachner (eds.), *Essays on Culture and Society in Modern Germany* (Arlington, TX, 1982), 122–66.
Stebbins, Robert A., *Amateurs, Professionals, and Serious Leisure* (Montreal, 1992).
Stedman Jones, Gareth, "The Determinist Fix: Some Obstacles to the Further Development of the Linguistic Approach to History in the 1990s," *History Workshop Journal* 42 (1996), 19–35.
Steedman, Carolyn, *Strange Dislocations: Childhood and the Idea of Human Interiority, 1780–1930* (London, 1995).
Steele, James, *Queen Mary* (London, 1995).
Steinmetz, Willibald, "Anbetung und Dämonisierung des Sachzwangs: Zur Archäologie einer deutschen Redeform," in Michael Jeismann (ed.), *Obsessionen: Beherrschende Gedanken im wissenschaftlichen Zeitalter* (Frankfurt, 1995), 293–333.
Stern, Fritz, *The Politics of Cultural Despair: A Study in the Rise of Germanic Ideology* (Berkeley, 1961).
Sternberger, Dolf, *Panorama oder Ansichten vom 19. Jahrhundert* (Hamburg, 1946).

Stewart, Susan, *On Longing: Narratives of the Miniature, the Gigantic, the Souvenir, the Collection* (Durham, NC, 1993).
Straub, Eberhard, *Albert Ballin: Der Reeder des Kaisers* (Berlin, 2001).
Street, Sarah, *British National Cinema* (London, 1997).
Supple, Barry, "Fear of Failing: Economic History and the Decline of Britain," in Peter Clarke and Clive Trebilcock (eds.), *Understanding Decline: Perceptions and Realities of British Economic Performance* (Cambridge, 1997), 9–29.
Syon, Guillaume de, *Zeppelin: Germany and the Airship, 1900–1939* (Baltimore, 2002).
Tabili, Laura, *We Ask for British Justice: Workers and Racial Difference in Late Imperial Britain* (Ithaca, 1994).
Tacke, Charlotte, *Denkmal im sozialen Raum: Nationale Symbole in Deutschland und Frankreich im 19. Jahrhundert* (Göttingen, 1995).
Taylor, John, "Kodak and the 'English' Market Between the Wars," *Journal of Design History* 7 (1994), 29–43.
Taylor, S. J., *The Great Outsiders: Northcliff, Rothermere and the Daily Mail* (London, 1996).
Terdiman, Richard, *Discourse/Counter-Discourse: The Theory and Practice of Symbolic Resistance in Nineteenth-Century France* (Ithaca, 1985).
*Present Past: Modernity and the Memory Crisis* (Ithaca, 1993).
Teubner, Andreas, "Justifying Risk," *Daedalus* 119 (1990), 235–54.
Theweleit, Klaus, *Male Fantasies*, 2 vols. (Cambridge, 1987–9).
Thurner, Ingrid, "'Grauenhaft. Ich muß ein Foto machen': Tourismus und Fotografie," *Fotogeschichte* 44 (1992), 23–42.
Tichi, Cecilia, *Shifting Gears: Technology, Literature, Culture in Modernist America* (Chapel Hill, 1987).
Tillyard, Stella K., *The Impact of Modernism, 1900–1920: Early Modernism and the Arts and Crafts Movement in Edwardian England* (London, 1988).
Tosh, John, *A Man's Place: Masculinity and the Middle-Class Home in Victorian England* (New Haven, 1999).
Trentmann, Frank, "Civilization and Its Discontents: English Neo-Romanticism and the Transformation of Anti-Modernism in Twentieth-Century Western Culture," *Journal of Contemporary History* 29 (1994), 583–625.
Trischler, Helmut, "Arbeitsunfälle und Berufskrankheiten im Bergbau 1851–1945," *Archiv für Sozialgeschichte* 28 (1988), 111–51.
"Self-Mobilization or Resistance? Aeronautical Research and National Socialism," in Monika Renneberg and Mark Walker (eds.), *Science, Technology and National Socialism* (Cambridge, 1994), 72–87.
Trommler, Frank, "The Rise and Fall of Americanism in Germany," in Frank Trommler and Joseph McVeigh (eds.), *America and the Germans: An Assessment of a Three-Hundred-Year History* (Philadelphia, 1985), 333–42.
Turner, Patricia A., *I Heard it Through the Grapevine: Rumor in African-American Culture* (Berkeley, 1993).
Turner, Victor, "Liminal to Liminoid in Play, Flow, and Ritual: An Essay in Comparative Symbology," in *From Ritual to Theatre: The Human Seriousness of Play* (New York, 1982), 20–60.

Usborne, Cornelie, "The New Woman and Generational Conflict: Perceptions of Young Women's Sexual Mores in the Weimar Republic," in Mark Roseman (ed.), *Generations in Conflict: Youth Revolt and Generation Formation in Germany, 1770–1968* (Cambridge, 1995), 137–63.

Vaughan, Diane, *The Challenger Launch Decision: Risky Technology, Culture, and Deviance at NASA* (Chicago, 1996).

Veldman, Meredith, *Fantasy, the Bomb and the Greening of Britain: Romantic Protest, 1945–1980* (Cambridge, 1994).

Vitou, Elisabeth, "Le métal dans le décor et le mobilier des transatlantiques," *Monuments Historiques* 130 (1984), 49–55.

Wachsmann, Nikolaus, "Marching Under the Swastika? Ernst Jünger and National Socialism, 1918–1933," *Journal of Contemporary History* 33 (1998), 573–89.

Walkowitz, Judith R., *City of Dreadful Delight: Narratives of Sexual Danger in Late-Victorian London* (London, 1992).

Walton, John K., *The English Seaside Resort: A Social History 1750–1914* (New York, 1983).

Ward, Paul, *Red Flag and Union Jack: Englishness, Patriotism and the British Left, 1881–1924* (Woodbridge, 1998).

Warwick, Paul, "Did Britain Change? An Inquiry in the Causes of National Decline," *Journal of Contemporary History* 20 (1985), 99–134.

Watkin, David, *Grand Hotel: The Golden Age of Palace Hotels: An Architectural and Social History* (London, 1984).

Weber, Max, "Science as a Vocation," in H. H. Gerth and C. Wright Mills (eds. and trans.), *Max Weber: Essays in Sociology* (Oxford, 1959), 129–56.

Wehler, Hans-Ulrich, *Deutsche Gesellschaftsgeschichte*, vol. IV: *Vom Beginn des Ersten Weltkriegs bis zur Gründung der beiden deutschen Staaten, 1914–1949* (Munich, 2003).

"Radikalnationalismus – erklärt er das 'Dritte Reich' besser als der Nationalsozialismus?" in *Umbruch und Kontinuität: Essays zum 20. Jahrhundert* (Munich, 2000), 47–64.

Weindling, Paul, *Health, Race and German Politics Between National Unification and National Socialism* (Cambridge, 1989).

Weisbrod, Bernd, "Gewalt in der Politik: zur politischen Kultur in Deutschland zwischen den Weltkriegen," *Geschichte in Wissenschaft und Unterricht* 16 (1990), 391–404.

Weise, Bernd, "Pressefotografie I: Die Anfänge in Deutschland," *Fotogeschichte* 31 (1989), 15–40.

"Pressefotografie II: Fortschritte der Fotografie- und Drucktechnik und Veränderungen des Pressemarktes im Deutschen Kaiserreich," *Fotogeschichte* 33 (1989), 27–61.

"Pressefotografie III: Das Geschäft mit dem aktuellen Foto," *Fotogeschichte* 37 (1990), 13–36.

Weiß, Hermann, "Die Ideologie der Freizeit im Dritten Reich: Die NS-Gemeinschaft 'Kraft durch Freude,'" *Archiv für Sozialgeschichte* 33 (1993), 289–303.

Welch, David, *Germany, Propaganda and Total War, 1914–1918* (London, 2000).
West, Nancy Martha, *Kodak and the Lens of Nostalgia* (Charlottesville, 2000).
Whalen, Robert Wheldon, *Bitter Wounds: German Victims of the Great War, 1914–1939* (Ithaca, 1984).
Wiener, Martin J., *English Culture and the Decline of the Industrial Spirit, 1850–1950* (Cambridge, 1981).
Wildenthal, Lora, *German Women for Empire, 1884–1945* (Durham, NC, 2001).
Williamson, Philip, *Stanley Baldwin: Conservative Leadership and National Values* (Cambridge, 1999).
Winner, Langdon, *Autonomous Technology: Technics-out-of-Control as a Theme in Political Thought* (Cambridge, MA, 1977).
Winter, Alison, "'Compasses All Awry:' The Iron Ship and the Ambiguities of Cultural Authority in Victorian Britain," *Victorian Studies* 38 (1994), 69–98.
Winter, James, *Secure from Rash Assault: Sustaining the Victorian Environment* (Berkeley, 1999).
Winter, Jay, *Sites of Memory, Sites of Mourning: The Great War in Cultural History* (Cambridge, 1995).
Withey, Lynne, *Grand Tours and Cook's Tours: A History of Leisure Travel, 1750–1915* (London, 1997).
Wohl, Robert, "Heart of Darkness: Modernism and Its Historians," *Journal of Modern History* 74 (2002), 573–621.
  *A Passion for Wings: Aviation and the Popular Imagination* (New Haven, 1994).
Wright, Patrick, *Tank: The Progress of a Monstrous War Machine* (London, 2000).
Wulf, Hans Albert, *"Maschinenstürmer sind wir keine": Technischer Fortschritt und sozialdemokratische Arbeiterbewegung* (Frankfurt, 1987).
Zimmermann, Patricia R., *Reel Families: A Social History of Amateur Film* (Bloomington, 1995).
Zischler, Hanns, *Kafka geht ins Kino* (Reinbeck, 1998).
Zitelmann, Rainer, "Die totalitäre Seite der Moderne," in Michael Prinz and Rainer Zitelmann (eds.), *Nationalsozialismus und Modernisierung* (Darmstadt, 1991), 1–20.
Žižek, Slavoj, *The Plague of Fantasies* (London, 1997).
  *The Sublime Object of Ideology* (London, 1989).

# Index

accidents, *see* physical risk
Adorno, Theodor W. 282
advertising and promotional campaigns 14, 72, 94–5, 134–6, 154, 159, 162–3, 201
aeroplane, *see under individual models*
AGFA 194, 196, 198, 201
airships, *see under individual models*
air war
    British attitudes to violence 264–6, 267–8, 270–1
    civil-defense measures 242, 251–3
    fears of 239–41
    German attitudes to violence 255–6, 257–8, 259–60, 261–3
    propaganda regimes 258, 267–8, 271
    technological constraints 254–5
    *See also under individual pilots*
Alcock, John 121, 130
*All Quiet on the Western Front* (1930) 108
amateur film
    antidote to the pressures of modernity 186, 207, 221
    clubs and societies 201, 219–20
    cost 196
    as cultural elitism 205
    and gender notions 195, 199, 213–16
    idealized family life 210–13
    and individual autonomy 202, 221, 278
    middle-class hobby 196, 203–5
    in Nazi Germany 218–21
    necessary skills 206–7, 209
    number of practitioners 201
    as personal history 207–9
    promotional strategies 201
    topics for movies 202, 209
    user-friendly designs 194–5
Anderson, John 241
Anglo-German rivalry 228–9
anti-Semitism 108–9, 110
*Aquitania* 158, 165, 172, 176
Arnheim, Rudolf 101, 218

Asquith, Herbert 97
Astor, John Jacob 60
*Atlantic* (1929) 23
Attlee, Clement 238
automobile 52, 158
aviation, *see* air war; pilots; physical risk; *and under individual models of aircraft*

Bailey, Peter 91
Baker, Josephine 105
Balazs, Bela 101
Baldwin, Stanley 238, 239
Ball, Albert 118, 264–5
Bangkok 150
Bartov, Omer 282, 283
Bassermann, Ernst 69
Bates, Percy 181
*Battleship Potemkin* (1926) 101
Bauhaus 178
Beardsell, Noel 217
Beck, Ulrich 83
Beinhorn, Elly 36, 130, 136, 141, 142
Benjamin, Walter 87
Beresford, Charles 68, 69
Berlin 62, 77, 98, 121, 154
*Berlin: Symphony of a City* (1927) 101
Berridge, Percy 210
Bewsher, Paul 265
Bishop, William 126, 265
Blériot, Louis 118, 236
Bölcke, Oswald 118, 255, 256, 257, 260
Boston 162
Brandenburg, Ernst von 262
Brandt, Rolf 73
*Bremen* 1, 178–80, 229
Bremen 161
Breuhaus de Groot, Fritz 179
*Bristol 142* 231
British Board of Film Censors 96
Bruce, S. M. 43
Brunel, Isambard Kingdom 160, 227

Buchan, Andrew 102
Bülow, Graf Bernhard von 243
Burke, Edmund 25
Butt, Archibald 60
Buxton, Noel 67

*Catherine the Great* (1934) 107
Chaplin, Charlie 102
Cheshire, Leonard 269, 270
Chislett, Charles 193, 212
Churchill, Winston 239
cinema, *see* film
cinema architecture 98–9
Cinematograph Films Act (1927) 106
Cinematograph Films Act (1937) 107
class 119–20, 148–9, 168–75, 183–4, 196, 203–5
Cobham, Alan 36, 43, 116, 117, 118, 120, 123, 125, 129, 231, 233
colonialism 36, 107, 147–8, 151–2, 233–5, 278
Columbus, Christopher 43, 70
Communist Party (KPD), *see* technology: wariness from the political Left
Cook, Thomas 184
Corn, Joseph 139
Courtenay, William 137
Croydon 68, 69
Cunard Line 22, 46, 74, 156, 158, 163, 165, 170, 171, 175, 228
Cunliffe Lister, Philip 106

Daedalus and Icarus 42
danger, *see* physical risk
*Dawn Patrol* (1939) 268
Deedes, William 241
*Deutschland* 243
Dietrich, Marlene 150
Dietrich, Otto 41
Dixon, Charles 149
*Do X* 31, 70, 72, 73
Dominicus, Alexander 244
Dornier 70, 73
*Drummond Castle* 57, 65
Duchess of Bedford 120, 139

Eckener, Hugo 61, 73, 245
*Edith, the White Slave* 92
Eisenstein, Sergej 101
Empire flying boats 154, 230
*Empress of Ireland* 57, 161
enthusiasm 20, 231
Epstein, Jacob 178
Ernemann 196
Etzdorf, Marga von 120, 127–8, 129, 139, 141, 144–6, 150
*Europa* 1

femininity 126–8, 139–42, 147–8, 151–2, 199
film
  American dominance 105–8
  as art 100–2
  British quota legislation 106
  censorship 93
  and entertainment culture 90, 99–100, 158
  not considered to be a technology 4, 88, 89, 111, 113–14, 277
  a moral danger 91–4
  National Socialist policies 110–12
  a political danger 103–5
  public moralists 90–1, 100
  a tool of deception 28–9, 39–40, 89, 103–4, 109, 112, 277
  *See also* advertising and promotional campaigns; amateur film; anti-Semitism; cinema architecture
Fisher, Philip 20
Flex, Walter 259
Focke-Wulf 154
Freedgood, Elaine 52
Friedrichshafen 245

*G.38* 38, 72, 73
Gaumont et cie 90
gender, *see* femininity; masculinity
George V, King 118
*George Washington* 162
Germany, distinctive attitudes to technology
  aggressive challenger 244–6, 250–1
  newcomer status 226, 243, 278
  *See also* air war; Anglo-German rivalry; National Socialism; national symbols
Gill, Eric 178
Gillis, John 211
Girouard, Mark 266
Glasgow 1
Goebbels, Joseph 100, 108, 110, 112
Goldwyn, Samuel 109
Gordon, Charles George 264
Göring, Hermann 34–5, 249, 251, 252, 260
*Graf Zeppelin* 72, 246, 248, 250
Grahame-White, Claude 129
gramophone 87
Graves, Robert 132
Great Britain, distinctive attitudes to technology
  anti-war sentiment 3, 11, 237–8, 241
  emphasis on national defense 235–7, 238–9, 269
  empire and technology 233–5, 278
  technological leadership 225–6
  *See also* air war, Anglo-German rivalry, national symbols

## Index

*Great Western* 160
Guggenheim, Benjamin 60

Haig, Douglas 249
Hamburg-American Line 22, 25, 156, 163, 165, 168, 185, 228
Harlan, Veit 111
Harvey, Lilian 150
Havelock, Henry 264
Havilland, Geoffrey de 116
Hawker, Harry 120, 132–3
Heath, Lady 120, 139–40
Heinkel, Ernst 41
Hendon 34, 268
Herf, Jeffrey 7, 281
*Der Herrscher* (1937) 111
Hillary, Richard 270, 271–3
*Hindenburg* 51, 59, 60, 76, 81–3, 250
Hindenburg, Paul von 120
Hirth, Hellmuth 128, 139
Hitler, Adolf 41, 109, 188, 238, 249, 261, 282
Holden, Charles 178
Hollywood 105, 107
Horkheimer, Max 282
*HP 42* 230
Hunter, Charles 209, 211
Hurricane, fighter plane 269

*Ile de France* 181
Immelmann, Max 118, 256, 257
*Imperator* 1, 25, 163, 164, 170, 229
Imperial Airways 73, 75, 77, 233, 234
imperialism, *see* colonialism
Ismay, Bruce 66

Jameson, Fredric 8
Johnson, Amy 120, 121, 125, 136, 141–4, 146–50, 151, 153, 231
*JU 87* 261
Jünger, Ernst 124, 244, 262
Junkers, company 38, 73, 80, 135
Junkers, Hugo 116, 118, 135

*Kaiser Wilhelm der Große* 1
*Kaiserin Auguste Victoria* 166
Klemperer, Victor 186
Knight, Laura 182
Kodak 194, 196, 197, 198, 201, 205
Köhl, Hermann 116, 117, 118, 120, 124, 131–2, 135, 154, 199, 247
*Kolportageroman* 91
Korda, Alexander 104
Kracauer, Siegfried 101
*Kraft durch Freude, see* Strength through Joy
Kraszna Krausz, Andor 218

*L1* 62
*L2* 58
Labour Party, *see* technology: wariness from the political Left
Lakehurst 65, 81
Lang, Fritz 101
Lawrence, T.E. 123
Leander, Zarah 150
Lee, Arthur 236
Lehmann, Ernst 59, 82
LeMahieu, Dan 6
Lindbergh, Charles 119
Liverpool 162, 189
Lloyd-George, David 264, 267
Loew, Marcus 109
Löfgren, Orvar 212
Loiperdinger, Martin 93
London 55, 65, 94, 97, 98, 163, 164, 231, 234, 240, 256
Ludendorf, Erich von 108
Lufthansa 55, 59, 65, 74, 245, 249
Lumière, Louis 88
*Lusitania* 1, 23, 26, 165, 175, 183, 228

McCudden, James 126, 265
Madeira 186, 187
Manchester 98
Mann, Thomas 186
masculinity 123–4, 126–8, 195, 255–6, 257–8, 259–60, 261–3, 264–6, 267–8, 270–1
*Mauretania* 1, 26, 163, 165
Melzer, Karl 219, 220
*Metropolis* (1926) 101, 114
Méwès, Charles Frédéric 164
Mickey Mouse 100
Milestone, Lewis 108
*Modern Times* (1936) 114
modernity
  and artistic modernism 8–9, 176–8
  contemporary attitudes 4, 10–11, 21, 91, 99–100, 121, 163–4, 243, 253, 274
  and modernization theory 9–10
  and National Socialism 8, 111, 113–14, 249, 280–1
  pilots as its paragons 125, 137, 146, 156
  theoretical models 8
  *See also* amateur film: antidote to the pressures of modernity
Mollison, James 123, 131, 154, 233
Moreck, Curt 99
Morgan, Stanley 235
Morris, B.W. 181
Morris Motor Works 74
Morris, William 177
Morrison, R.C. 107

## Index

*Motor Mad* (1909) 92
music halls 91
Muswell Hill 99

National Socialism 7–8, 40–2, 80–3, 110–12, 186–90, 218–21, 249, 250, 279
national symbols 226, 229–30, 252
Neagle, Anna 148
Neptune 42
New York City 162
Newcastle 163
Niezoldi und Krämer 196
*Normandie* 181
North German Lloyd 74, 178, 228, 245
Northcliffe, Lord 172, 175
Nuremberg 35

Orwell, George 271
Osnabrück 98
Overy, Richard 239

Paris 55, 164
passenger liners
 cruises 185–6, 189–90
 health-enhancing qualities 168
 modernist design 176–83
 on-board luxury 164–6, 277
 re-creations of urban spaces 163–4
 seasickness 161–2
 social segregation 168–71
 sociability 166–7, 187
 traditionally seen as unsafe 161–2
 working conditions 171–5, 183–4
 as works of art 175–6
 *See also* Strength through Joy
Pathé 196
Pathé-Frères 90
photography 14, 198, 205
physical risk
 accounts of heroism 59–61
 celebration of danger in aviation 125, 137–8, 155
 fears of 54–5
 and innovation processes 66–70, 75–6, 77–9
 limits of public acceptance 53, 79–80, 84
 National Socialist attitudes to 80–3
 public shock 55–7, 61–2
 safety arrangements 73–5
 sensationalism in media coverage 71–3
 *See also* film, pilots
Pick, Frank 178
pilots
 female pilots' self-assertiveness 139–42
 female pilots as women of empire 147–8, 151–2
 first flight as initiation rite 126–8
 gender notions 123–4, 127–8, 138–9, 142–6, 148–9, 150–1, 152–3
 as media celebrities 136–7
 mental qualities 121–4
 physical qualities 121
 social background 119–20, 148–9
 transcendence 128–34, 156, 277–80
 *See also under individual pilots*, physical risk, air war
Priestley, J. B. 172
*Private Life of Henry VIII, The* (1933) 104, 107
productivism 7, 8, 42, 175, 188, 280
public sphere 3, 13–14, 71–3, 136–7
Pudovkin, Vsevolod 102

*Queen Mary* 1, 2, 17, 180–3, 184, 193

*R101* 36, 56, 57, 58, 61, 66, 67, 76, 77, 79
radio 13, 87, 198
Rappaport, Erika 167
*Reichsfilmkammer* 219
*Reichsluftschutzbund* 251
Reitsch, Hanna 152–3
Remarque, Erich Maria 108
Reuleaux, Franz 226
Richthofen, Manfred von 118, 127, 256, 260
Riefenstahl, Leni 112, 150, 152
Ritter, Karl 112
Ritz, César 164
*Robert Ley* 189
Roberts, Mary Louise 138
Rolls, Charles 119, 129
Rosenberg, Alfred 40
Rotha, Paul 89, 101, 102
Rothermere, Lord 231
Ruskin, John 177
Ruttmann, Walter 101

Saint-Exupéry, Antoine de 137
Salford 98
*Sanders of the River* (1935) 107
Sassoon, Philip 65
Scheidemann, Philipp 258
Schivelbusch, Wolfgang 244
serial novels 91
ships, *see* passenger liners; *and under individual vessels*
*Singing Fool* (1929) 24
Social Democracy, *see* technology: wariness from the political Left
social fantasies 16–17, 157, 158, 192, 194, 277
Söderbaum, Kristina 150
Southampton 71, 163, 186
Southgate 99

Spengler, Oswald 35, 36–7, 114
Spitfire 269
*Stahlhelm* 247
Stanley, Oliver 100, 107
Strasser, Alex 218
Strength through Joy 186–90
*Suevic* 71
Surbiton 99

tabloid press 3
technology
  aesthetics of 23–4, 175–83
  ambivalence towards 1–2, 16, 21, 25, 276
  designs for non-experts 194–5
  and European superiority 36
  film not conceived as a technology 4, 88, 89, 111, 113–14
  and gender 17, 276
  knowledge problems 2, 26, 29–32, 64–6, 87–8, 194–5, 276
  and the media 13–14, 26–8
  and modernity 3, 11, 32–4, 42–5
  National Socialist attitudes 7–8, 40–2, 246, 249, 250, 279
  a reified category 5, 34–5, 47
  and science 37–9
  a secular miracle 20–1, 22–5
  and social fantasies 16–18
  wariness from the political Left 45–6, 75–6, 77–9, 148–9, 183–4, 231–2, 246–7, 258

  *See also under individual technologies;* air war, amateur film, modernity, National Socialism; pilots; social fantasies
Telford, Thomas 227
Thomson, Lord 62, 147, 148
Tirpitz, Alfred 67, 69
*Titanic* 1, 23, 43, 51–2, 54, 60–1, 66, 67, 68, 76, 77–8, 161
Tokyo 121
Trenchard, Hugh 264

*Vaterland* 163, 164, 170
Versailles Treaty 41, 152, 244, 245
Vickers 74

war, *see* air war
Weber, Max 24
White Star Line 43, 66, 67, 71, 236
*Wilhelm Gustloff* 189
William II, Kaiser 97
*Wings* (1928) 259, 267
Wolfe Murray, Stella 125
Women's League of the German Colonial Association 152
Wright, Patrick 6

Zeiss-Ikon 196
Zeppelin, company 72, 73, 245
Zeppelin, Count Ferdinand von 232, 248–9
Zinkeisen, Doris 182
*ZR3* 245
Zukor, Adolph 109

# NEW STUDIES IN EUROPEAN HISTORY

*Books in the series*

Royalty and Diplomacy in Europe, 1890–1914
RODERICK R. McLEAN

Catholic Revival in the Age of the Baroque
Religious Identity in Southwest Germany, 1550–1750
MARC R. FORSTER

Helmuth von Moltke and the Origins of the First World War
ANNIKA MOMBAUER

Peter the Great
The Struggle for Power, 1671–1725
PAUL BUSHKOVITCH

Fatherlands
State Building and Nationhood in Nineteenth-Century Germany
ABIGAIL GREEN

The French Second Empire
An Anatomy of Political Power
ROGER PRICE

Origins of the French Welfare State
The Struggle for Social Reform in France, 1914–1947
PAUL V. DUTTON

Ordinary Prussians
Brandenburg Junkers and Villagers, 1500–1840
WILLIAM W. HAGEN

Liberty and Locality in Revolutionary France
Six Villages Compared
PETER JONES

Vienna and Versailles
The Courts of Europe's Dynastic Rivals, 1550–1780
JEROEN DUINDAM

From *Reich* to State
The Rhineland in the Revolutionary Age, 1780–1830
MICHAEL ROWE

Re-Writing the French Revolutionary Tradition
ROBERT ALEXANDER

Provincial Power and Absolute Monarchy
JULIAN SWANN

People and Politics in France, 1848–1870
ROGER PRICE

Nobles and Nation in Central Europe
WILLIAM D. GODSEY, JR.

*The Russian Roots of Nazism*
White Emigrés and the Making of National Socialism, 1917–1945
MICHAEL KELLOGG

Technology and the Culture of Modernity in Britain and Germany, 1890–1945
BERNHARD RIEGER

For EU product safety concerns, contact us at Calle de José Abascal, 56–1°,
28003 Madrid, Spain or eugpsr@cambridge.org.

www.ingramcontent.com/pod-product-compliance
Ingram Content Group UK Ltd.
Pitfield, Milton Keynes, MK11 3LW, UK
UKHW042149130625
459647UK00011B/1255